CONTENTS

CONTENTS

6. VOLUME AND CAPACITY

7. MASS

8. TIME

9. POSITION

ANSWERS

ABOUT CATCH-UP MATHS

The Catch-Up Maths series enables students to start from scratch when they are struggling with their year level maths. Each book takes maths back to the foundation and ensures that all basic concepts are consolidated before moving forwards. Lots of revision and opportunities to practise and build confidence are provided before moving on to new topics.

Each new strand and sub-strand of the primary maths curriculum is introduced clearly with simple explanations, examples and trial questions (with answers), before children move to the Practice section. To ensure concepts are understood fully, videos of the author working through and explaining new concepts are included in every chapter.

SCAN to watch video

A QR code on the topic page provides access to the subtitled video.

This book has 9 chapters divided between the Year 5 Measurement and Space strands of the Australian Maths Curriculum. The chapters are:

1 Length
2 Angles
3 2D Shapes
4 3D Objects
5 Area
6 Volume & Capacity
7 Mass
8 Time
9 Position

★ A Review section that can be used as assessment and to check students' progress is included at the end of each chapter.

★ Answers are at the back of the book.

How to use this book

Children can work through the pages from front to back, or choose individual topics to reinforce areas where they are struggling.

The topics are introduced with:

- clear instructions, using simple language
- completed examples and incomplete examples for students to tackle before moving on to the **Your Turn** section
- a video linked by QR code that shows the incomplete examples and has the author giving extra instruction about the page.

Each **Your Turn** section contains a SELF CHECK for students to reflect and give self-assessment on their understanding.

HOW TO USE THE QR CODES IN CATCH-UP MATHS

A unique aspect of the **Catch-Up Maths** series is the **instructional video** created by the author for every new topic.

Access the video using a QR code reader app

SCAN to watch video

The videos give a step-by-step explanation of the examples and work through them to provide a helpful lesson for the student. The videos are simply accessed via the QR code on the same page and can be watched on a phone, tablet, computer or interactive whiteboard.

Each video shows the page from the book. The author talks through the concepts and examples, and demonstrates what students need to do. Her words are shown as easy-to-read captions so that the audio can be turned down in noisy classrooms, and for students with hearing difficulties. The solutions to the examples are presented before students are expected to tackle the 'Your Turn' section. This careful instruction ensures that students can confidently move on to the following Practice questions. Those assisting students should encourage them to check their 'Your Turn' answers before moving on.

Over 60 instructional videos!

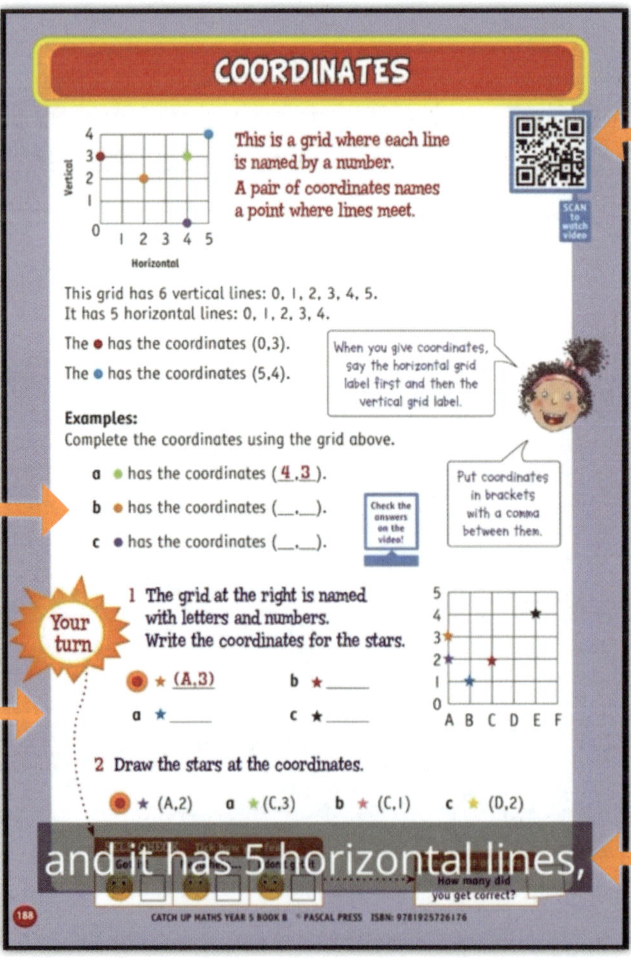

Scan this to access the video.

The answers to these examples can only be found on the video.

After watching the video, students can confidently complete the **Your turn** section.

Clear and easy-to-read captions

COORDINATES

This is a grid where each line is named by a number.
A pair of coordinates names a point where lines meet.

This grid has 6 vertical lines: 0, 1, 2, 3, 4, 5.
It has 5 horizontal lines: 0, 1, 2, 3, 4.

The ● has the coordinates (0,3).
The ● has the coordinates (5,4).

When you give coordinates, say the horizontal grid label first and then the vertical grid label.

Examples:
Complete the coordinates using the grid above.

a ● has the coordinates (4,3).
b ● has the coordinates (_,_).
c ● has the coordinates (_,_).

Check the answers on the video!

Put coordinates in brackets with a comma between them.

Your turn

1 The grid at the right is named with letters and numbers. Write the coordinates for the stars.

★ (A,3) b ★____
a ★____ c ★____

2 Draw the stars at the coordinates.

★ (A,2) a ★ (C,3) b ★ (C,1) c ★ (D,2)

and it has 5 horizontal lines,

AUSTRALIAN CURRICULUM CORRELATIONS

Australian Curriculum: Mathematics F–6 Version 9.0

ACARA CODE	Content Description	Pages
1. Length		**PAGES 1–27**
AC9M3M01 Year 3	identify which metric units are used to measure everyday items; use measurements of familiar items and known units to make estimates	1, 2, 3
AC9M3M02 Year 3	measure and compare objects using familiar metric units of length, mass and capacity, and instruments with labelled markings	4, 5, 6, 7, 8, 9, 10, 11
AC9M4M01 Year 4	interpret unmarked and partial units when measuring and comparing attributes of length, mass, capacity, duration and temperature, using scaled and digital instruments and appropriate units	1, 2, 3, 4, 5, 6, 7, 8, 9, 10, 11, 16, 17, 18, 19
AC9M4M02 Year 4	recognise ways of measuring and approximating the perimeter and area of shapes and enclosed spaces, using appropriate formal and informal units	12, 13, 14, 15
AC9M5M02 Year 5	solve practical problems involving the perimeter and area of regular and irregular shapes using appropriate metric units	13, 14, 15, 98, 99
AC9M6M01 Year 6	convert between common metric units of length, mass and capacity; choose and use decimal representations of metric measurements relevant to the context of a problem	16, 17
2. Angles		**PAGES 28–47**
AC9M4M04 Year 4	estimate and compare angles using angle names including acute, obtuse, straight angle, reflex and revolution, and recognise their relationship to a right angle	28, 29, 30, 31, 32, 33, 34, 35
AC9M5M04 Year 5	estimate, construct and measure angles in degrees, using appropriate tools including a protractor, and relate these measures to angle names	36, 37, 38, 39, 40, 41, 42

ACARA CODE	Content Description	Pages
3. 2D Shapes		**PAGES 48–71**
AC9M1SP01 Year 1	make, compare and classify familiar shapes; recognise familiar shapes and objects in the environment, identifying the similarities and differences between them	48, 49, 50, 51, 52, 53, 54, 55, 56, 57, 58, 59
AC9M2M05 Year 2	identify, describe and demonstrate quarter, half, three-quarter and full measures of turn in everyday situations	60, 61, 62, 63
AC9M2SP01 Year 2	recognise, compare and classify shapes, referencing the number of sides and using spatial terms such as "opposite", "parallel", "curved" and "straight"	50, 51, 54, 55, 56, 57, 58, 59
AC9M4SP03 Year 4	recognise line and rotational symmetry of shapes and create symmetrical patterns and pictures, using dynamic geometric software where appropriate	64, 65
AC9M5SP03 Year 5	describe and perform translations, reflections and rotations of shapes, using dynamic geometric software where appropriate; recognise what changes and what remains the same, and identify any symmetries	60, 61, 62, 63
4. 3D Objects		**PAGES 72–89**
AC9M3SP01 Year 3	make, compare and classify objects, identifying key features and explaining why these features make them suited to their uses	72, 73, 74, 75, 76, 77, 82, 83
AC9M3SP02 Year 3	interpret and create two-dimensional representations of familiar environments, locating key landmarks and objects relative to each other	82, 83
AC9M5SP01 Year 5	connect objects to their nets and build objects from their nets using spatial and geometric reasoning	80, 81
AC9M6SP01 Year 6	compare the parallel cross-sections of objects and recognise their relationships to right prisms	78, 79
5. Area		**PAGES 90–107**
AC9M4M02 Year 4	recognise ways of measuring and approximating the perimeter and area of shapes and enclosed spaces, using appropriate formal and informal units	90, 91, 92, 93, 94, 95, 96, 97, 100, 101
AC9M6M02 Year 6	establish the formula for the area of a rectangle and use it to solve practical problems	90, 91, 98, 99

Australian Curriculum Correlations continued

ACARA CODE	Content Description	Pages
6. Volume and Capacity		**PAGES 108–127**
AC9M1SP01 Year 1	make, compare and classify familiar shapes; recognise familiar shapes and objects in the environment, identifying the similarities and differences between them	108, 109
AC9M3M01 Year 3	identify which metric units are used to measure everyday items; use measurements of familiar items and known units to make estimates	110, 111
AC9M3M02 Year 3	measure and compare objects using familiar metric units of length, mass and capacity, and instruments with labelled markings	108, 109, 110, 111, 112, 113, 114, 115, 116, 117, 118, 119, 120
AC9M4M01 Year 4	interpret unmarked and partial units when measuring and comparing attributes of length, mass, capacity, duration and temperature, using scaled and digital instruments and appropriate units	112, 113, 114, 115, 116, 117, 118
AC9M5M01 Year 5	choose appropriate metric units when measuring the length, mass and capacity of objects; use smaller units or a combination of units to obtain a more accurate measure	108, 109
7. Mass		**PAGES 128–147**
AC9M3M01 Year 3	identify which metric units are used to measure everyday items; use measurements of familiar items and known units to make estimates	128, 130, 138, 140, 141
AC9M3M02 Year 3	measure and compare objects using familiar metric units of length, mass and capacity, and instruments with labelled markings	128, 129, 131, 132, 134, 135, 138, 139
AC9M4M01 Year 4	interpret unmarked and partial units when measuring and comparing attributes of length, mass, capacity, duration and temperature, using scaled and digital instruments and appropriate units	132, 135, 137
AC9M5M01 Year 5	choose appropriate metric units when measuring the length, mass and capacity of objects; use smaller units or a combination of units to obtain a more accurate measure	129
AC9M6M01 Year 6	convert between common metric units of length, mass and capacity; choose and use decimal representations of metric measurements relevant to the context of a problem	129, 132, 133, 136, 137, 138, 139

ACARA CODE	Content Description	Pages
8. Time		**PAGES 148–176**
AC9M1M03 Year 1	describe the duration and sequence of events using years, months, weeks, days and hours	166, 167, 168, 169
AC9M2M03 Year 2	identify the date and determine the number of days between events using calendars	158, 159, 160, 161
AC9M2M04 Year 2	recognise and read the time represented on an analog clock to the hour, half-hour and quarter-hour	148, 149, 150, 151
AC9M3M03 Year 3	recognise and use the relationship between formal units of time including days, hours, minutes and seconds to estimate and compare the duration of events	148, 149, 150, 151, 152, 153
AC9M3M04 Year 3	describe the relationship between the hours and minutes on analog and digital clocks, and read the time to the nearest minute	148, 149, 150, 151, 152, 153
AC9M4M01 Year 4	interpret unmarked and partial units when measuring and comparing attributes of length, mass, capacity, duration and temperature, using scaled and digital instruments and appropriate units	150, 151
AC9M4M03 Year 4	solve problems involving the duration of time including situations involving "am" and "pm" and conversions between units of time	154, 155
AC9M5M03 Year 5	compare 12- and 24-hour time systems and solve practical problems involving the conversion between them	154, 155, 156, 157
AC9M6M03 Year 6	interpret and use timetables and itineraries to plan activities and determine the duration of events and journeys	162, 163, 164, 165
9. Position		**PAGES 177–196**
AC9M1SP02 Year 1	give and follow directions to move people and objects to different locations within a space	177, 178, 179, 180, 181
AC9M2M05 Year 2	identify, describe and demonstrate quarter, half, three-quarter and full measures of turn in everyday situations	177, 178, 179, 180, 181
AC9M2SP02 Year 2	locate positions in two-dimensional representations of a familiar space; move positions by following directions and pathways	177, 178, 179, 180, 181
AC9M3SP01 Year 3	make, compare and classify objects, identifying key features and explaining why these features make them suited to their uses	182, 183, 184, 189, 190
AC9M4SP02 Year 4	create and interpret grid reference systems using grid references and directions to locate and describe positions and pathways	182, 183, 184, 185, 186, 189, 190
AC9M5SP02 Year 5	construct a grid coordinate system that uses coordinates to locate positions within a space; use coordinates and directional language to describe position and movement	185, 186, 187, 188, 189

METRES

A metre (m) is a unit of measurement used to measure big things.
One metre is one-thousandth of a kilometre.
There are 100 cm in one metre.

Cars and motorbikes are measured in metres.

1 m — 100 cm

A skateboard and a newborn baby are around 50 cm long.

$\frac{1}{2}$ m — 50 cm

This book and a shoe are about 25 cm long.

$\frac{1}{4}$ m — 25 cm

Examples: Colour the boxes of the items that would measure less than one metre in blue and more than one metre in red.

a ■ a train **c** ☐ a bridge **e** ☐ a laptop

b ☐ a fingernail **d** ☐ a tricycle **f** ☐ a road

Check the answers on the video!

Your turn

List five things that measure each length.

1 metre	$\frac{1}{2}$ m	$\frac{1}{4}$ m
a big dog	a night stand	my ruler

SELF CHECK Tick how you feel

Got it! ☐	Need help... ☐	I don't get it ☐

Check your answers
How many did you get correct? ☐

  ISBN: 9781925726176

PRACTICE

Write these lengths in centimetres.

- 1 m 27 cm 127 cm
- a 5 m 28 cm ______
- b 3 m 17 cm ______
- c 2 m 31 cm ______
- d 8 m 59 cm ______
- e 6 m 44 cm ______
- f 3 m 9 cm ______
- g 7 m 10 cm ______
- h 9 m 20 cm ______
- i 6 m 99 cm ______
- j 4 m 4 cm ______
- k 8 m 8 cm ______

Write in metres and centimetres.

- 243 cm 2 m 43 cm
- a 637 cm ______
- b 601 cm ______
- c 952 cm ______
- d 481 cm ______
- e 545 cm ______
- f 890 cm ______
- g 330 cm ______
- h 206 cm ______
- i 660 cm ______
- j 107 cm ______
- k 715 cm ______

3 Complete the tables.

	Centimetres	Metres and centimetres
	169 cm	1 m 69 cm
a		2 m 10 cm
b	865 cm	
c		9 m 89 cm
d	400 cm	
e		6 m 37 cm
f	302 cm	

	Centimetres	Metres and centimetres
g		5 m 20 cm
h	429 cm	
i		6 m 3 cm
j	646 cm	
k		7 m 11 cm
l	801 cm	
m		9 m 90 cm

  ISBN: 9781925726176

4 Write the lengths in question 3 in order from shortest (1) to longest (14).

1 1 m 69 cm, 2 ________, 3 ________, 4 ________,

5 ________, 6 ________, 7 ________, 8 ________,

9 ________, 10 ________, 11 ________, 12 ________,

13 ________, 14 ________

5 Draw lines to match the items with the correct measurements.

●	length of a car	$\frac{1}{4}$ metre
a	length of a motorbike	3 metres
b	length of an aeroplane	$2\frac{1}{2}$ metres
c	length of a cruise ship	45 metres
d	length of a mountain bike	5 metres
e	length of a toy truck	2 metres
f	length of a child's ride-on car	300 metres
g	length of a kayak	1 metre

6 How many centimetres?

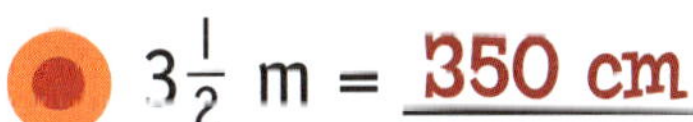

● $3\frac{1}{2}$ m = 350 cm

a $6\frac{1}{4}$ m = ________

b $2\frac{3}{4}$ m = ________

c $1\frac{1}{4}$ m = ________

d $5\frac{1}{2}$ m = ________

e $7\frac{1}{4}$ m = ________

f $9\frac{3}{4}$ m = ________

g $10\frac{1}{4}$ m = ________

h $1\frac{1}{2}$ m = ________

i $12\frac{1}{4}$ m = ________

j $4\frac{1}{4}$ m = ________

k $8\frac{1}{2}$ m = ________

CENTIMETRES

A centimetre (cm) is a unit of measurement used to measure smaller things. One centimetre is one-hundredth of a metre.

This ruler is 15 centimetres (15 cm) long.

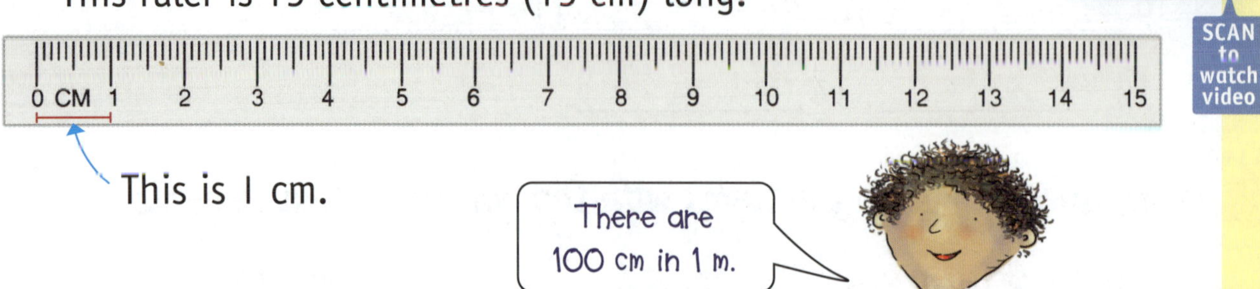

Examples: Tick the items you would measure in centimetres.

a ☑ a shoe
b ☐ a bed
c ☐ a pen
d ☐ a bus
e ☐ a toy car
f ☐ a backpack

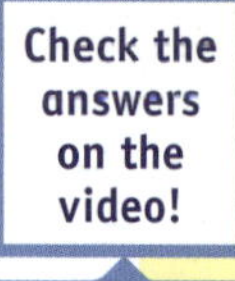

Start at 0 when you measure with a ruler.

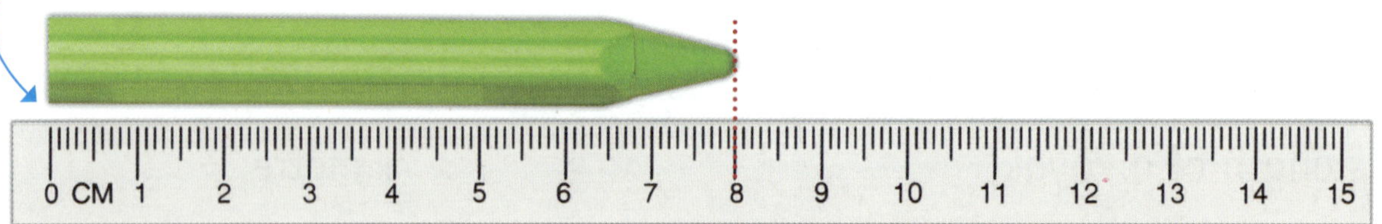

This crayon is 8 cm long.

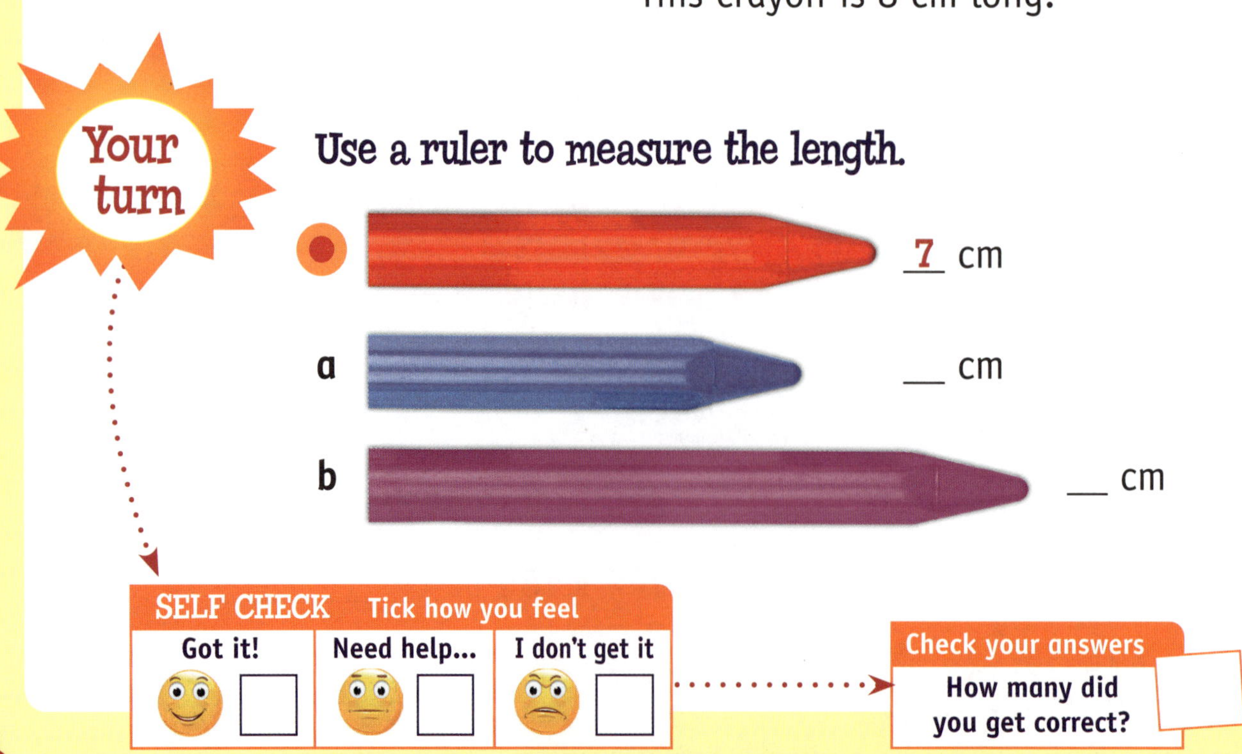

Use a ruler to measure the length.

● 7 cm

a ___ cm

b ___ cm

SELF CHECK Tick how you feel

Got it!	Need help...	I don't get it
☐	☐	☐

Check your answers

How many did you get correct? ☐

CATCH UP MATHS YEAR 5 BOOK B © PASCAL PRESS ISBN: 9781925726176

PRACTICE

1 Draw lines of these lengths.

- 13 cm ____________________

a 4 cm

b 8 cm

c 10 cm

d 2 cm

2 Measure the lines and write the length.

- 6 cm ____________

a ______

b ______

c ______

d ______

3 Measure the pencils and write the length.

- 10 cm

a ______

b ______

c ______

4 Use the pencils in Question 3 to answer the following.

a What is the colour of the shortest pencil? ______________

b What is the colour of the longest pencil? ______________

c Write the colours of the pencils in order from longest to shortest.

____

  ISBN: 9781925726176

MILLIMETRES

Millimetres (mm) are used to measure very small lengths. One millimetre is one-tenth ($\frac{1}{10}$) of a centimetre.

This ruler is 15 cm long. 15 cm = 150 mm

The small marks on the ruler are millimetres.

This is 16 mm.

Example: Write these measurements in millimetres.

a 4 cm 40 mm

b 8 cm ____ mm

c 1 cm ____ mm

d 3 cm ____ mm

e 12 cm ____ mm

f 20 cm ____ mm

Your turn

1 Write the lengths marked on the ruler.

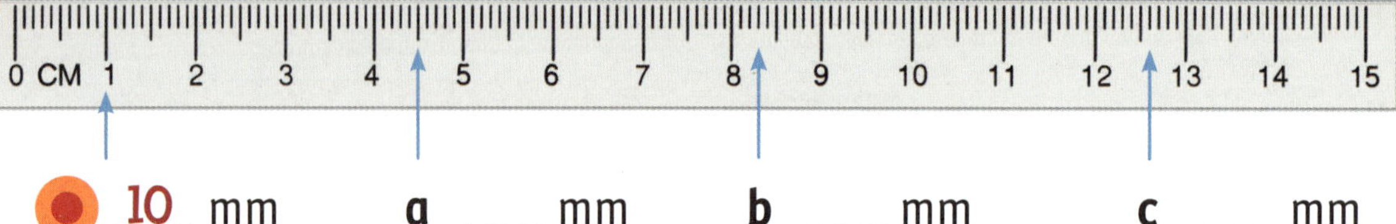

● 10 mm **a** ___ mm **b** ___ mm **c** ___ mm

2 Mark these lengths on the ruler.

● 7 mm **a** 19 mm **b** 58 mm **c** 110 mm

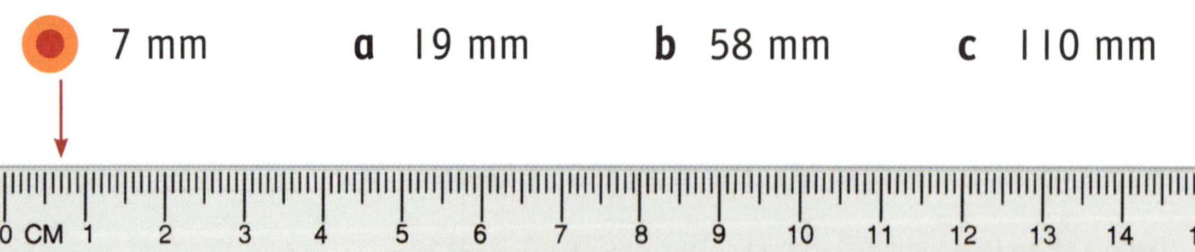

SELF CHECK Tick how you feel

Got it!	Need help...	I don't get it
☐	☐	☐

Check your answers

How many did you get correct? ☐

CATCH UP MATHS YEAR 5 BOOK B © PASCAL PRESS ISBN: 9781925726176

PRACTICE

1 Name four things that would be measured in millimetres.

- an ant ____________

2 Would these things be measured in mm or cm?
Write the units to complete the meaurements.

- pen: 9 cm

b book: 25 ____

d staple: 12 ____

a thumbtack: 8 ____

c fingernail: 15 ____

e sticky note: 10 ____

3 Measure and write the length in millimetres.

- 72 mm

a ________

b ________

c ________

4 Write the lengths in Question 3 from shortest to longest.

13 mm ____________

5 Complete the tables.

	cm and mm	mm
●	4 cm 2 mm	42 mm
a		71 mm
b		50 mm
c	6 cm 9 mm	

	cm and mm	mm
d		83 mm
e	9 cm 1 mm	
f	14 cm 4 mm	
g		28 mm

6 How many mm?

- 4 cm 40 mm

b 11 cm ________

d 20 cm ________

a 8 cm ________

c 2 cm ________

e 80 cm ________

  ISBN: 9781925726176

MILLIMETRES, CENTIMETRES AND METRES

Length can be measured in millimetres (mm), centimetres (cm) and metres (m).

Millimetres (mm) are used to measure very small things, like a toenail.

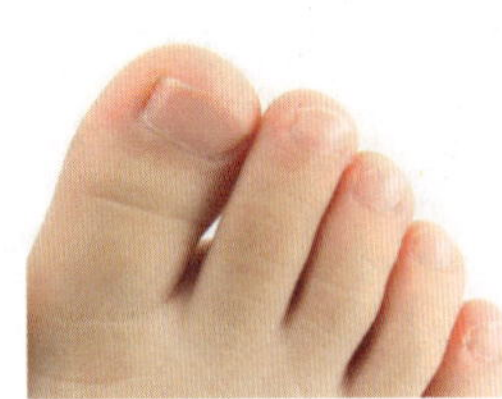

Centimetres (cm) are used to measure small things, like your hand.

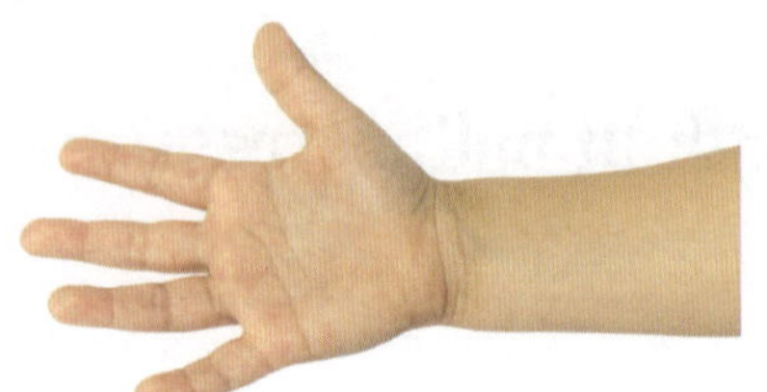

Metres (m) are used to measure large things, like your height.

Example: Use red to circle the items you would measure in millimetres, use blue for centimetres and green for metres.

a	a tablet (circled)	**e**	a small dog
b	an eraser	**f**	a netball court
c	a mouse's eye	**g**	an ant
d	the height of a house	**h**	a train

Circle the best estimate for the length.

●	a remote-control car	30 cm (circled)	30 mm	30 m
a	a calculator	20 mm	20 m	20 cm
b	a fence	30 cm	30 m	30 mm
c	the width of a baby's toe	13 mm	13 cm	13 m
d	a pen	10 m	10 cm	10 mm

SELF CHECK Tick how you feel

Got it!	Need help...	I don't get it
☐	☐	☐

Check your answers
How many did you get correct? ☐

CATCH UP MATHS YEAR 5 BOOK B © PASCAL PRESS ISBN: 9781925726176

PRACTICE

1 Circle the correct set of unit abbreviations.

a mm = metre, cm = centimetre, m = millimetre

b mm = millimetre, cm = cubic metre, m = metre

c mm = millimetre, cm = centimetre, m = metre

2 Convert these measurements.

● 50 mm = 5 cm

a 3 m = ______ cm

b 5 cm = ______ mm

c 120 mm = ______ cm

d 40 mm = ______ cm

e 100 mm = ______ cm

f 800 cm = ______ m

g 170 mm = ______ cm

h 60 mm = ______ cm

i 13 m = ______ cm

3 Are these measurement conversions correct? Write Yes or No.

● 30 mm = 3 cm Yes

a 5 metres = 50 cm ______

b 190 mm = 19 cm ______

c 700 cm = 7 m ______

d 300 cm = 3 m ______

e 200 mm = 2 cm ______

f 18 m = 1800 cm ______

g 5 mm = 50 cm ______

h 40 cm = 4 m ______

i 60 mm = 6 cm ______

4 Complete these unit conversions.

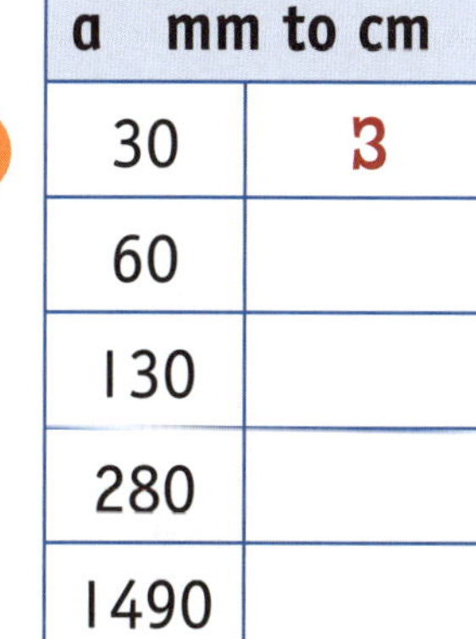

a mm to cm	
30	3
60	
130	
280	
1490	

b cm to mm	
5	50
14	
67	
133	
552	

c cm to m	
50	0.5
200	
550	
600	
1400	

d m to cm	
8	800
24	
56	
740	
153	

KILOMETRES

Long distances are measured in kilometres (km).
One kilometre is equal to 1000 metres.

The length of this road would be measured in kilometres.

Examples:
Tick the items you would measure in kilometres.

a ☑ The distance to the next suburb

b ☐ The length of a hair ribbon

c ☐ The distance to Canada from Australia

d ☐ The distance you would walk in one day

e ☐ The distance around your house

f ☐ The distance from Sydney to Perth

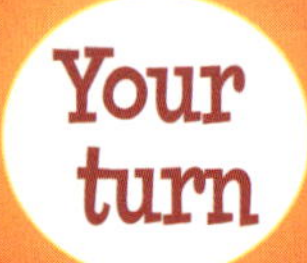

1 Write in metres.

- ● 5 km 5000 m
- a 7 km ________
- b 12 km ________
- c 14 km ________
- d 26 km ________
- e 315 km ________

2 Write in kilometres.

- ● 6000 m 6 km
- a 9000 m ________
- b 16 000 m ________
- c 34 000 m ________
- d 25 000 m ________
- e 537 000 m ________

SELF CHECK Tick how you feel

Got it!	Need help...	I don't get it
☐	☐	☐

Check your answers
How many did you get correct? ☐

CATCH UP MATHS YEAR 5 BOOK B © PASCAL PRESS ISBN: 9781925726176

PRACTICE

1 Name five things that would be measured using kilometres.

a river

2 Write these lengths in metres.

- 2 km 354 m = 2354 metres
- **a** 5 km 25 m = ______ metres
- **b** 7 km 47 m = ______ metres
- **c** 1 km 9 m = ______ metres
- **d** 4 km 653 m = ______ metres
- **e** 6 km 814 m = ______ metres
- **f** 24 km 102 m = ______ metres
- **g** 96 km 12 m = ______ metres
- **h** 10 km 146 m = ______ metres
- **i** 38 km 31 m = ______ metres

3 Write these lengths in kilometres and metres.

- 36 485 m = 36 km 485 m
- **a** 2829 m = ______ km ______ m
- **b** 5945 m = ______ km ______ m
- **c** 6003 m = ______ km ______ m
- **d** 9012 m = ______ km ______ m
- **e** 14 388 m = ______ km ______ m
- **f** 21 510 m = ______ km ______ m
- **g** 35 007 m = ______ km ______ m
- **h** 46 030 m = ______ km ______ m
- **i** 71 010 m = ______ km ______ m
- **j** 74 136 m = ______ km ______ m
- **k** 847 528 m = ______ km ______ m

4 Complete the table.

	cm	m	km
●	400 cm	4 m	0.004 km
a		6.25 m	
b			2 km
c		3 m	
d	914 cm		
e		53 740 m	

PERIMETER

Perimeter (P) is the distance around a 2D shape.

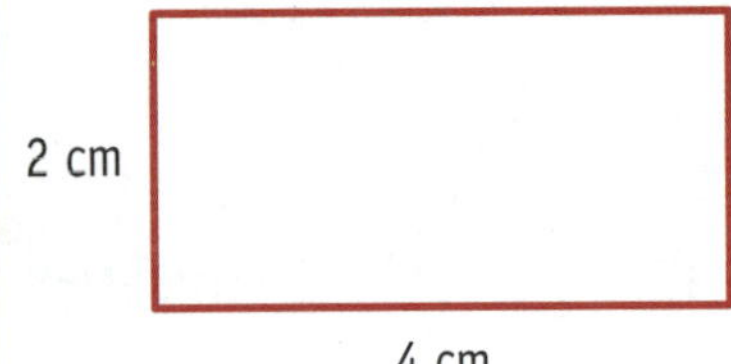

Perimeter = 2 cm + 4 cm + 2 cm + 4 cm

P = 12 cm

The perimeter (P) of the rectangle is 12 cm.

2 cm

Perimeter = 2 cm + 2 cm + 2 cm + 2 cm

P = 8 cm

The perimeter (P) of the square is 8 cm.

Examples: Find the perimeter.

a (3 m, 1 m)

P = 1 m + 3 m + 1 m + 3 m

= __8__ m

Check the answers on the video!

b

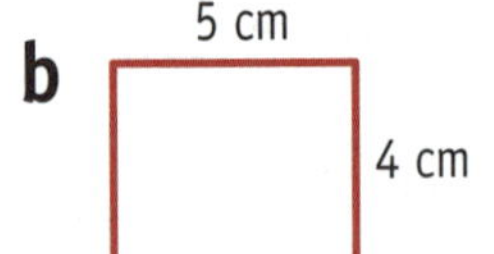

P = 5 cm + 4 cm + 5 cm + 4 cm

= ___ cm

Calculate the perimeter.

P = __1__ + __1__ + __1__ + __1__

= __4__ cm

a (5 cm, 1 cm)

P = ___ + ___ + ___ + ___

= ___ cm

b (2 m, 1 m, 3 m, 4 m)

P = ___ + ___ + ___ + ___

= ___ m

Check your answers

How many did you get correct?

CATCH UP MATHS YEAR 5 BOOK B © PASCAL PRESS ISBN: 9781925726176

PRACTICE

Calculate the perimeter of these 2D shapes.

6 m

P = 6 m + 6 m + 6 m + 6 m

= 24 m

a 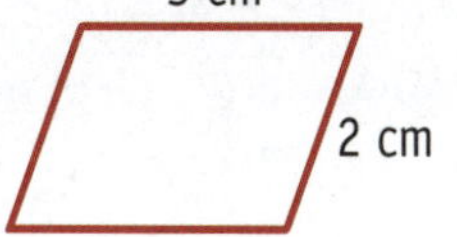

P = ______

= ______

b 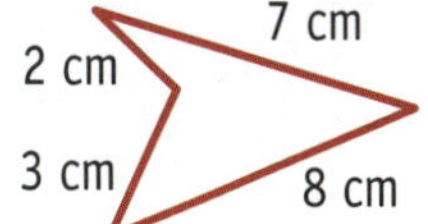

P = ______

= ______

c 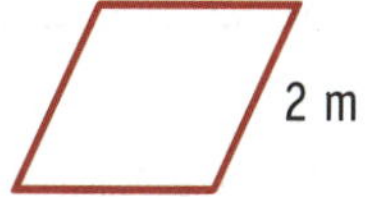

P = ______

= ______

d 

P = ______

= ______

e 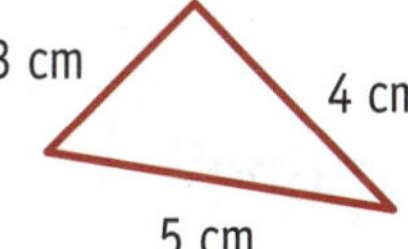

P = ______

= ______

f 

P = ______

= ______

2 Draw three different shapes with a perimeter of 24 cm.

  ISBN: 9781925726176

Use this diagram to answer the following questions.

Flower Centre Scale: 1 cm = 1 m

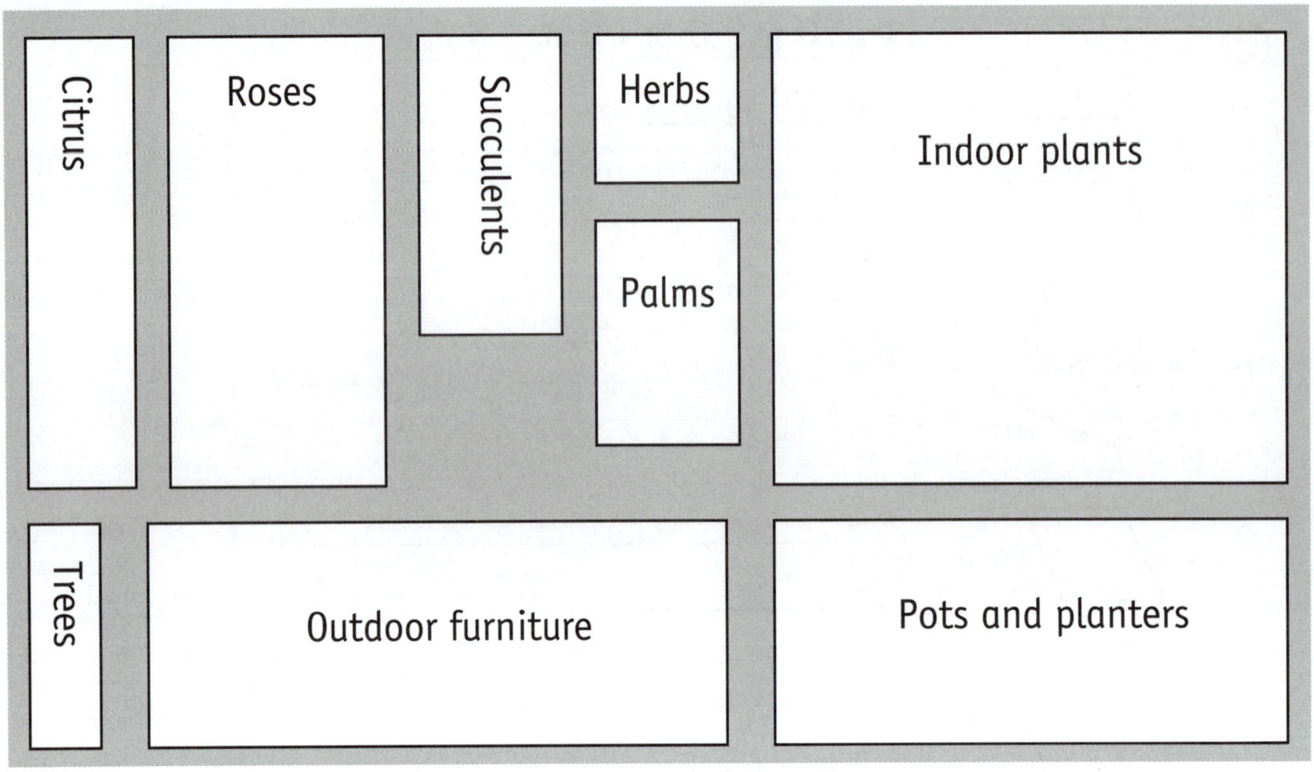

Citrus
Roses
Succulents
Herbs
Palms
Indoor plants
Trees
Outdoor furniture
Pots and planters

3 Peter wants to fence some of the enclosures.
How many metres of fencing will these enclosures require?

- Palms 10 m
- a Indoor plants ______
- b Pots and planters ______
- c Roses ______

4 Did Peter order enough fencing for these enclosures? Write Yes or No.

- Citrus: 15 m Yes
- a Succulents: 10 m ______
- b Herbs: 8 m ______
- c Outdoor furniture: 22 m ______
- d Trees: 6 m ______

5 Order the perimeters from smallest (1) to largest (8).

- [ ] Citrus
- [ ] Succulents
- [1] Herbs
- [ ] Outdoor Furniture
- [1] Trees
- [ ] Palms
- [ ] Indoor plants
- [ ] Pots and planters
- [ ] Roses

CATCH UP MATHS YEAR 5 BOOK B © PASCAL PRESS ISBN: 9781925726176

6 Complete the tables for rectangles with these dimensions.

	Height	Width	Perimeter
	22 cm	13 cm	70 cm
a		5 m	18 m
b	7 mm	8 mm	
c	8 m		32 m

	Height	Width	Perimeter
d	3 cm	6 cm	
e	2 m	2 m	
f	32 m		100 m
g		2 mm	38 mm

7 Work out the perimeter.

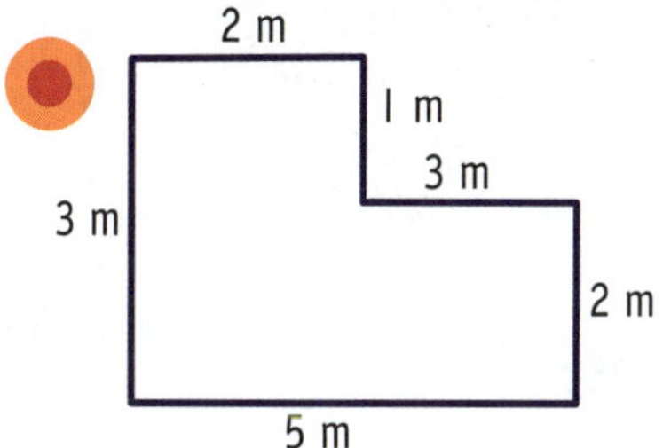

P = 2 + 1 + 3 + 2 + 5 + 3 m

= 16 m

a

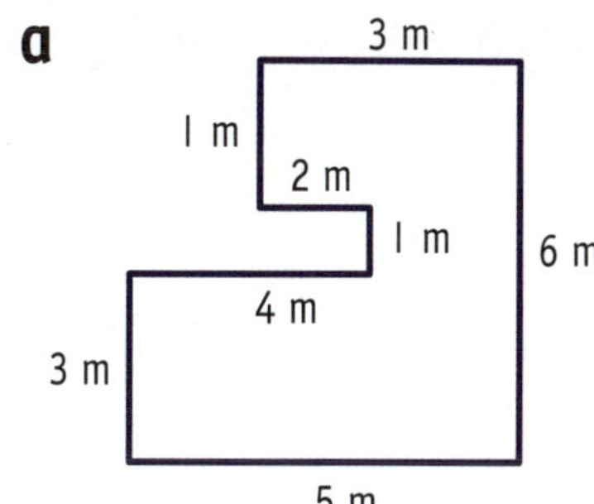

P = ____________________

= __________

b

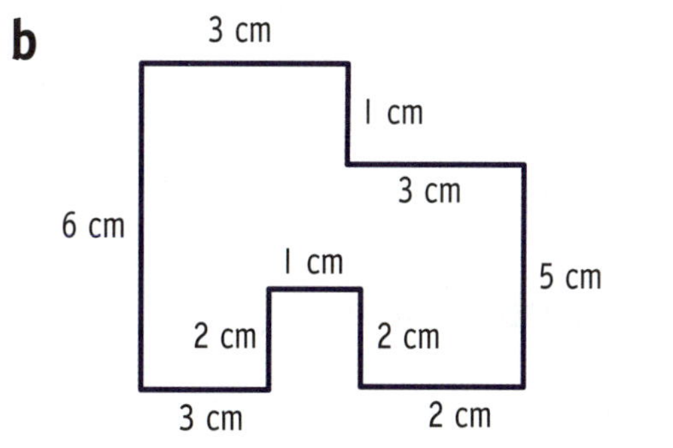

P = ____________________

= __________

c

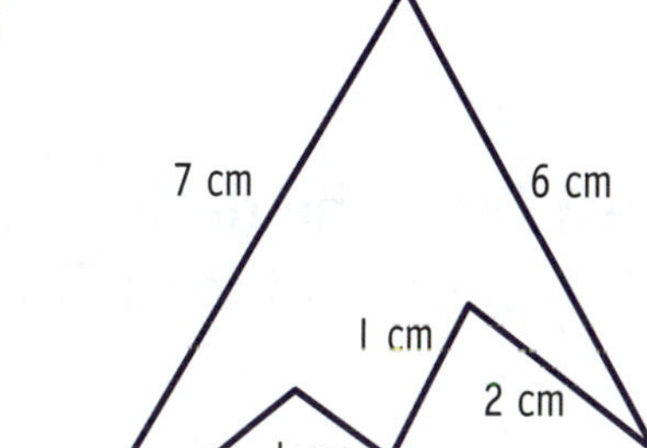

P = ____________________

= __________

d

5 cm
1 cm
4 cm
3 cm
2 cm
2 cm
4 cm
4 cm
1 cm
2 cm
6 cm
6 cm

P = ____________________

= __________

DECIMAL NOTATION

There are 100 centimetres in 1 metre.
Centimetres are hundredths of a metre.

6 metres and 54 centimetres = 6.54 m

5 whole metres

524 cm = 5.24 m

Whole numbers (metres) — Tenths (of a metre) — Hundredths (of a metre)

The decimal point separates the whole number from the fractional part.

Examples: Draw the decimal point in the correct position for each measurement in metres.

a 487 cm = 4.87 m

b 203 cm = 203 m

c 100 cm = 100 m

d 815 cm = 815 m

e 320 cm = 320 m

f 599 cm = 599 m

Your turn

Complete using decimal notation.

● 606 cm = 6.06 m

a 634 cm = ______ m

b 102 cm = ______ m

c 400 cm = ______ m

d 520 cm = ______ m

e 135 cm = ______ m

f 118 cm = ______ m

g 485 cm = ______ m

h 94 cm = ______ m

i 14 253 cm = ________ m

j 27 595 cm = ________ m

k 43 682 cm = ________ m

SELF CHECK Tick how you feel

Got it!	Need help...	I don't get it
 ☐	 ☐	 ☐

Check your answers
How many did you get correct? 

CATCH UP MATHS YEAR 5 BOOK B © PASCAL PRESS ISBN: 9781925726176

PRACTICE

1 Write these lengths in metres using decimal form.

- 4 metres 32 centimetres = 4.32 m

a 1 metre 58 centimetres = ______

b 6 metres 92 centimetres = ______

c 1 metre 9 centimetres = ______

d 9 m 20 cm = ______

e 3 m 1 cm = ______

f 30 m 30 cm = ______

g 16 m 41 cm = ______

h 39 m 87 cm = ______

i 62 m 3 cm = ______

j 43 cm = ______

k 64 cm = ______

2 Rewrite in metres (m) and centimetres (cm).

- 2.07 m = 2 m 7 cm

a 0.16 m = ____ m ____ cm

b 3.79 m = ____ m ____ cm

c 5.20 m = ____ m ____ cm

d 4.05 m = ____ m ____ cm

e 8.17 m = ____ m ____ cm

f 13.06 m = ____ m ____ cm

g 52.8 m = ____ m ____ cm

h 0.08 m = ____ m ____ cm

i 0.02 m = ____ m ____ cm

3 Complete the table.

	Centimetres	Metres and Centimetres	Decimal
●	605 cm	6 m 5 cm	6.05 m
a	216 cm		
b		8 m 10 cm	
c			36.83 m
d	15 453 cm		
e		247 m 38 cm	

TEMPERATURE

SCAN to watch video

Temperature is the measurement of how hot or cold something is. You use a thermometer to measure temperature. The units for measuring temperature are degrees Celsius (°C).

The temperature on this thermometer is 25 °C.

The higher the level, the hotter it is.

Examples:

Write the letters **a** to **h** in the correct places on the thermometer scale.

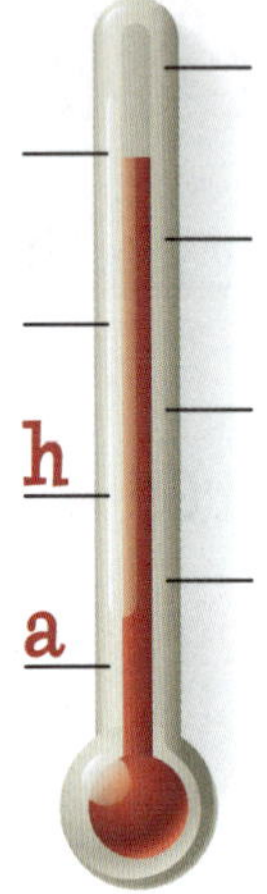

a freezing
b cool
c warm
d hot
e very cold
f boiling
g very hot
h cold

Write the temperatures in order from coldest to hottest.

● 35 °C, 20 °C, 10 °C, 75 °C, 49 °C

10 °C, 20 °C, 35 °C, 49 °C, 75 °C

a 23 °C, 82 °C, 53 °C, 21 °C, 13 °C

____

b 401 °C, 910 °C, 300 °C, 650 °C, 330 °C

____

c 9 °C, 0 °C, 102 °C, 90 °C, 37 °C

____

Check your answers
How many did you get correct?

CATCH UP MATHS YEAR 5 BOOK B © PASCAL PRESS ISBN: 9781925726176

PRACTICE

1 Write the temperatures in numerals with °C.

- sixty-four degrees Celsius 64 °C

a forty-two degrees Celsius _______

b ninety-nine degrees Celsius _______

c sixteen degrees Celsius _______

d one hundred and seventy-three degrees Celsius _______

2 Write these temperatures in words.

- 29 °C twenty-nine degrees Celsius

a 15 °C ______________________________

b 98 °C ______________________________

c 61 °C ______________________________

d 404 °C ______________________________

3 Write the temperature shown on the thermometer.

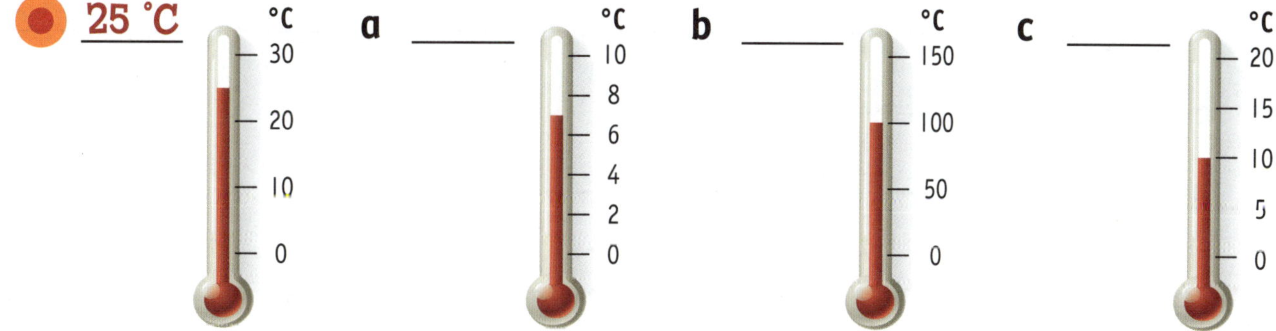

4 Colour the thermometer to show the temperature.

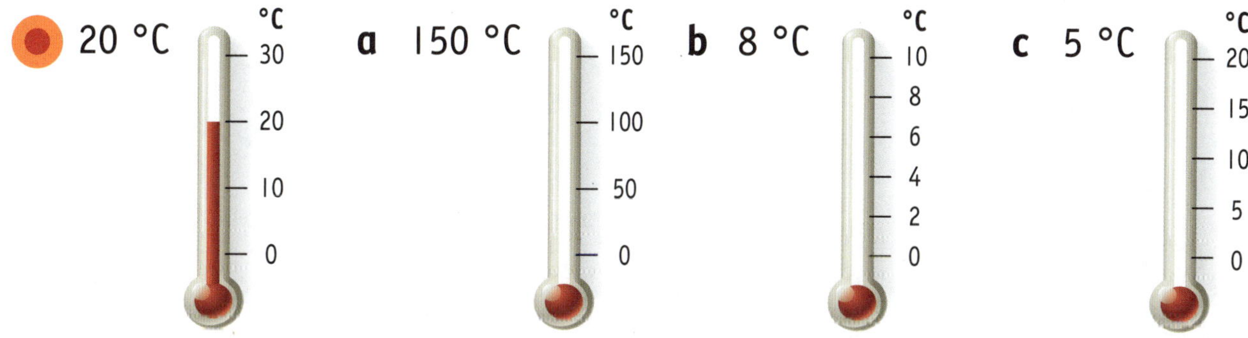

LENGTH REVIEW

1 Complete.

a 1 metre = ____ cm b $\frac{1}{2}$ metre = ____ cm c $\frac{1}{4}$ metre = ____ cm

2 Write these lengths in centimetres.

a 2 m 53 cm = ______ cm

b 8 m 7 cm = ______ cm

c 6 m 16 cm = ______ cm

d 8 m 8 cm = ______ cm

e 15 m 16 cm = ______ cm

f 24 m 85 cm = ______ cm

3 Write in metres and centimetres.

a 357 cm = __ m ____ cm

b 805 cm = __ m ____ cm

c 670 cm = __ m ____ cm

d 479 cm = __ m ____ cm

e 38 cm = __ m ____ cm

f 115 cm = __ m ____ cm

4 Measure and write the lengths in centimetres.

a ____ cm ________________________________

b ____ cm ____

c ____ cm ______

d ____ cm __________________________

5 Write the lengths in Question 4 in order from shortest to longest.

____

6 Draw lines of these lengths.

a 12 cm

b 9 cm

c 1 cm

  ISBN: 9781925726176

d 4 cm

e 7 cm

f 13 cm

7 **Would these things be measured in mm or cm? Write the units to complete the measurements.**

a pencil length: 10 ___

b pin length: 4 ___

c computer keyboard key width: 9 ___

d mobile phone length: 15 ___

e thickness of a fingernail: 1 ___

f length of an ant: 2 ___

8 **Write the lengths marked on the ruler.**

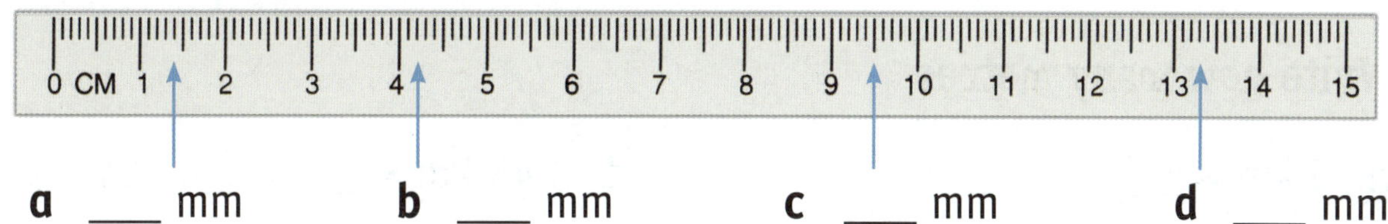

a ___ mm b ___ mm c ___ mm d ___ mm

Mark these measurements on the ruler.

a 8 mm b 26 mm c 58 mm d 142 mm

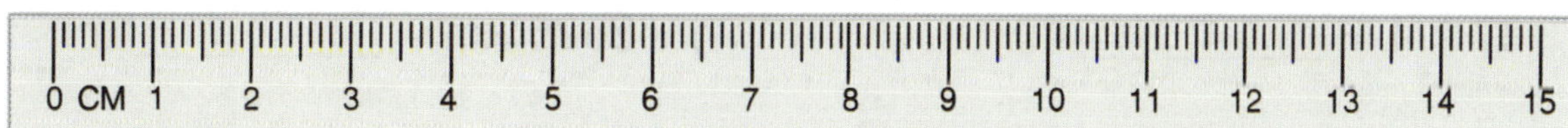

10 **Circle the answer that is the best estimate.**

a	length of a glue stick	11 mm	11 cm	11 m
b	width of a road	6 m	6 mm	6 cm
c	length of a train	100 m	100 cm	100 mm
d	length of a pin head	1 cm	1 mm	1 m
e	length of a lolly packet	25 mm	25 m	25 cm

REVIEW

11 Convert these measurements.

a 70 mm = ______ cm

b 4 cm = ______ mm

c 20 m = ______ cm

d 100 cm = ______ m

e 6 m = ______ cm

f 136 mm = ______ cm

g $\frac{1}{4}$ m = ______ cm

h 1500 cm = ______ m

i 56 cm = ______ mm

j 290 mm = ______ cm

k 140 mm = ______ cm

l 172 m = ______ cm

m 800 cm = ______ m

n 1570 mm = ______ cm

o 5935 mm = ______ cm

p 1250 cm = ______ m

q 670 cm = ______ m

r 690 m = ______ cm

12 Write how many metres.

a 6 km = __________ m

b 8 km = __________ m

c 17 km = __________ m

d 24 km = __________ m

e 37 km = __________ m

f 243 km = __________ m

g 146 km = __________ m

h 549 km = __________ m

i 6847 km = __________ m

j 8625 km = __________ m

13 Write how many kilometres.

a 7000 m = ________ km

b 11 000 m = ________ km

c 2000 m = ________ km

d 13 000 m = ________ km

e 43 000 m = ________ km

f 612 000 m = ________ km

g 33 000 m = ________ km

h 81 000 m = ________ km

i 549 000 m = ________ km

j 642 000 m = ________ km

  ISBN: 9781925726176

14 Complete the table.

	cm	m	km
a	500 cm		
b		2.25 m	
c		5000 m	
d			0.0627 km
e		33 520 m	

15 Find the perimeters.

a

3 m

P = ______________________

= __________

b

P = ______________________

= __________

c

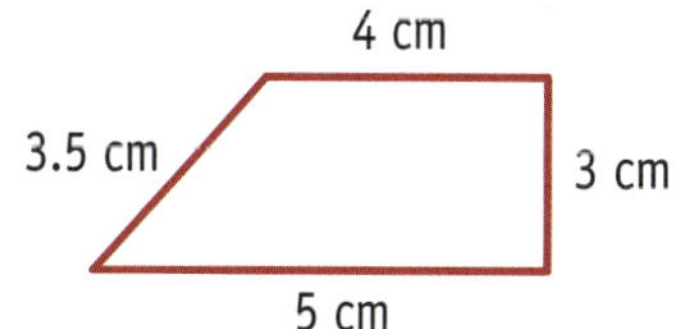

P = ______________________

= __________

d

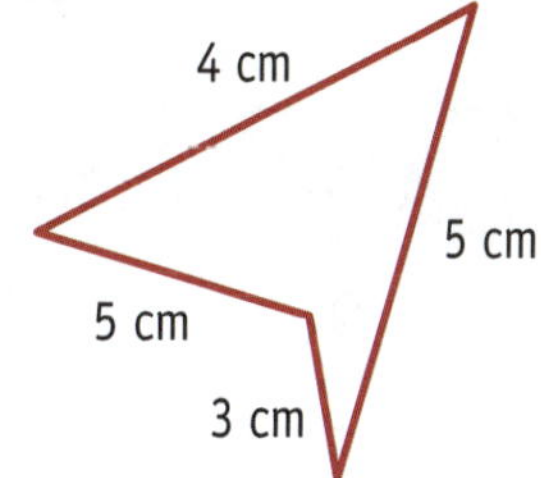

P = ______________________

= __________

e

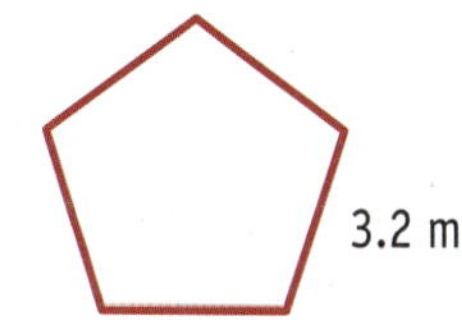

P = ______________________

= __________

f

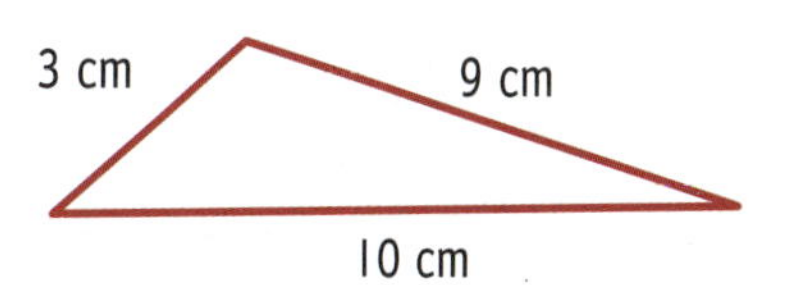

P = ______________________

= __________

  ISBN: 9781925726176

REVIEW

16 Draw three different shapes with a perimeter of 18 cm.

17 Complete the tables for rectangles with these dimensions.

	Height	Width	Perimeter
a		9	40
b	8	6	
c		14	60
d		8	36
e	9		44

	Height	Width	Perimeter
f	66		156
g	7	28	
h		15	74
i	9		38
j	16	35	

18 Work out the perimeter.

a

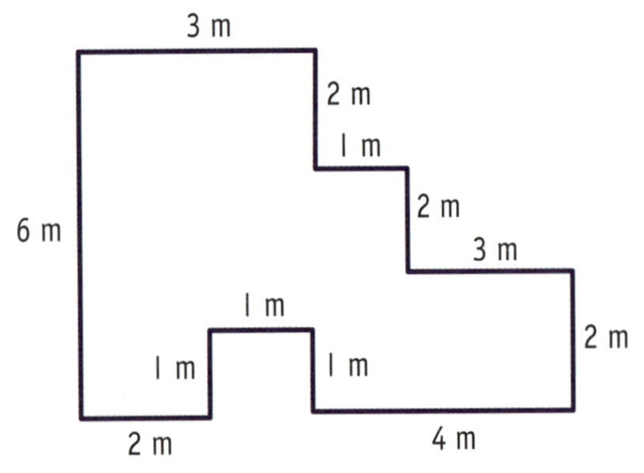

P = ______________________

= __________

b

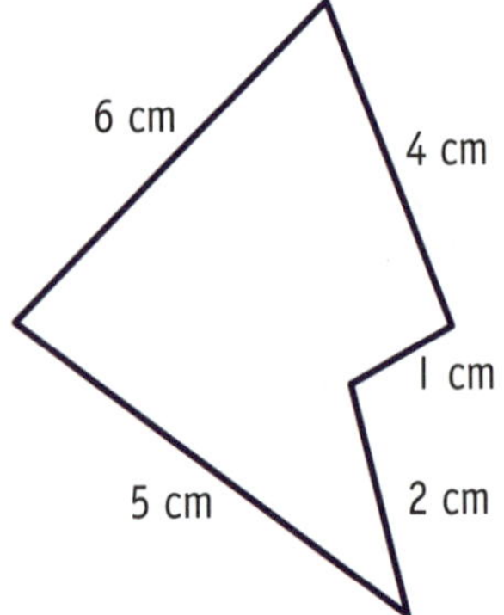

P = ______________________

= __________

CATCH UP MATHS YEAR 5 BOOK B © PASCAL PRESS ISBN: 9781925726176

19 Draw the decimal point in the correct position for each measurement in metres.

a 749 cm = 749 m

b 307 cm = 307 m

c 200 cm = 200 m

d 995 cm = 995 m

e 382 cm = 382 m

f 58 cm = 58 m

g 11 cm = 11 m

h 6 cm = 6 m

i 470 cm = 470 m

j 41 532 cm = 41532 m

k 55 595 cm = 55595 m

l 24 930 cm = 24930 m

20 Complete using decimal notation.

a 709 cm = ______ m

b 493 cm = ______ m

c 207 cm = ______ m

d 893 cm = ______ m

e 500 cm = ______ m

f 660 cm = ______ m

g 83 cm = ______ m

h 72 cm = ______ m

i 3 cm = ______ m

j 8 cm = ______ m

k 23 265 cm = __________ m

l 83 451 cm = __________ m

21 Write the lengths as metres (m) and centimetres (cm).

a 7.03 m = ___ m _____ cm

b 8.30 m = ___ m _____ cm

c 2.19 m = ___ m _____ cm

d 1.00 m = ___ m _____ cm

e 0.08 m = ___ m _____ cm

f 60.36 m = ____ m _____ cm

g 0.53 m = ___ m _____ cm

h 153.45 m = _____ m _____ cm

i 1.04 m = ___ m _____ cm

j 245.63 m = ______ m _____ cm

k 346.20 m = ______ m _____ cm

l 657.06 m = ______ m _____ cm

m 908.42 m = ______ m _____ cm

n 421.87 m = ______ m _____ cm

22 Rewrite using decimal form.

a 8 m 33 cm = _______ m

b 4 m 56 cm = _______ m

c 7 m 57 cm = _______ m

d 20 centimetres = _______ m

e 59 centimetres = _______ m

f 6 centimetres = _______ m

23 Rewrite using decimal form.

a 1 metre 85 centimetres = _______ m

b 17 metres 62 centimetres = _______ m

c 50 metres 15 centimetres = _______ m

24 Order these temperatures from warmest to coldest.

a 21 °C, 3 °C, 34 °C, 10 °C, 8 °C

____

b 43 °C, 5 °C, 24 °C, 13 °C, 32 °C

____

c 0 °C, 7 °C, 14 °C, 2 °C, 19 °C

____

d 18 °C, 11 °C, 22 °C, 17 °C, 4 °C

____

e 41 °C, 15 °C, 63 °C, 45 °C, 10 °C

____

25 Write the temperatures in numerals and °C.

a nineteen degrees Celsius _______

b twenty-six degrees Celsius _______

c seven hundred and forty-eight degrees Celsius _______

CATCH UP MATHS YEAR 5 BOOK B © PASCAL PRESS ISBN: 9781925726176

26 Write the temperatures in words.

a 6 °C ______________________

b 83 °C ______________________

c 410 °C ______________________

27 Write the temperature shown on each thermometer.

a ____________

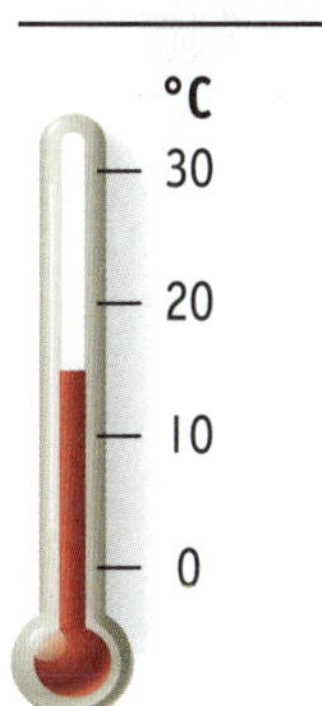

b ____________

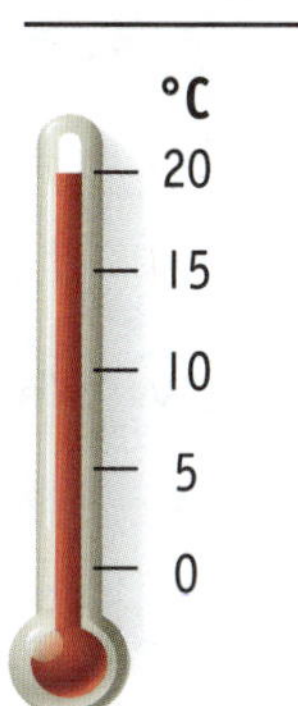

c ____________

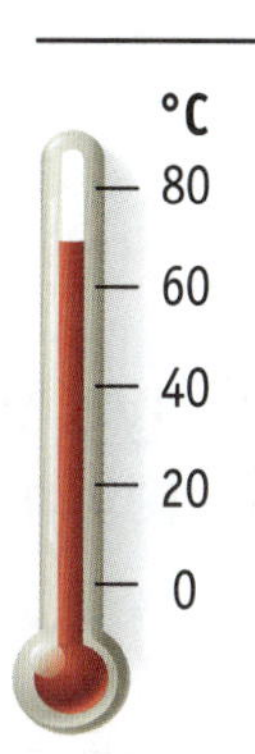

d ____________

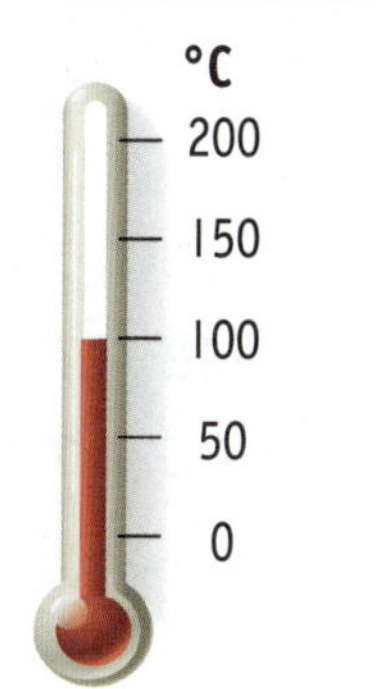

28 Colour each thermometer to show the temperature.

a 30 °C

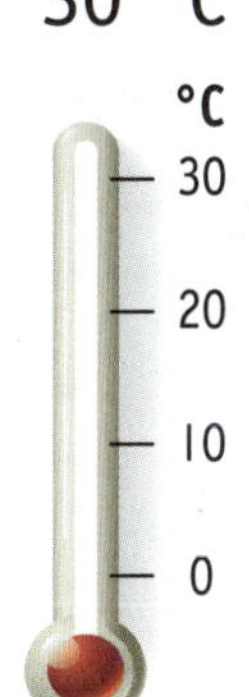

c 30 °C

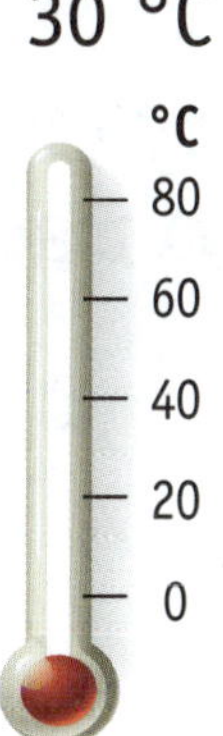

e 25 °C

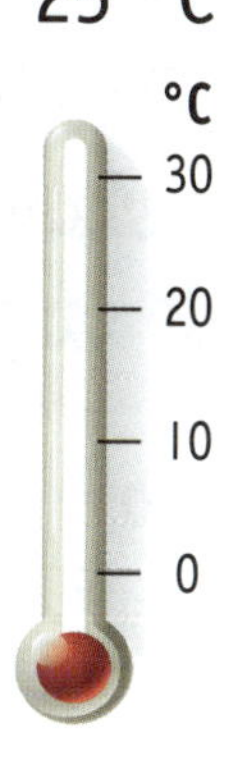

g 100 °C

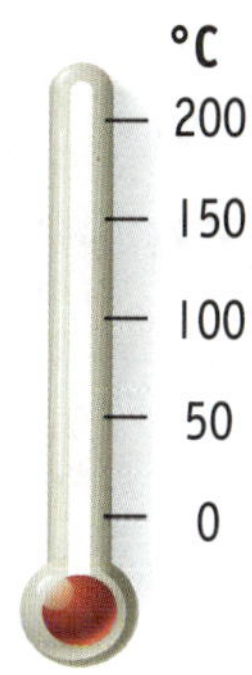

b 10 °C

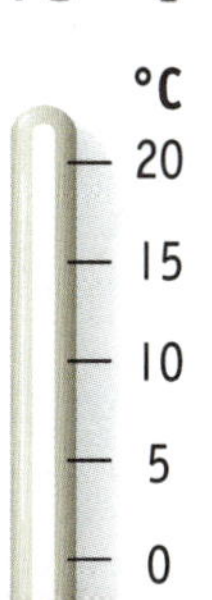

d 0 °C

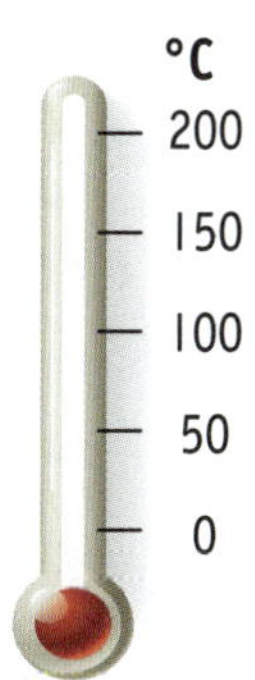

f 10 °C

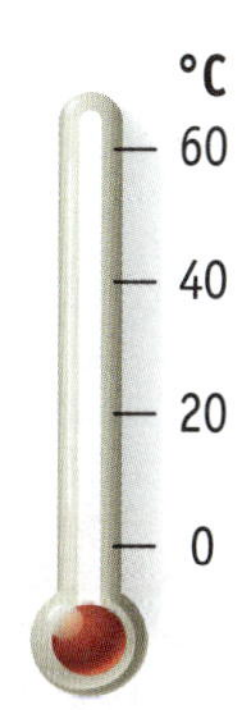

h 20 °C

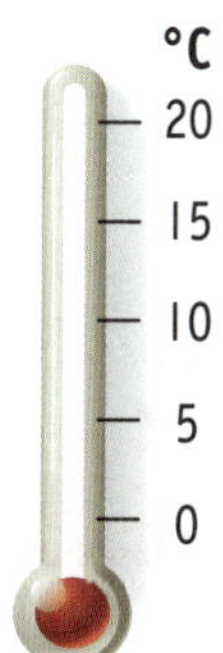

ANGLES

An angle is made by measuring the amount of turn between the two arms. The two arms (rays) meet at a point (vertex).

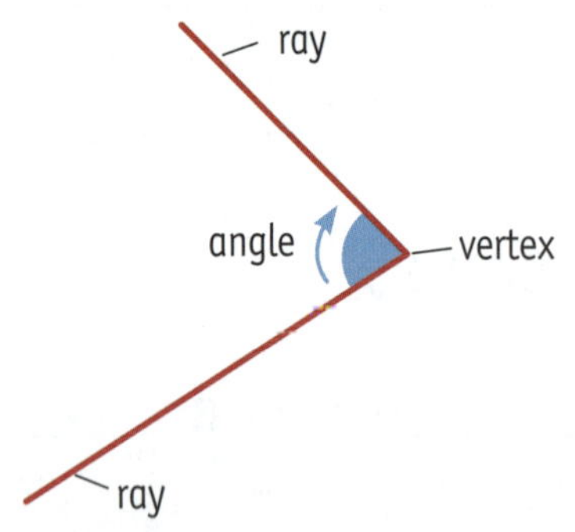

One arm or ray of an angle can be shorter than the other.

Examples:
Use red to trace the rays, use blue to show the angle and draw a green circle on the vertex.

a

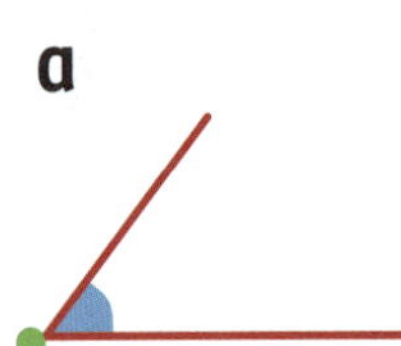

b

c

d

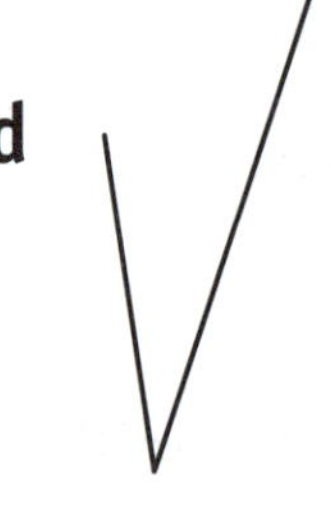

Check the answers on the video!

Make each line into an angle by drawing a ray in red. Use blue to show the angle you made and draw a green circle on the vertex.

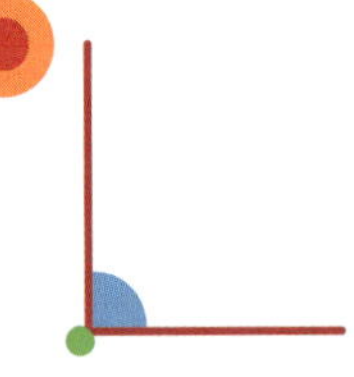

b

d

f

a

c

e

g

SELF CHECK Tick how you feel

Got it!	Need help...	I don't get it

Check your answers
How many did you get correct?

CATCH UP MATHS YEAR 5 BOOK B © PASCAL PRESS ISBN: 9781925726176

PRACTICE

1 Draw four objects in your environment that have angles in them.

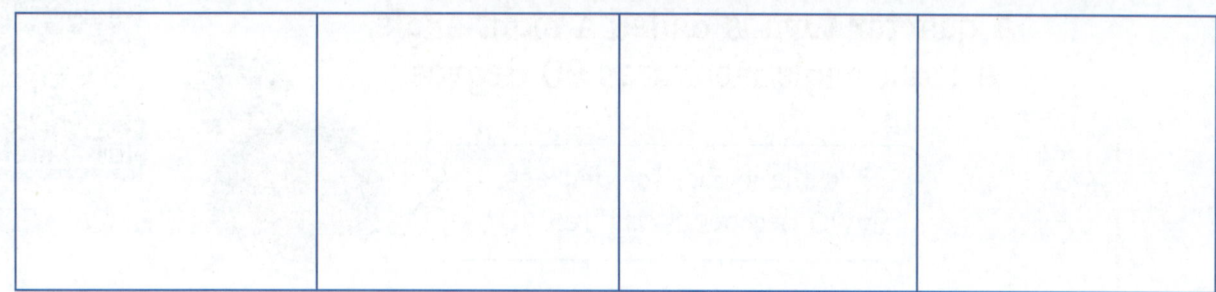

2 Use red to trace the rays, use blue to show the angle and draw a green circle on the vertex.
Then number the angles to order them from smallest (1) to largest (5).

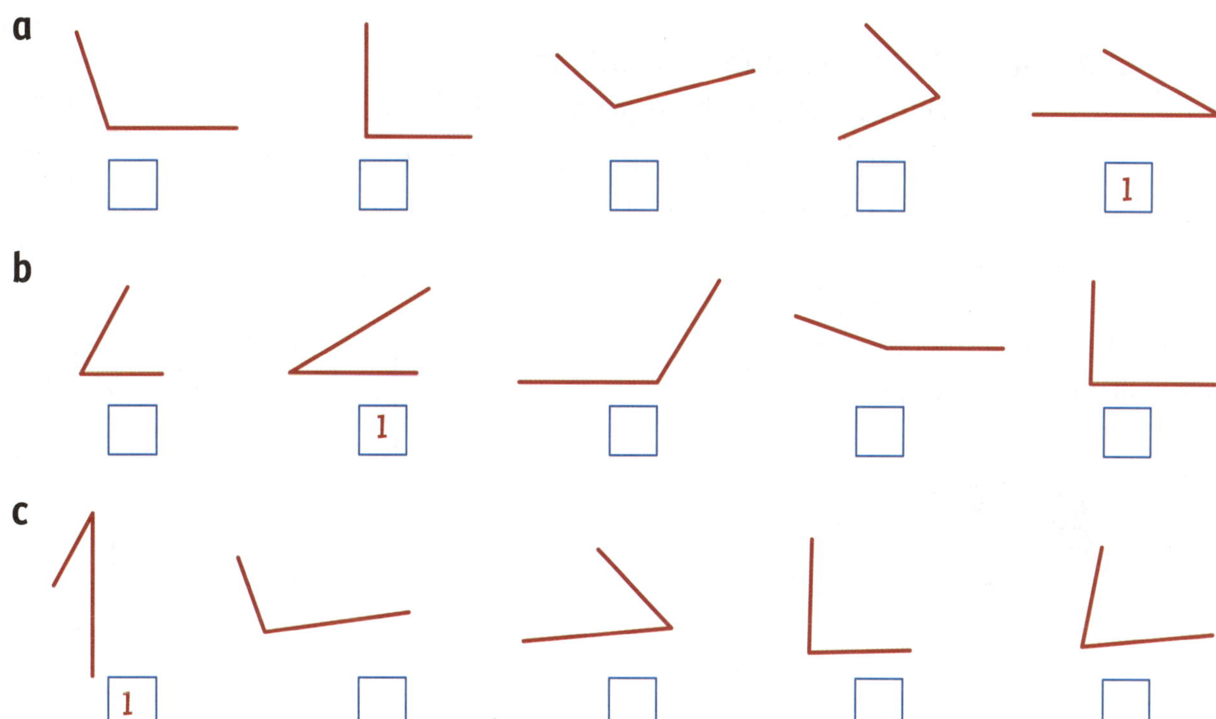

3 Circle the angles in the picture.

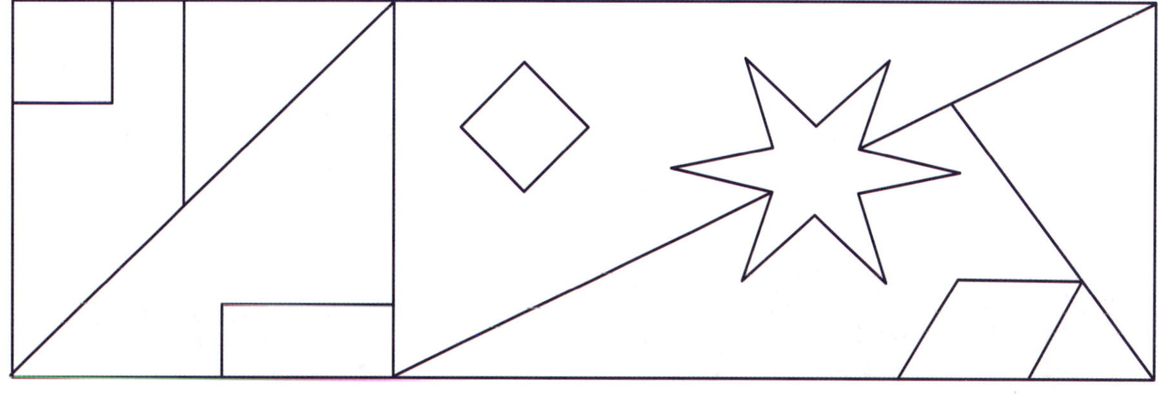

RIGHT ANGLES

A quarter turn is called a right angle.
A right angle measures 90 degrees.

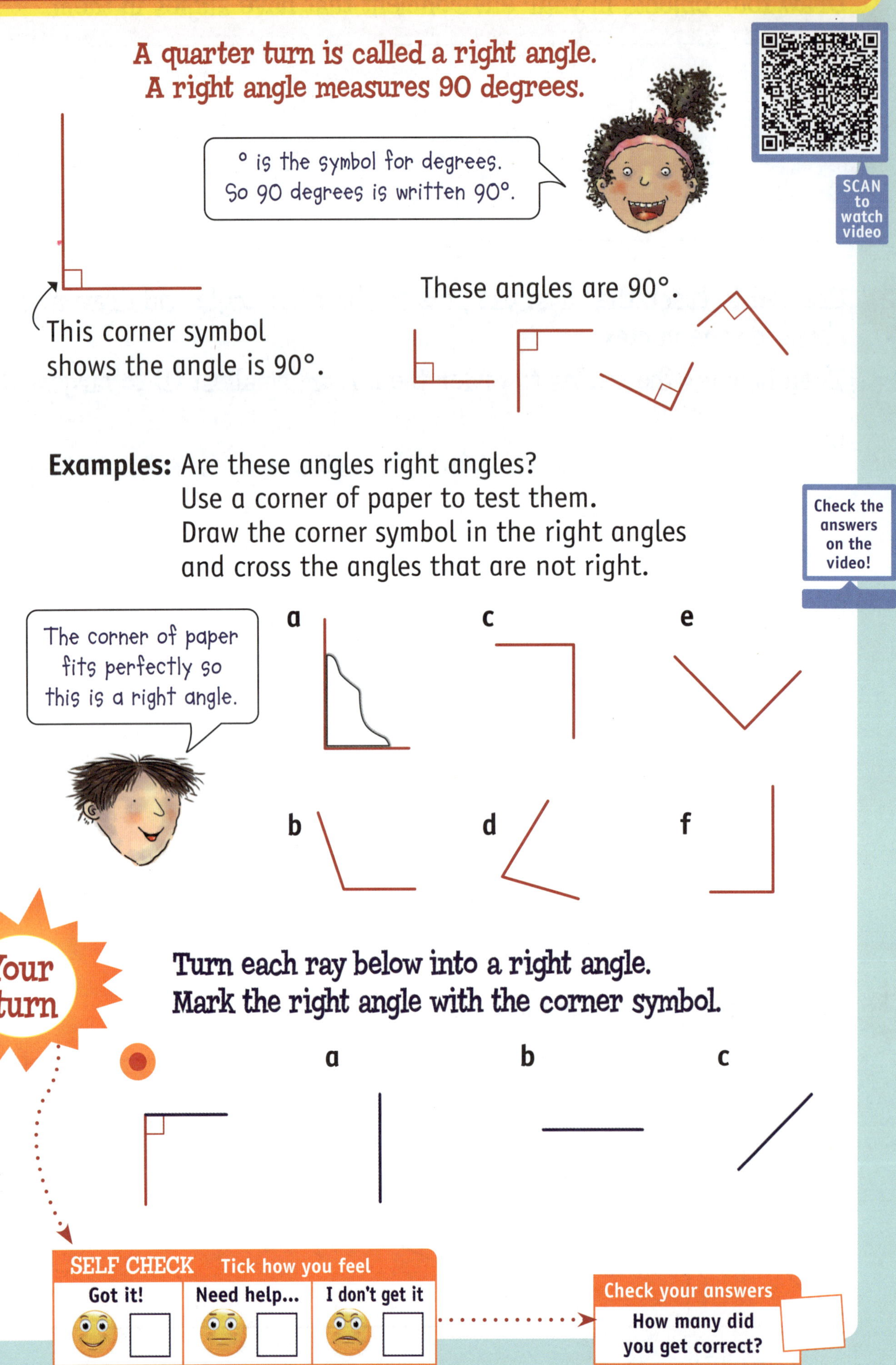

Examples: Are these angles right angles?
Use a corner of paper to test them.
Draw the corner symbol in the right angles and cross the angles that are not right.

Your turn

Turn each ray below into a right angle.
Mark the right angle with the corner symbol.

SELF CHECK Tick how you feel		
Got it! ☐	Need help... ☐	I don't get it ☐

Check your answers
How many did you get correct? ☐

  ISBN: 9781925726176

PRACTICE

1 Draw four angles in each box that match the description.

Smaller than 90°	90°	Larger than 90°

2 List five things in your environment that have right angles.

picture frame

3 Label the angles as equal to (=), less than (<) or greater than (>) a right angle.

>

b 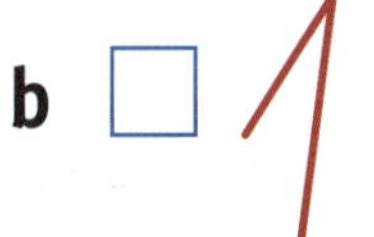

d 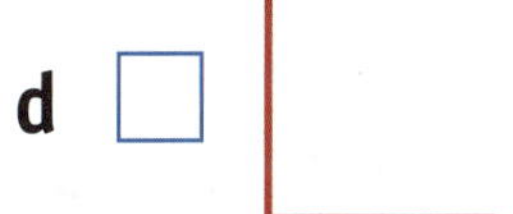

f 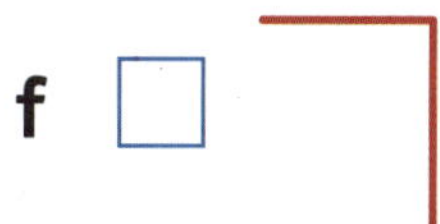

a

c

e 

g

4 Trace over the right angles in each shape.

b 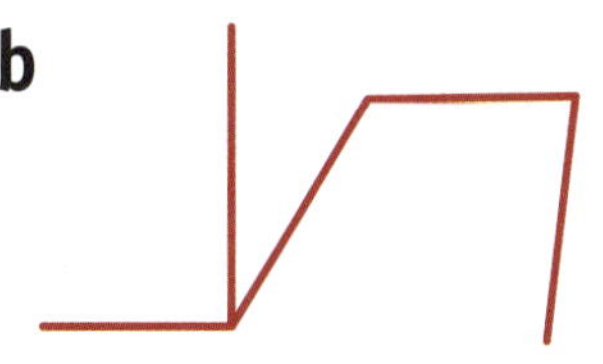

d 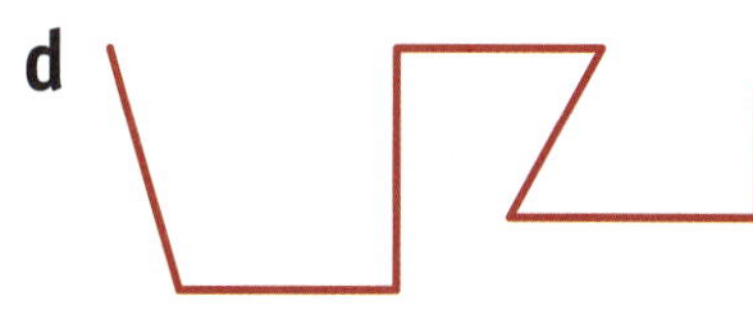

a

c 

e 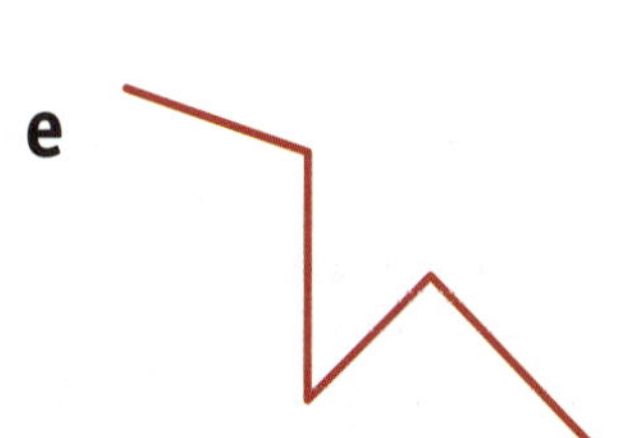

ACUTE AND OBTUSE ANGLES

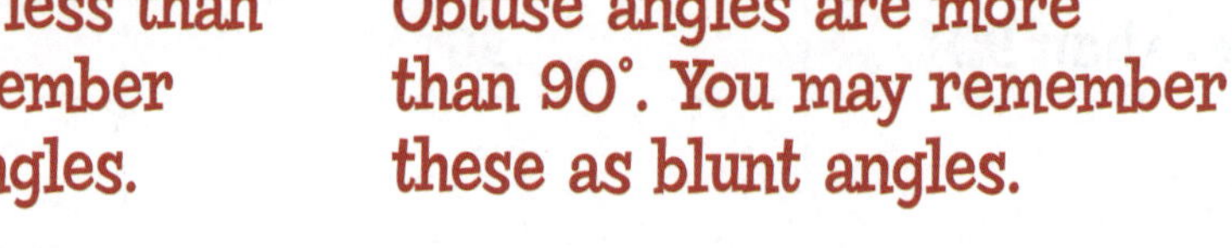

Acute angles are less than 90°. You may remember these as sharp angles.

Obtuse angles are more than 90°. You may remember these as blunt angles.

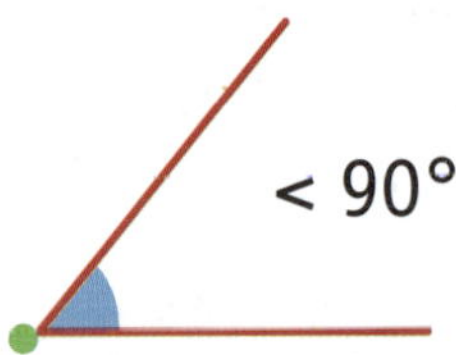

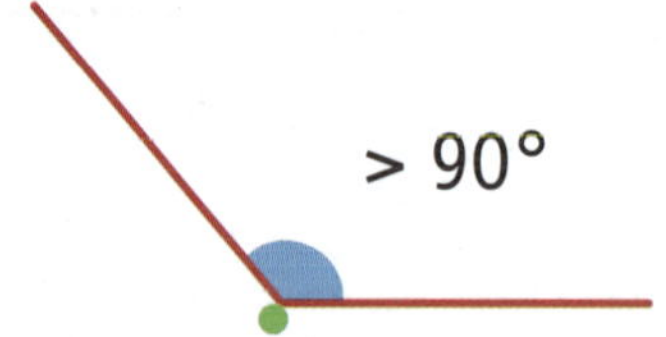

Examples:

Circle the acute angles in orange and the obtuse angles in green.

a 

c 

e 

g 

i 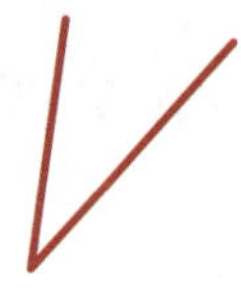

b 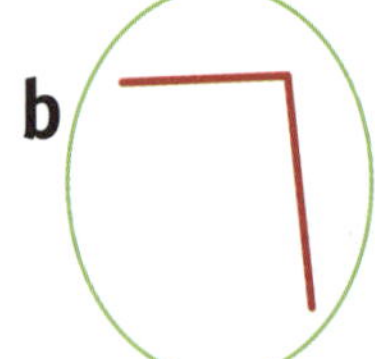

d 

f 

h 

1 Draw another ray to turn each line into an acute angle.

a

b

c 

d 

e 

2 Draw another ray to turn each line into an obtuse angle.

a 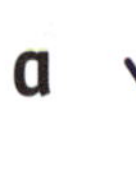

b

c

d 

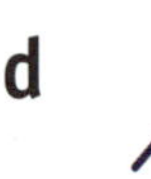

e

SELF CHECK Tick how you feel

Got it!	Need help...	I don't get it

Check your answers

How many did you get correct?

CATCH UP MATHS YEAR 5 BOOK B © PASCAL PRESS ISBN: 9781925726176

PRACTICE

We name angles using three letters.
Put the letter at the vertex in the middle and use ∠ for the angle sign.

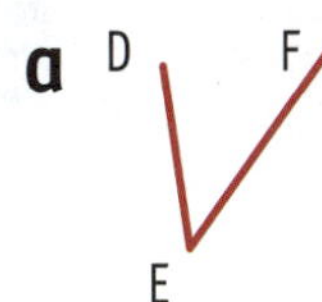

Name: ∠ABC or ∠CBA
Type: Acute

c 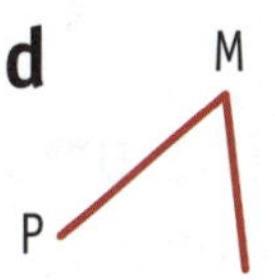

Name: ______________
Type: __________

a 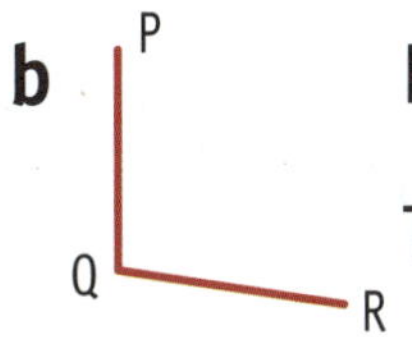

Name: ______________
Type: __________

d 

Name: ______________
Type: __________

b

Name: ______________
Type: __________

e 

Name: ______________
Type: __________

Order these angles from sharpest (1) to bluntest (5).

a

☐ ☐ ☐ 1 ☐

b

 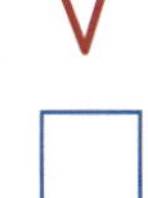     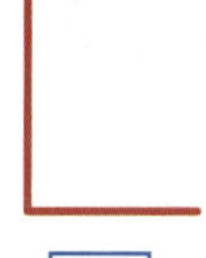

☐ 1 ☐ ☐ ☐

c

☐ ☐ ☐ 1 ☐

Shade the acute angles orange and the obtuse angles green.

a 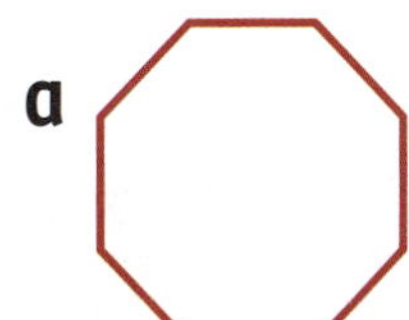

b 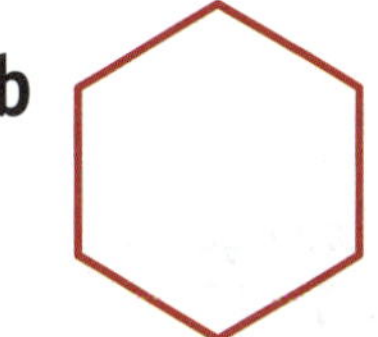

c 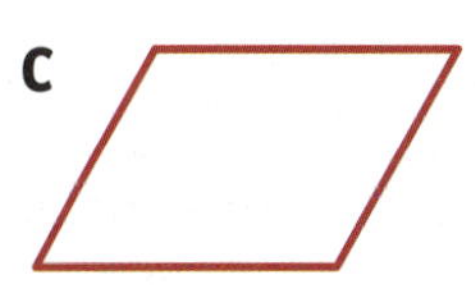

d 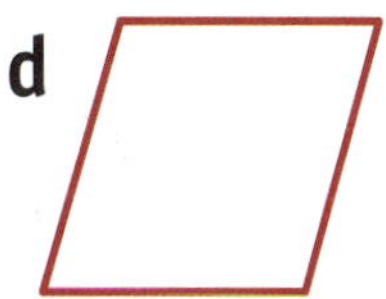

STRAIGHT, REFLEX AND REVOLUTION ANGLES

Straight angles

Always 180°

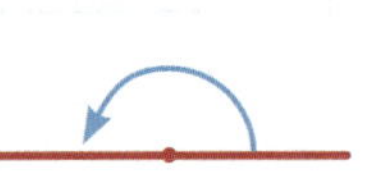

Looks like a straight line.

Reflex angles

Between 180° and 360°

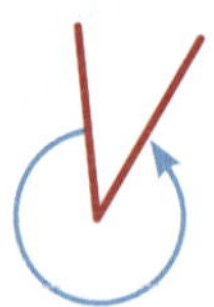

The arrow on the outside shows it is a reflex angle.

Revolution (full turn)

Always 360°

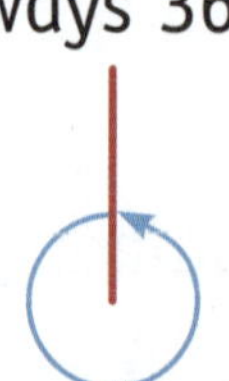

The two arms are together.

Examples: Write Reflex, Straight or Revolution for each angle.

a Revolution

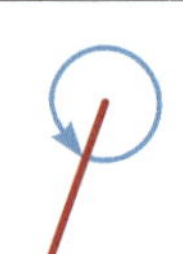

b __________

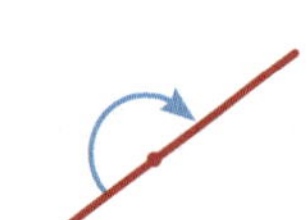

c __________

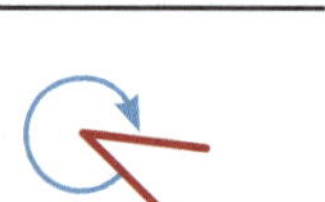

d __________

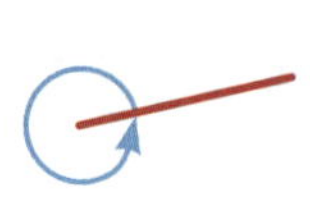

Check the answers on the video!

Your turn

1 Mark in the angle to show straight angles.

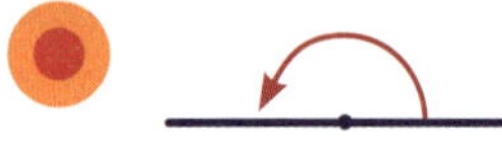

a

b

2 Mark in the angle to show revolutions.

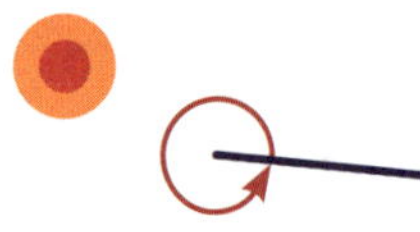

a

b

3 Mark in the angle to show reflex angles.

a

b

SELF CHECK Tick how you feel

Got it!	Need help...	I don't get it

Check your answers

How many did you get correct?

  ISBN: 9781925726176

PRACTICE

1 Use yellow to circle the reflex angles, use black to circle the straight angles and use purple to circle the revolutions.

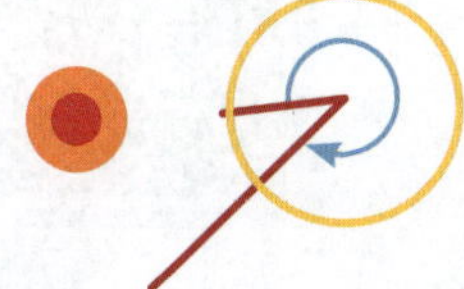

b 

d

f 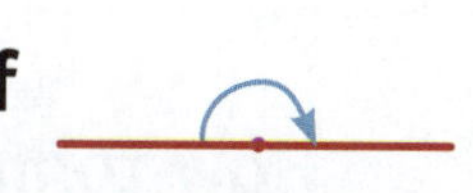

a 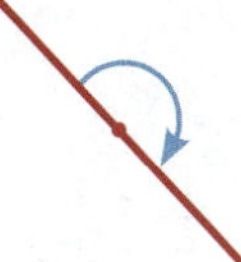

c 

e

g 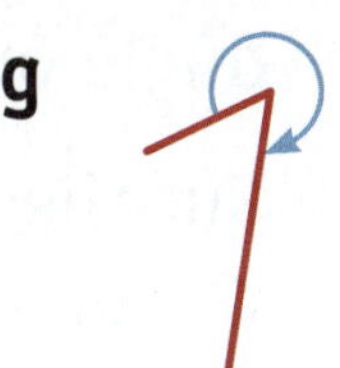

2 Draw five of each type of angle.

Straight Angles	Revolutions	Reflex Angles

3 Write the letter of the angle that matches each type.

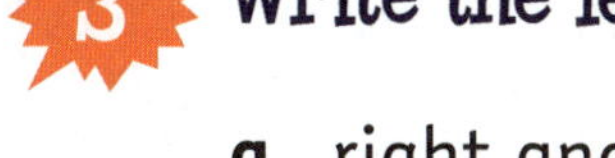

a right angle __

b reflex angle __

c acute angle __

d obtuse angle __

e straight angle __

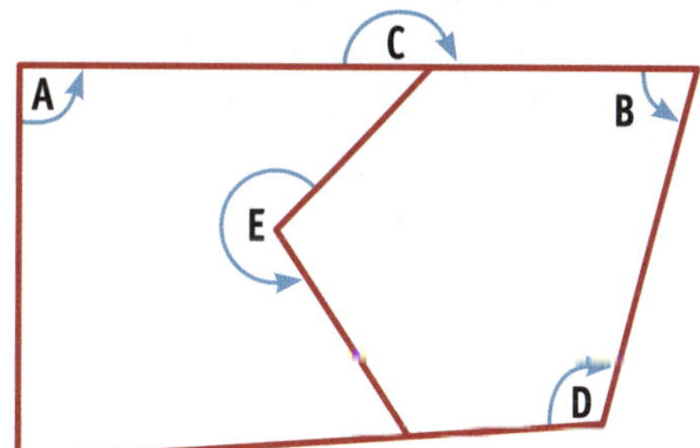

4 Find three more acute angles in the shape in Question 3. Label them F, G and H.

5 Find three more obtuse angles in the shape in Question 3. Label them I, J and K.

USING A PROTRACTOR

A protractor is used to measure or draw angles. It measures the amount of turn.

SCAN to watch video

How to measure an angle

1 Put the lower arm of the angle on the base line.

2 Put the vertex of the angle at the centre point.

3 Find 0° on the lower arm to work out if you use the inside or outside scale.

4 Read the scale.

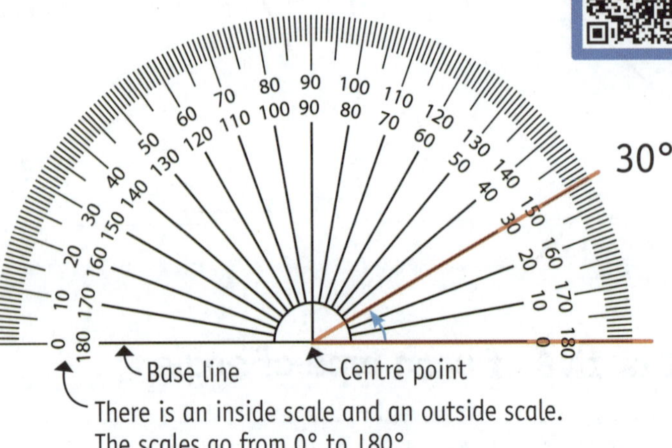

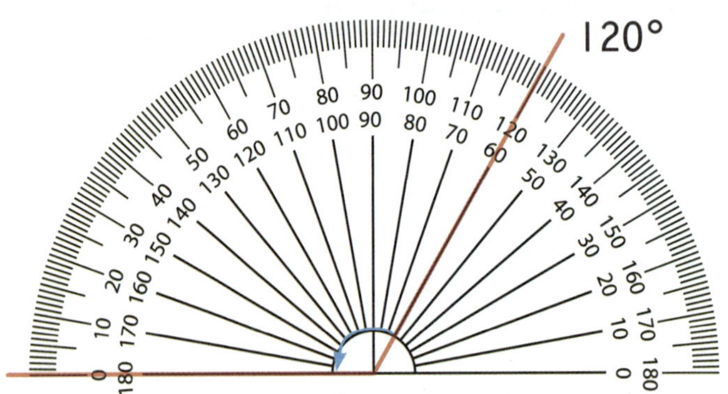

Example:

Label the protractor above with these features.

a Inside numbers

b Base line

c Centre point

d Outside numbers

Check the answers on the video!

Your turn

Should you use the inside numbers (I) or the outside numbers (O) to measure these angles?

● O a ☐ b ☐ c ☐

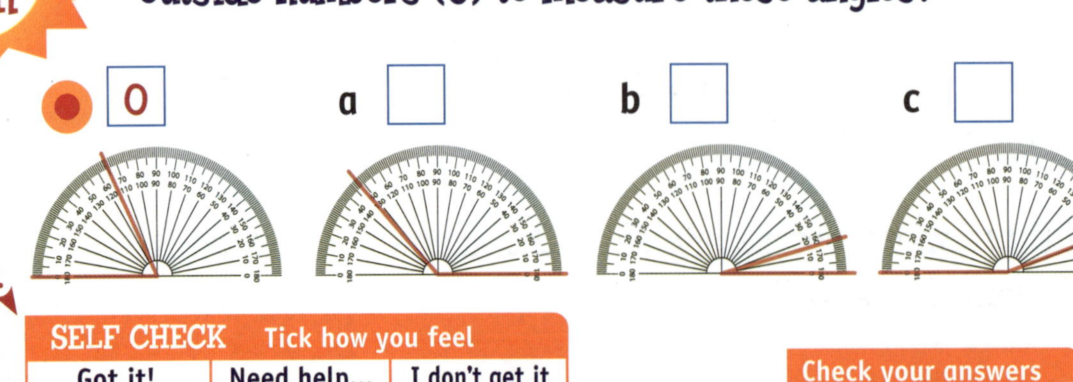

SELF CHECK Tick how you feel

Got it!	Need help...	I don't get it
☐	☐	☐

Check your answers

How many did you get correct? ☐

CATCH UP MATHS YEAR 5 BOOK B © PASCAL PRESS ISBN: 9781925726176

1 Write the size of each angle.

● 40°

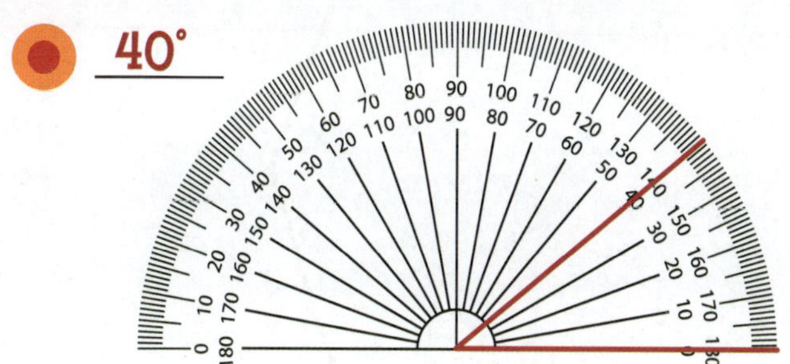

c ______

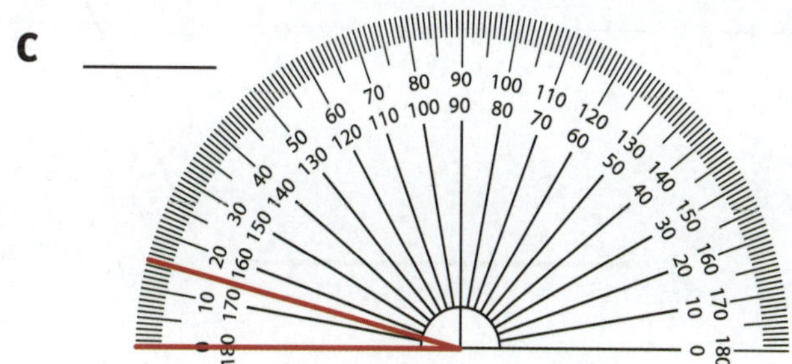

a ______

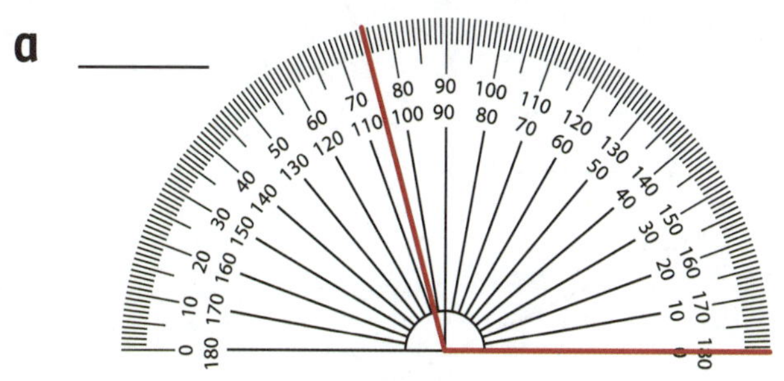

d ______

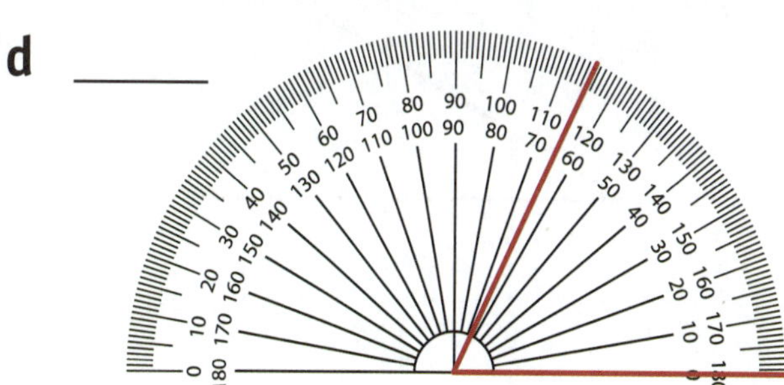

b ______

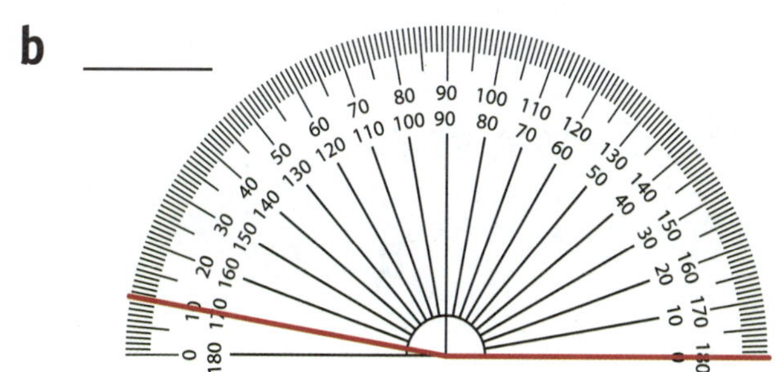

e ______

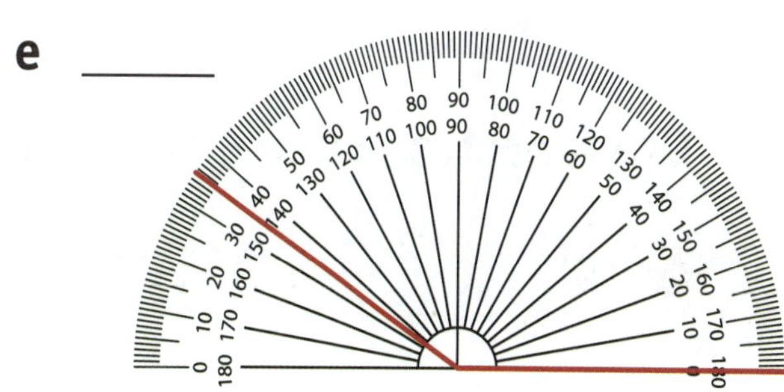

2 Measure these angles using a protractor.

● 50°

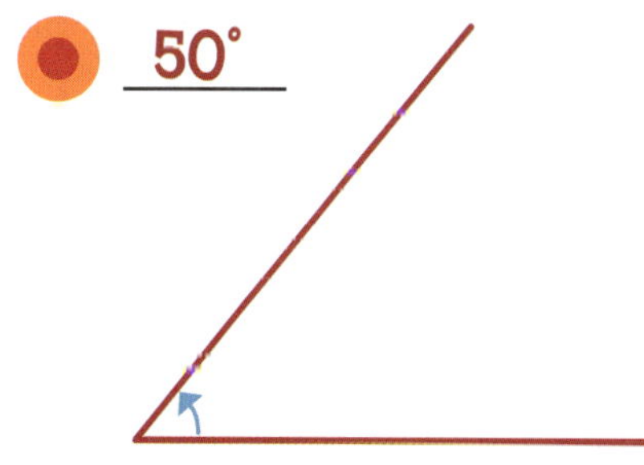

b ______

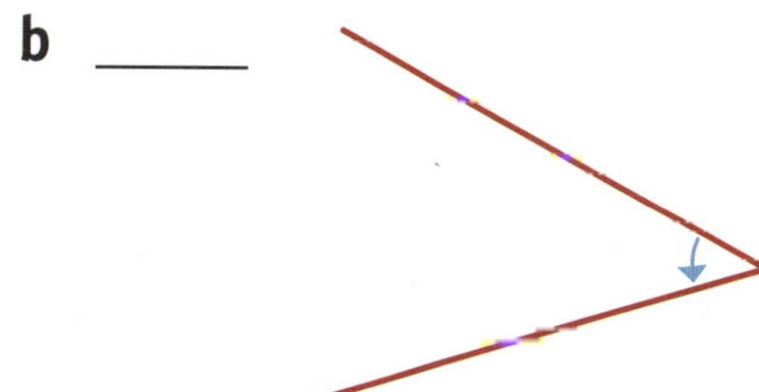

a ______

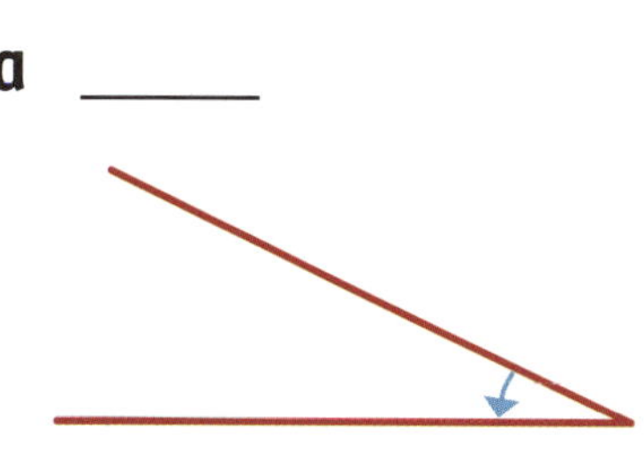

c ______

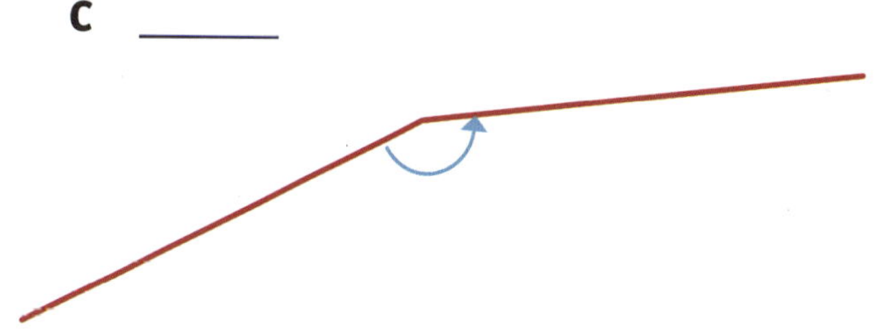

d ______

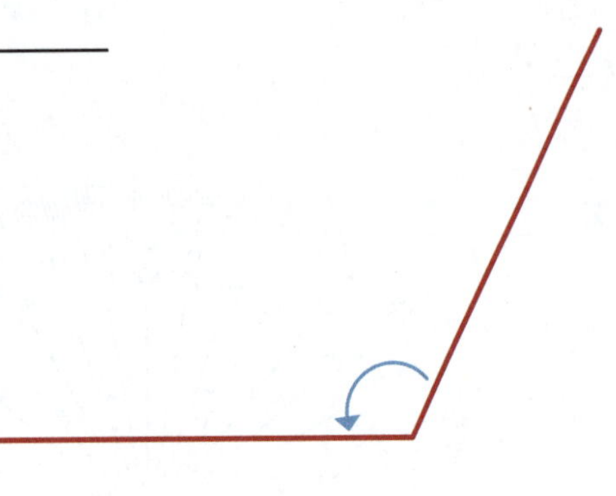

e ______

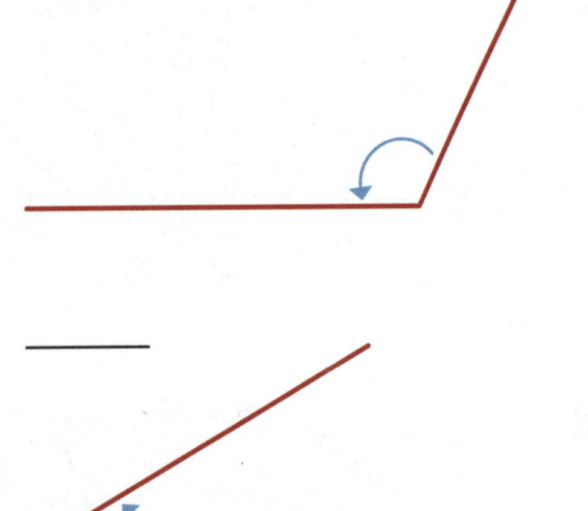

f ______

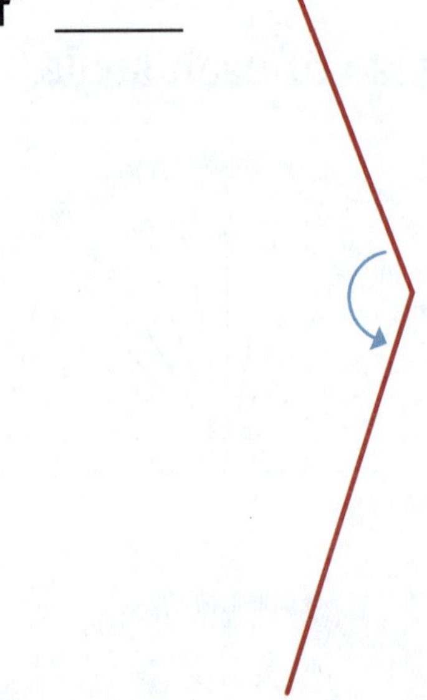

g ______

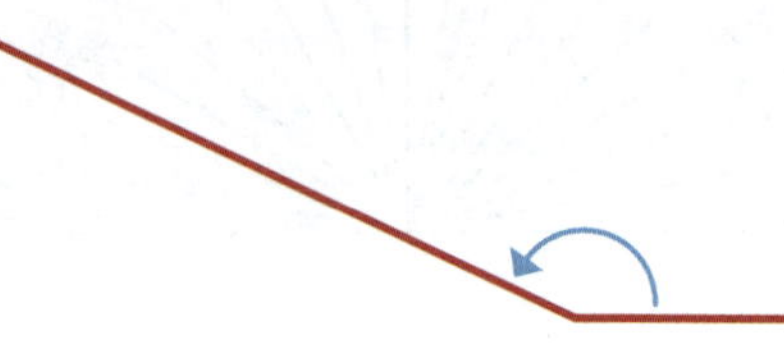

Use a protractor to draw an angle of the given size.

10°

b 105°

d 170°

a 60°

c 20°

e 26°

CATCH UP MATHS YEAR 5 BOOK B © PASCAL PRESS ISBN: 9781925726176

ANGLE SUM OF A TRIANGLE

The three angles in any triangle always add up to 180°.

SCAN to watch video

Equilateral triangle

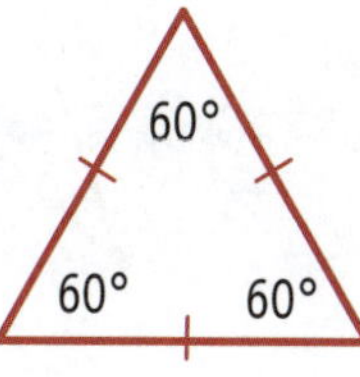

60° + 60° + 60°
= 180°

Right-angle triangle

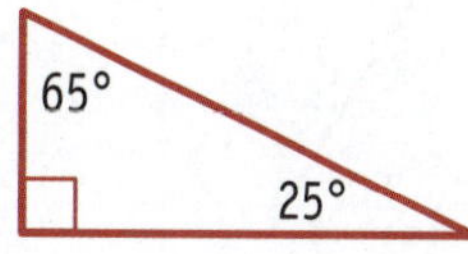

65° + 25° + 90°
= 180°

Isosceles triangle

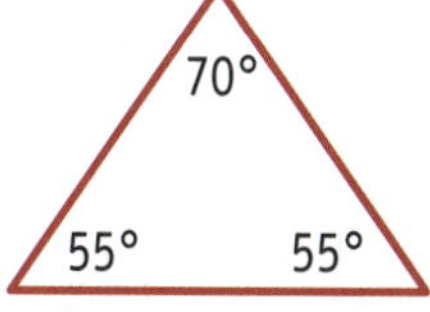

70° + 55° + 55°
= 180°

Scalene triangle

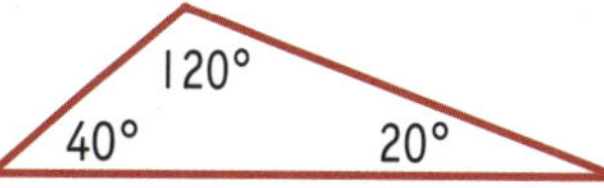

120° + 40° + 20°
= 180°

Examples: Write the type of triangle.

a scalene

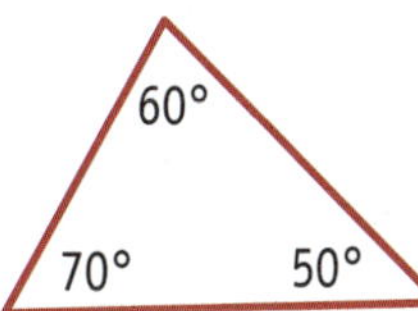

b ______

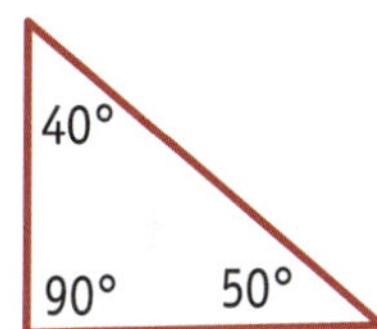

c ______

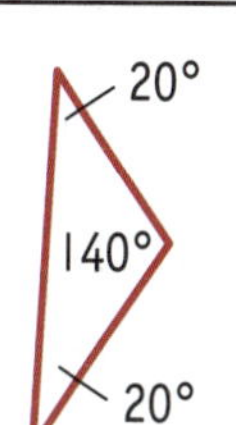

d ______

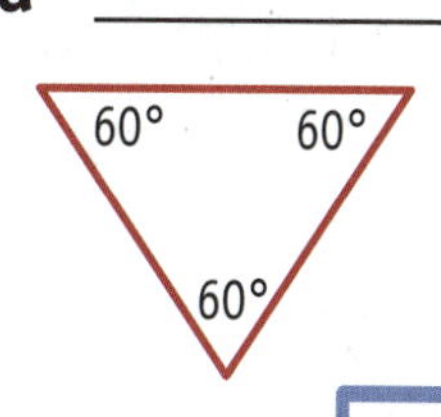

Check the answers on the video!

Your turn

Write the size of the missing angles.

60°

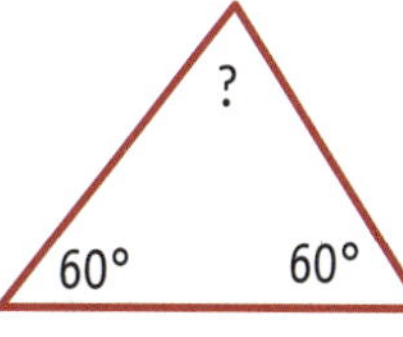

a ____

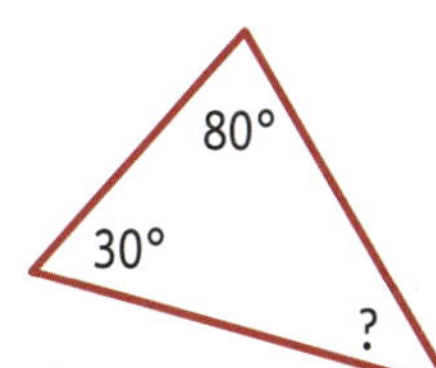

b ____

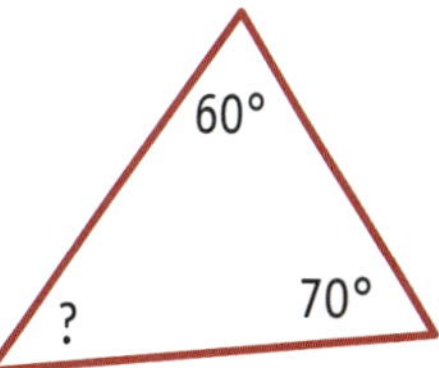

c ____

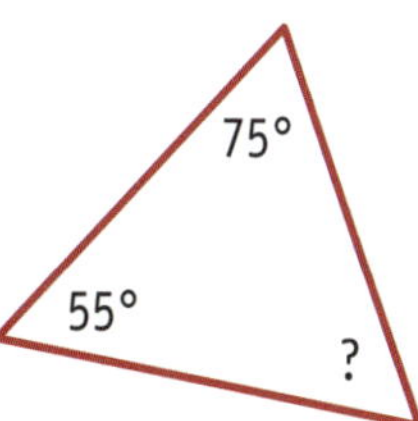

SELF CHECK Tick how you feel

Got it!	Need help...	I don't get it

Check your answers

How many did you get correct?

PRACTICE

1 Find the size of the missing angle and then write the triangle type.

● (example)

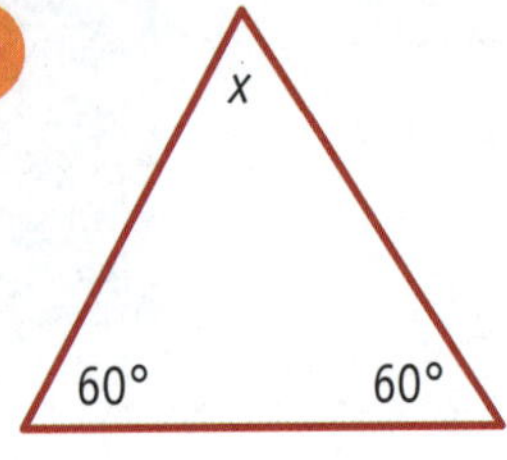

x = 60°

equilateral

b

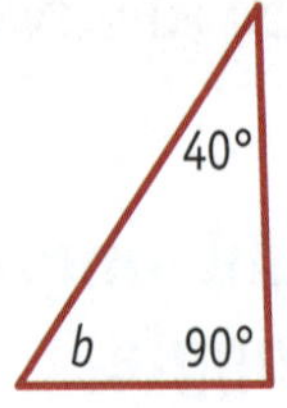

b = ____

____

d

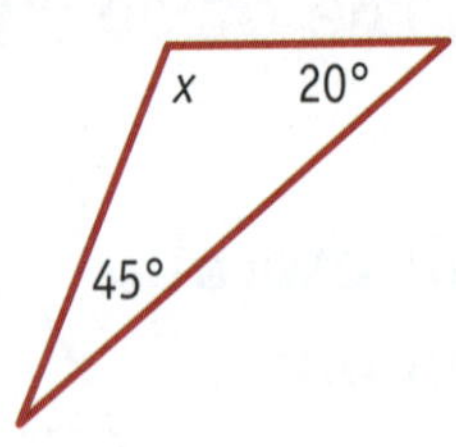

x = ____

____

a

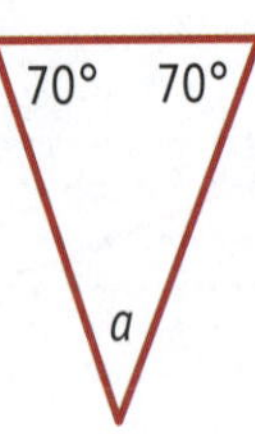

a = ____

____

c

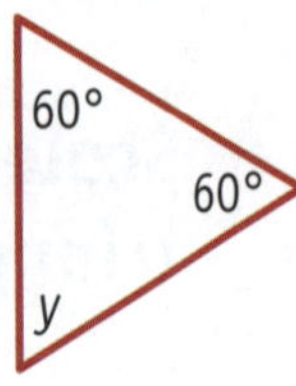

y = ____

____

e

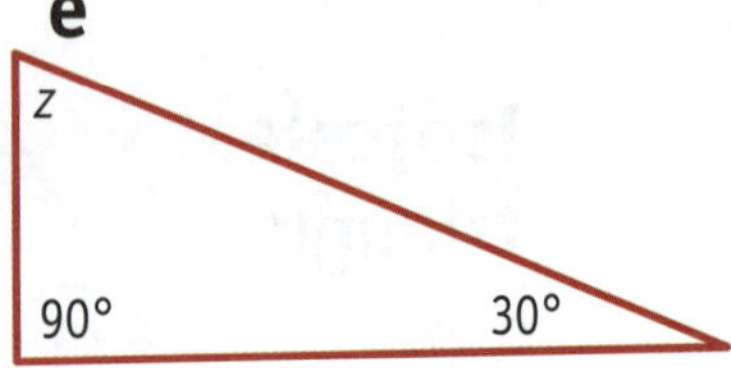

z = ____

____

2 Annie worked out the angle sizes of these triangles.
Write the total on the line inside the triangle.
Tick the box if Annie is correct.

● ☒

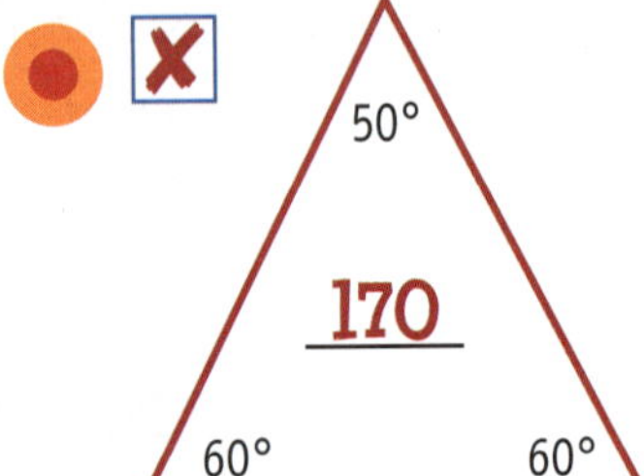

b ☐

30°
____
120°
40°

d ☐

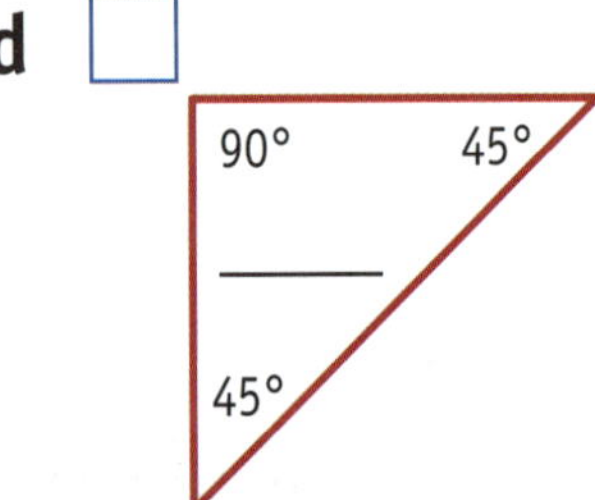

a ☐

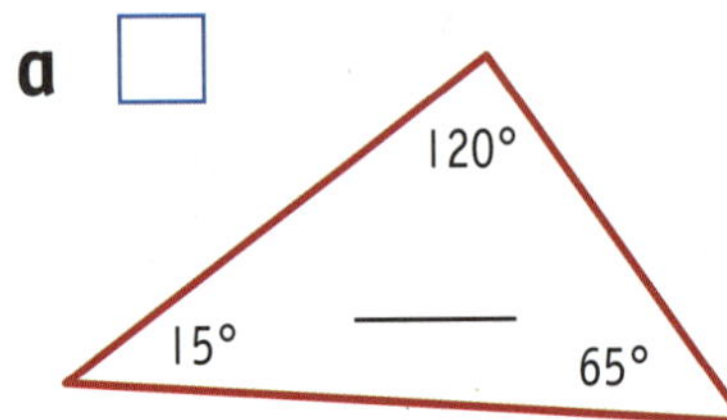

c ☐

40°
____
10°
130°

e ☐

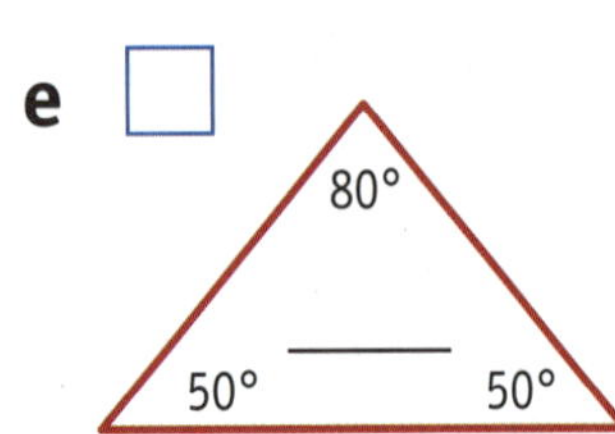

CATCH UP MATHS YEAR 5 BOOK B © PASCAL PRESS ISBN: 9781925726176

ANGLE SUM OF QUADRILATERALS

The four angles in any quadrilateral always add up to 360°.

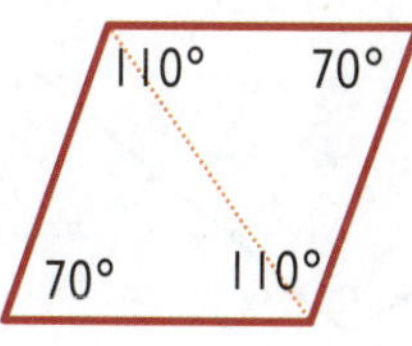

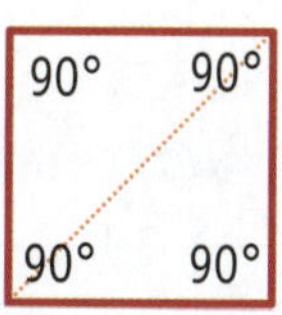

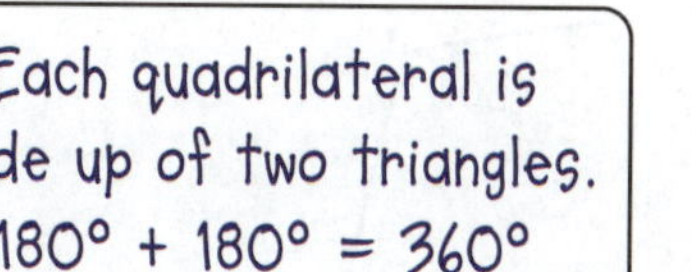

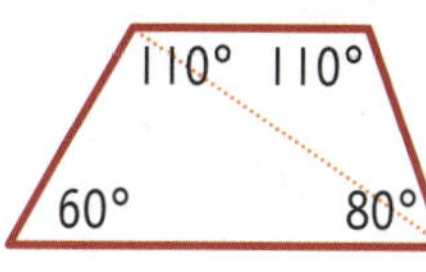

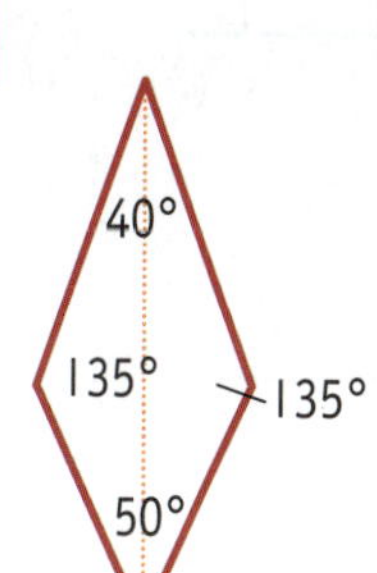

Examples: Divide each quadrilateral into two triangles.

a

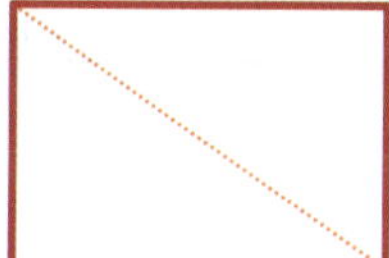

b

c

d

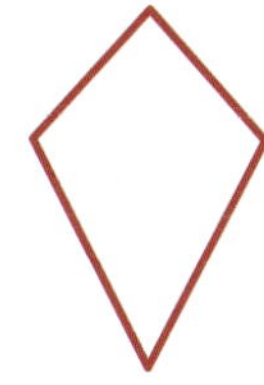

Check the answers on the video!

Your turn

Find the size of the missing angle in each quadrilateral.

x = 90°

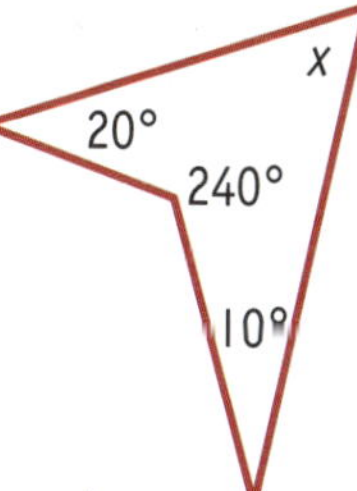

b b = ______

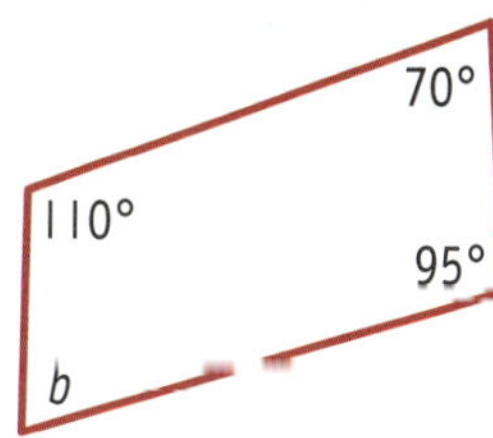

d d = ______

65°, d, 105°, 110°

a a = ______

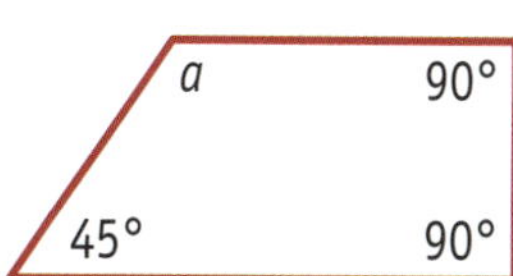

c c = ______

90°, 90°, 90°, c

e e = ______

105°, 75°, e, 105°

SELF CHECK Tick how you feel

Got it!	Need help...	I don't get it
☐	 ☐	 ☐

Check your answers

How many did you get correct? ☐

  ISBN: 9781925726176

PRACTICE

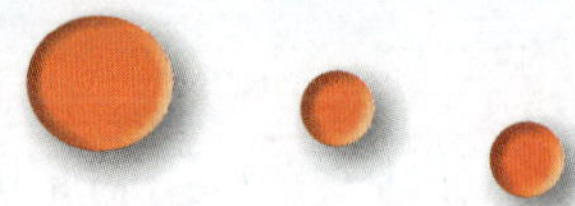

1 Find the size of the missing angles.

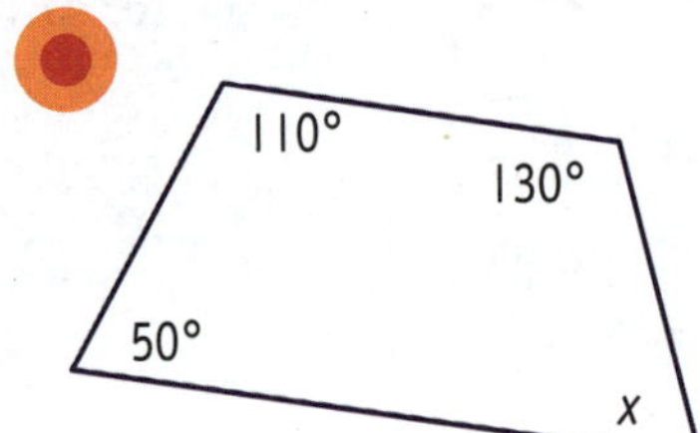

$x =$ 70°

d

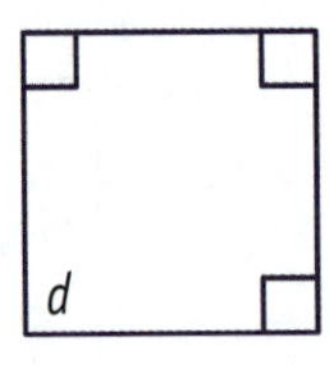

$d =$ ______

h

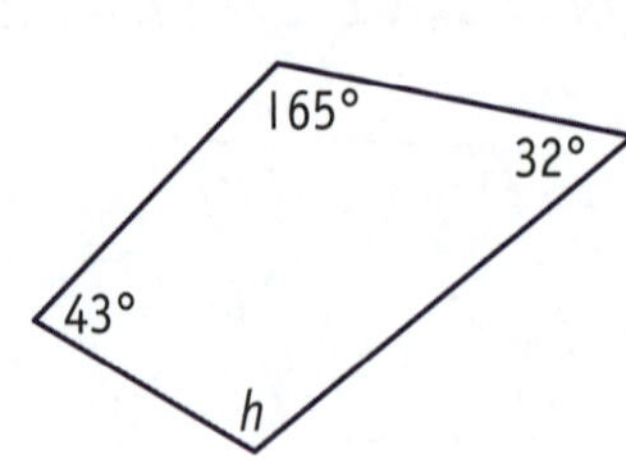

$h =$ ______

a

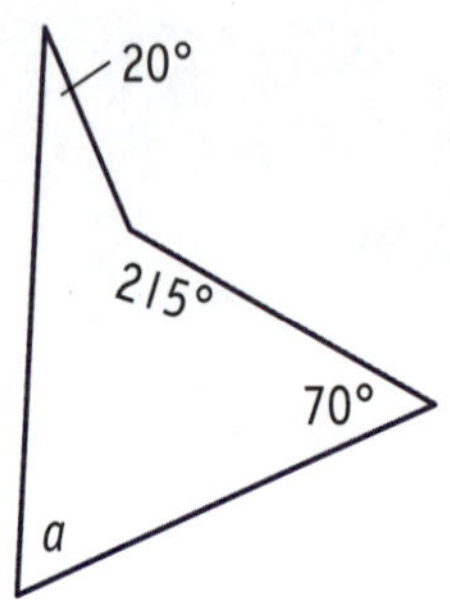

$a =$ ______

e

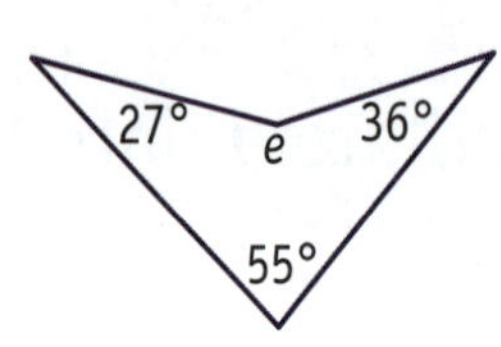

$e =$ ______

i

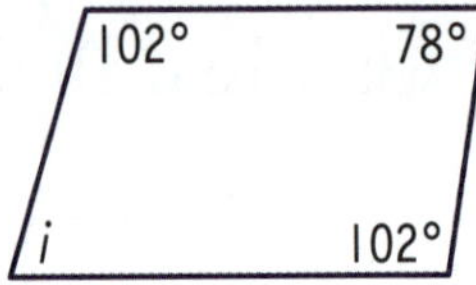

$i =$ ______

b

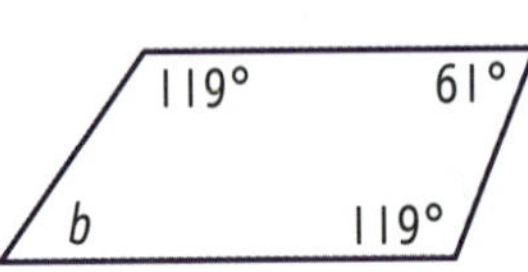

$b =$ ______

f

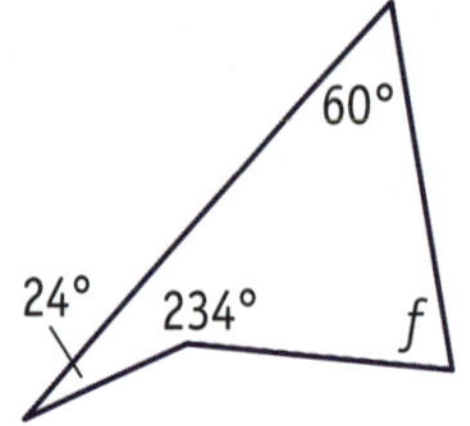

$f =$ ______

j

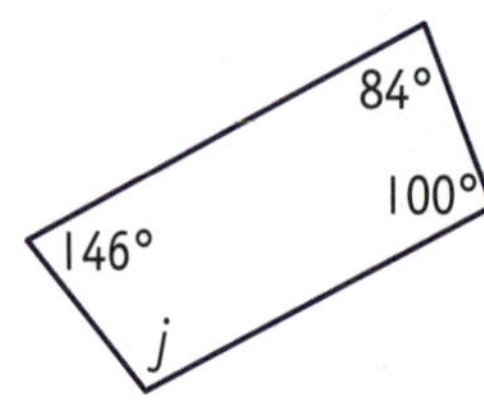

$j =$ ______

c

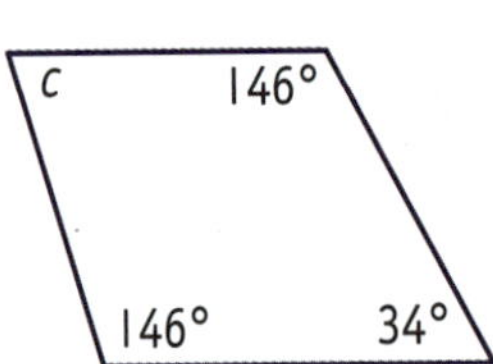

$c =$ ______

g

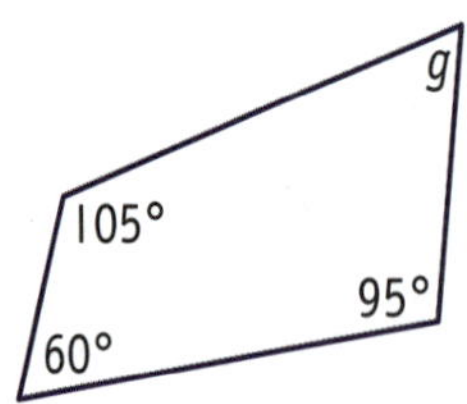

$g =$ ______

k

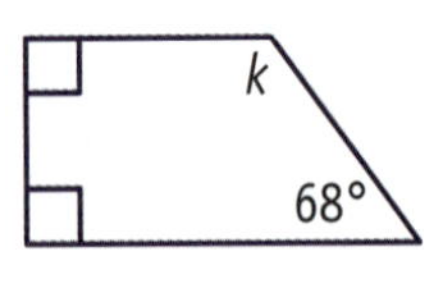

$k =$ ______

  ISBN: 9781925726176

ANGLES REVIEW

Use these words to complete the statements: ninety, arms, rays, right, turn.

a An angle is made by measuring the amount of ________ between the two ________ or ________.

b A quarter turn is called a ________ angle. It measures ________ degrees.

Trace over the rays in red and the angle made in blue, and draw a green circle on the vertex.

a

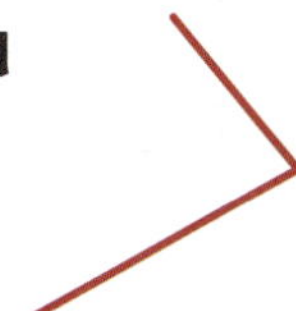

b

c

d

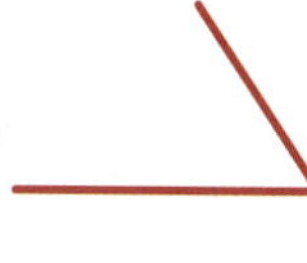

3 **Turn each ray into a right angle.**

a

b

c

d

4 **Trace over the right angles in each shape.**

a

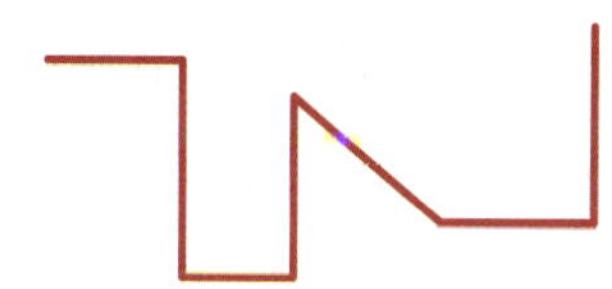

c

e

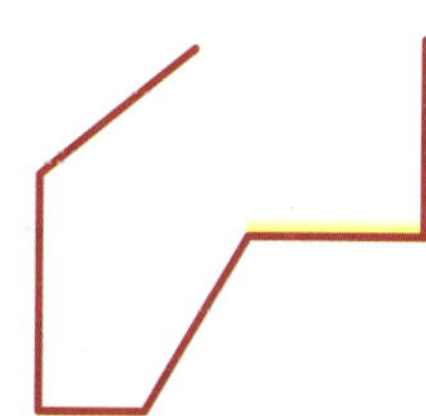

b

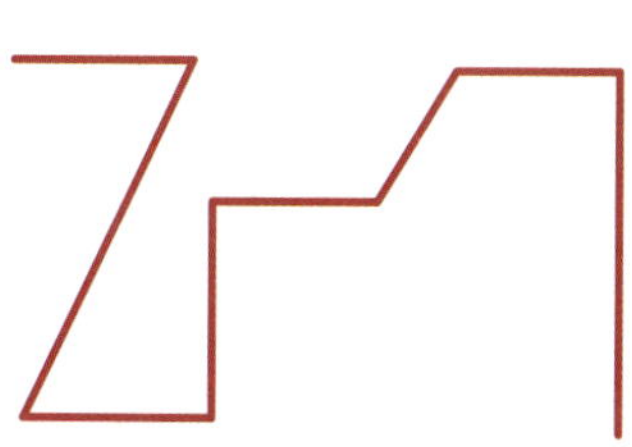

d

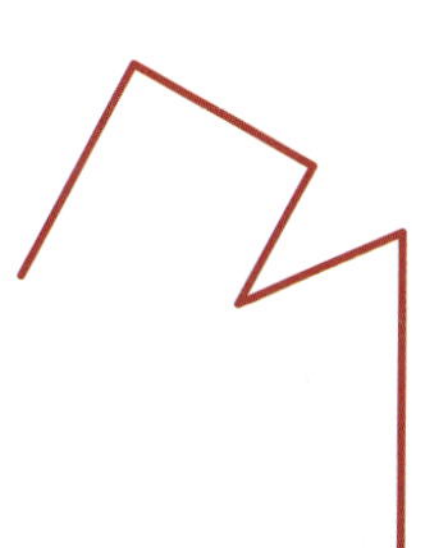

f

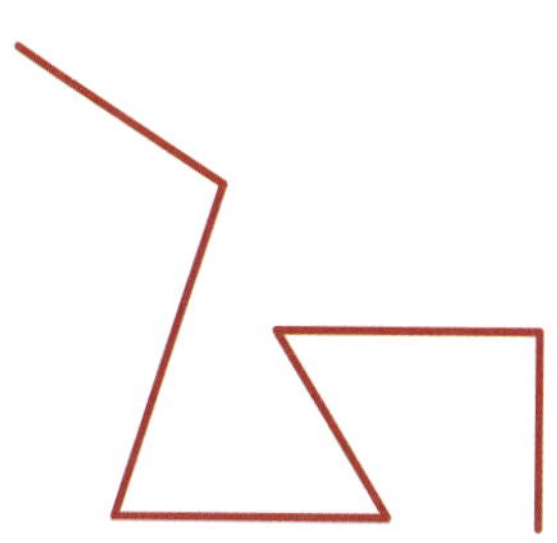

REVIEW

5 Draw five of each type of angle.

Acute angles	Obtuse angles

6 Name and label each angle.

a A B C

Name: ____________

Type: __________

b

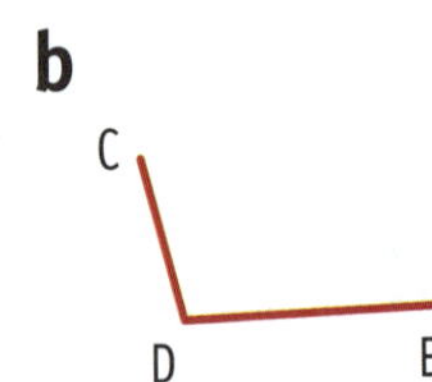

Name: ____________

Type: __________

c

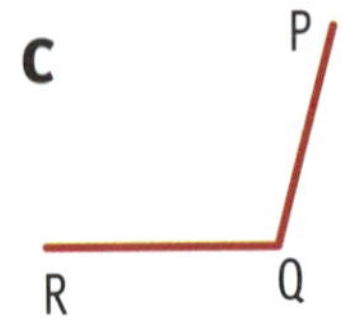

Name: ____________

Type: __________

d F G H

Name: ____________

Type: __________

e

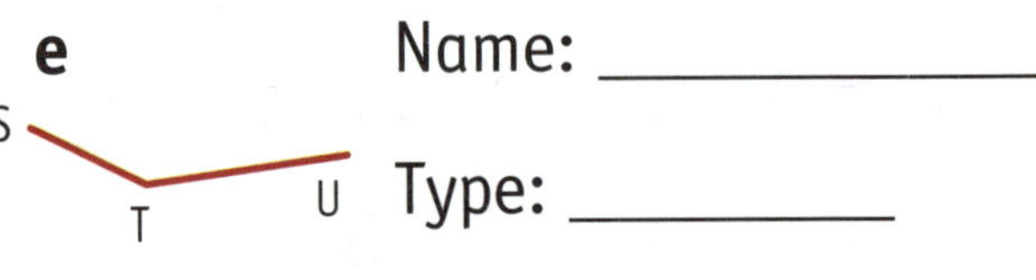

Name: ____________

Type: __________

f

Name: ____________

Type: __________

7 Use orange to circle the acute angles, green to circle the obtuse angles, red to circle the reflex angles, blue to circle the straight angles, and yellow to circle the revolutions.

a

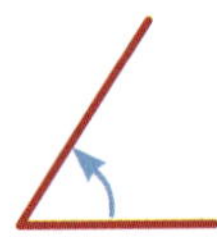

c

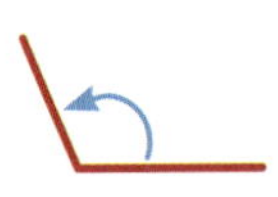

e

g

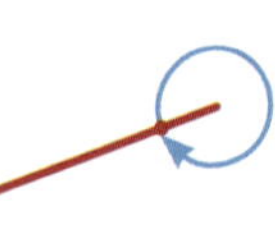

b

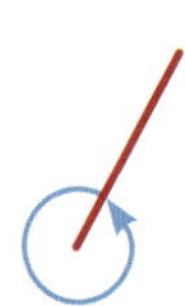

d

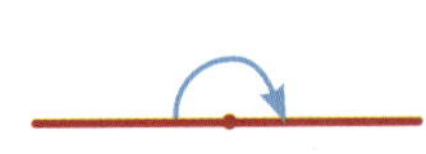

f

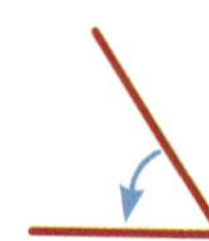

h

CATCH UP MATHS YEAR 5 BOOK B © PASCAL PRESS ISBN: 9781925726176

8 Write the letter of the angle that matches each type.

a reflex angle __

b obtuse angle __

c acute angle __

d straight angle __

e right angle __

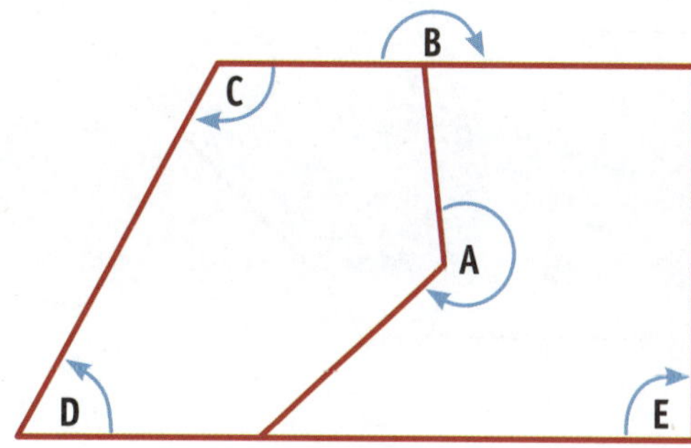

9 Find two more acute angles in the shape in Question 8. Label them F and G.

10 Find three more obtuse angles in the shape in Question 8. Label them H, I and J.

11 Draw another right angle in the shape in Question 8. Label it K.

12 Write the size of these angles.

a ______

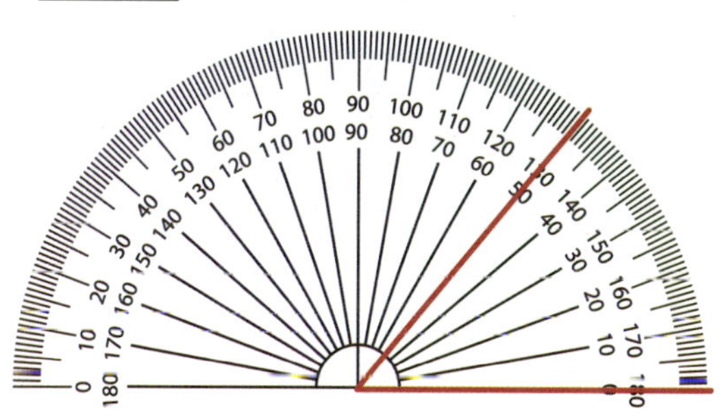

c ______

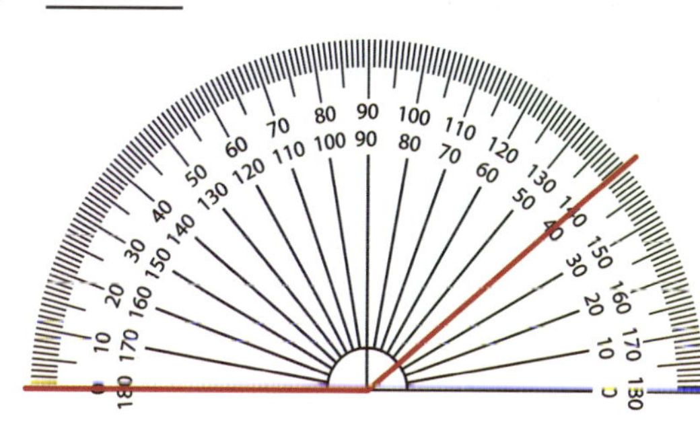

b ______

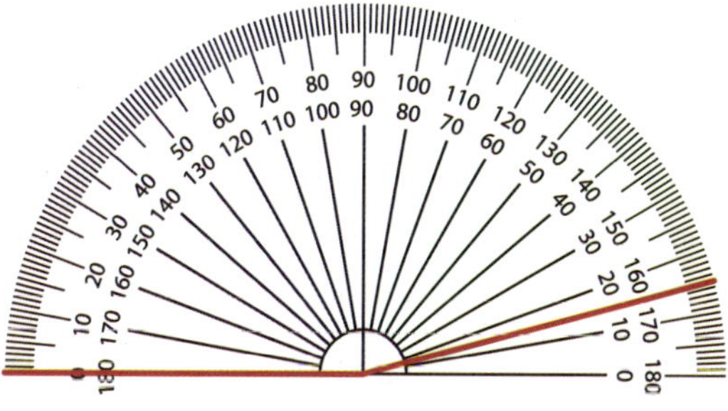

d ______

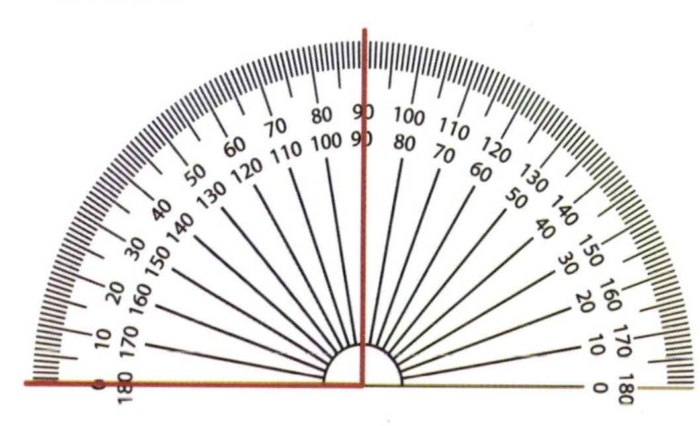

  ISBN: 9781925726176

REVIEW

13 Measure these angles using a protractor.

a ______

d ______

b ______

e ______

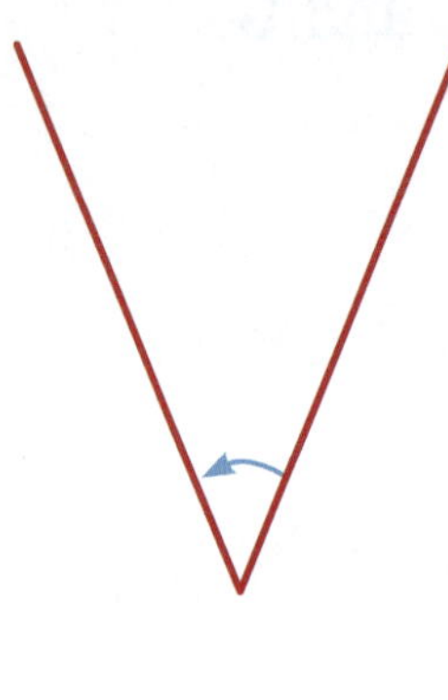

c ______

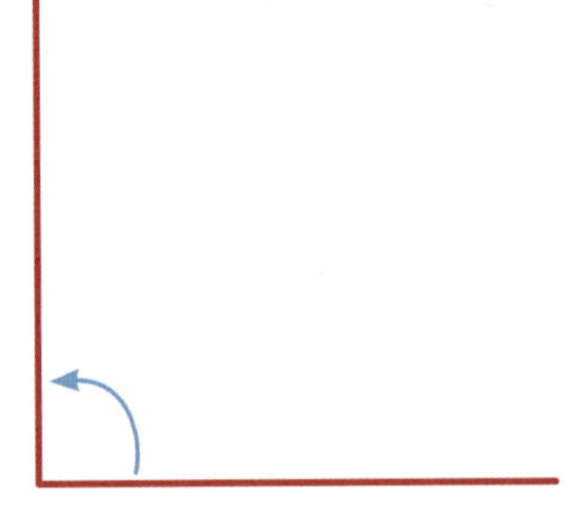

14 Use a protractor to draw an angle of the given size.

a 15°

b 72°

c 133°

d 243°

e 122°

f 84°

CATCH UP MATHS YEAR 5 BOOK B © PASCAL PRESS ISBN: 9781925726176

Find the size of the missing angle and then write the triangle type.

a

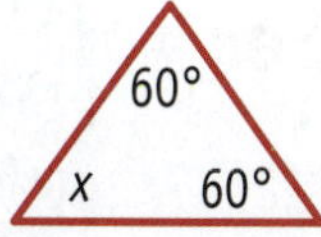

x = ____

____

b

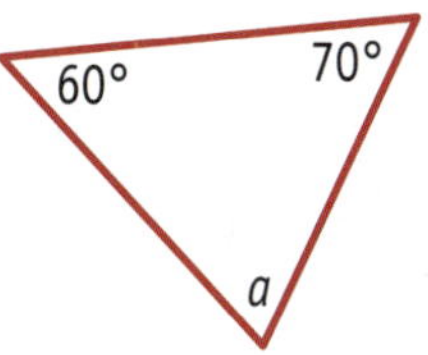

a = ____

____

c

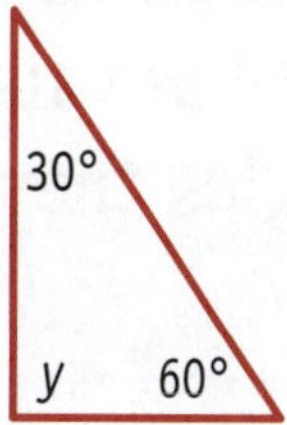

y = ____

____

d

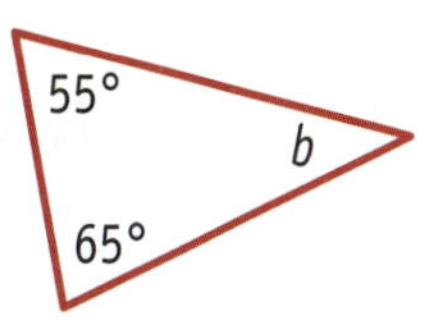

b = ____

____

e

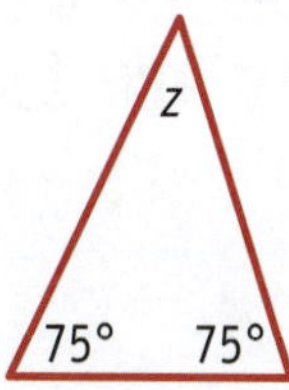

z = ____

____

f

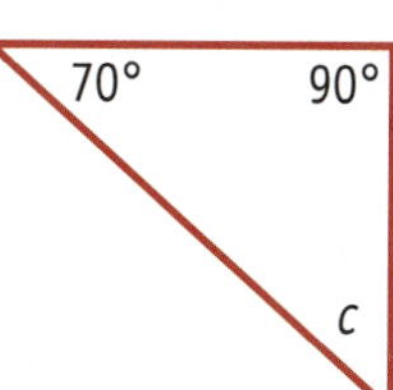

c = ____

____

16 Find the size of the missing angles in these quadrilaterals.

a

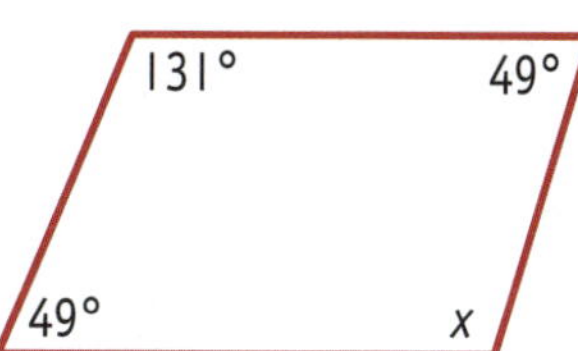

x = ____

b

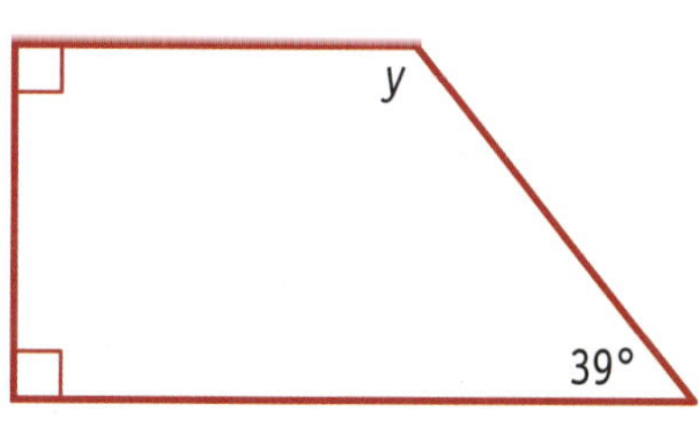

y = ____

c

b = ____

d

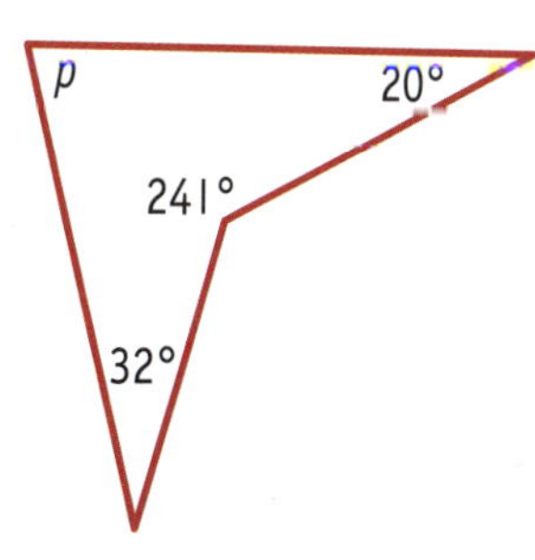

p = ____

e

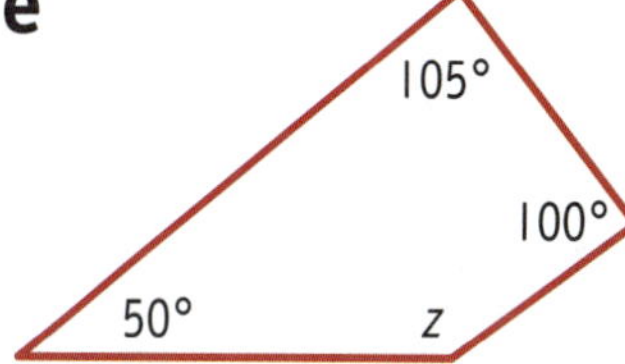

z = ____

f

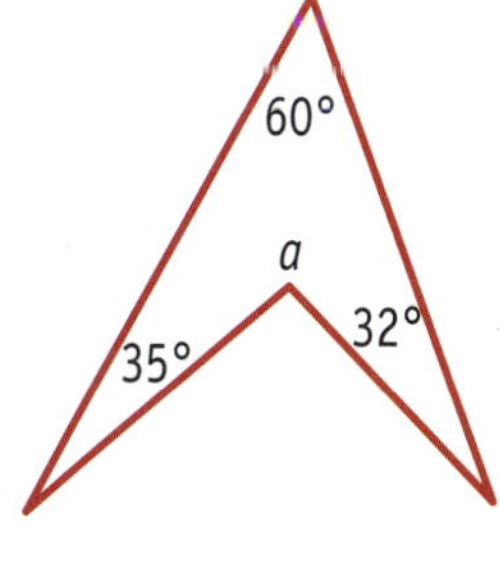

a = ____

2D SHAPES

2D shapes (polygons) have two dimensions: length and width. They are flat. Some 2D shapes are named by the number of sides and angles they have.

SCAN to watch video

triangle:
3 angles,
3 sides

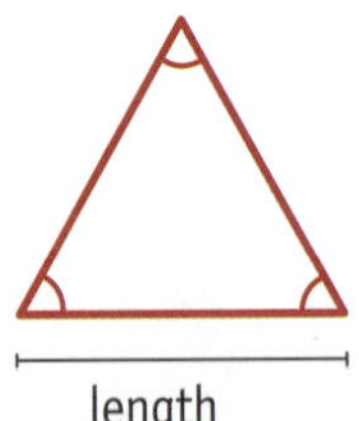

width (height)

length

hexagon:
6 sides,
6 angles

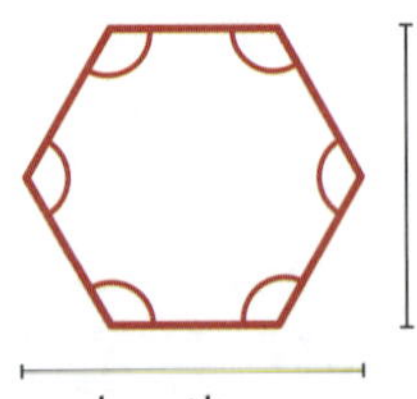

width (height)

length

In **regular** 2D shapes, all angles and sides are the same size.

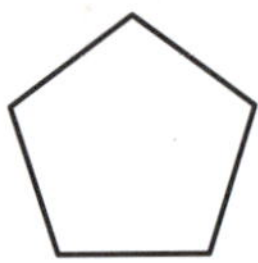

regular pentagon

square (regular quadrilateral)

regular octagon

Irregular 2D shapes have sides and angles of different sizes.

irregular pentagon

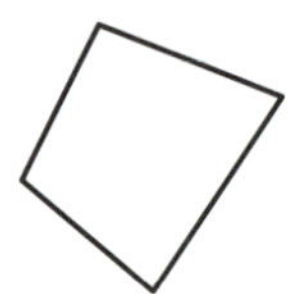

irregular quadrilateral

irregular octagon

Examples: Colour the regular 2D shapes red and the irregular 2D shapes blue.

a 

b 

c 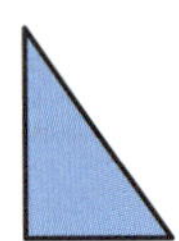

d 

e 

f

Check the answers on the video!

Your turn

Write the names of the 2D shapes in the Examples.

a square ____________

b ____________

c ____________

d ____________

e ____________

f ____________

SELF CHECK Tick how you feel

Got it!	Need help...	I don't get it

Check your answers
How many did you get correct?

CATCH UP MATHS YEAR 5 BOOK B © PASCAL PRESS ISBN: 9781925726176

PRACTICE

1 Trace the regular shapes with red and the irregular shapes with blue. Then write the name of each shape.

● irregular pentagon

a ____________________

b ____________________

c ____________________

d ____________________

e ____________________

2 Complete the chart.

	Polygon	Name	Letters that help identify shape	No. angles	No. sides	No. corners
●		quadrilateral	quad	4	4	4
a						
b						
c						
d		heptagon				
e						
f		nonagon				
g		decagon				

TYPES OF LINES

Many shapes have lines that are vertical, horizontal, parallel and perpendicular.

Vertical lines go up and down.

Parallel lines never meet and are equal distances apart.

Horizontal lines go left to right.

Perpendicular lines meet at right angles.

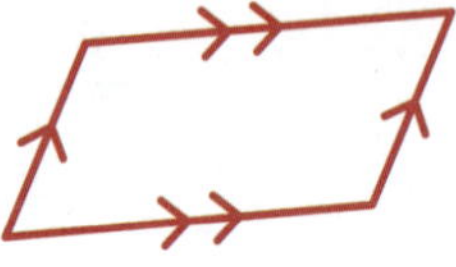

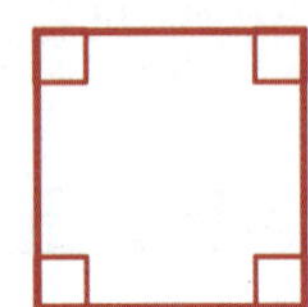

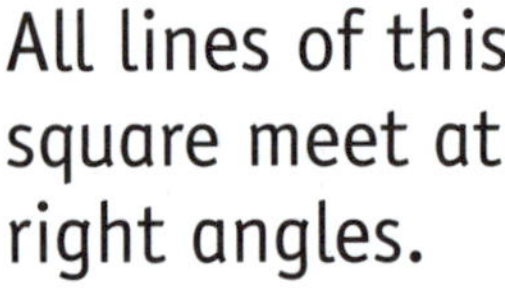

This parallelogram has two sets of parallel lines.

This right-angled triangle has two lines that meet at right angles.

All lines of this square meet at right angles.

Examples:

Label the lines as vertical, horizontal, parallel or perpendicular.

a parallel

b ____________

c ____________

d ____________

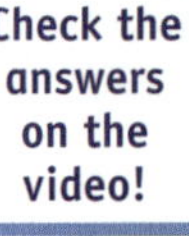

Trace the vertical lines with blue, horizontal lines with orange, parallel lines with green and perpendicular lines with purple.

b d f h

a c e g i

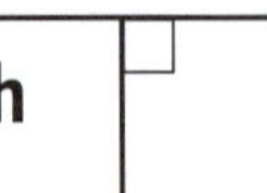

SELF CHECK Tick how you feel		
Got it! ☐	Need help... ☐	I don't get it ☐

Check your answers

How many did you get correct? ☐

CATCH UP MATHS YEAR 5 BOOK B © PASCAL PRESS ISBN: 9781925726176

PRACTICE

Draw five examples of each type of line.

Horizontal	Parallel	Perpendicular	Vertical

List five things in your environment that have parallel lines.

train tracks

Trace over the lines that match the line type.

a perpendicular

b parallel

c horizontal

d vertical

CIRCLES

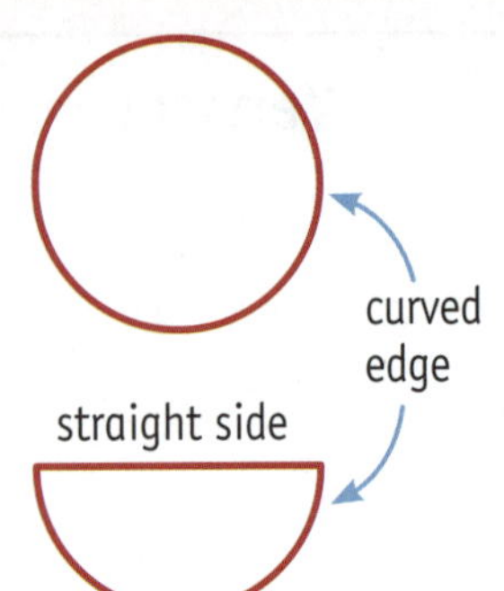

A circle is a regular shape with one curved edge.
It is two dimensional and has no corners.

A semicircle is half a circle.
It has one curved edge and one straight side.

These shapes are irregular circles.

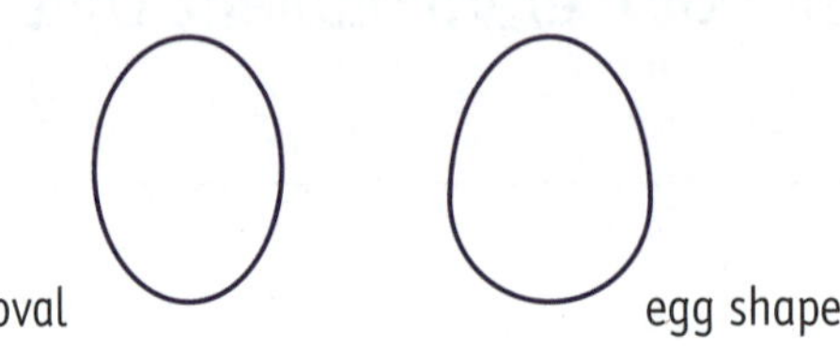

Examples:
Trace to complete the shape and then label it.

a 
circle

b 
____

c 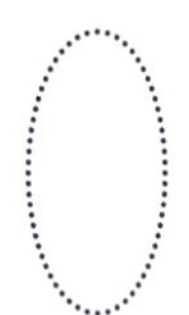
____

d
____

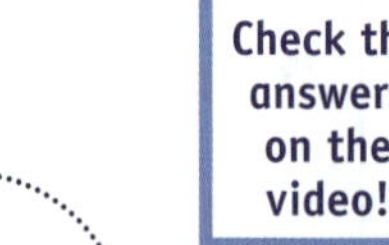

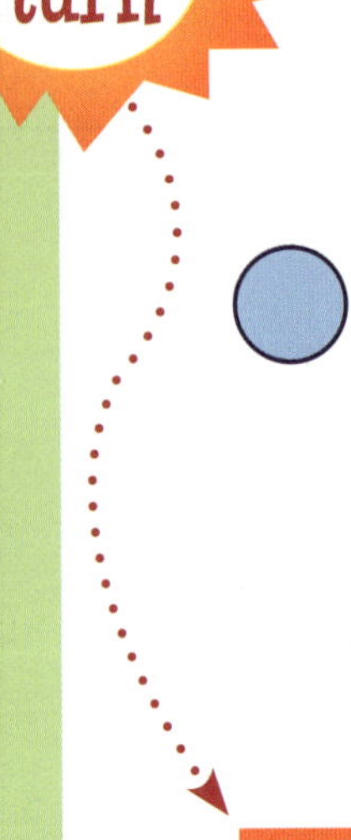

Colour the regular circles blue and the irregular circles red.

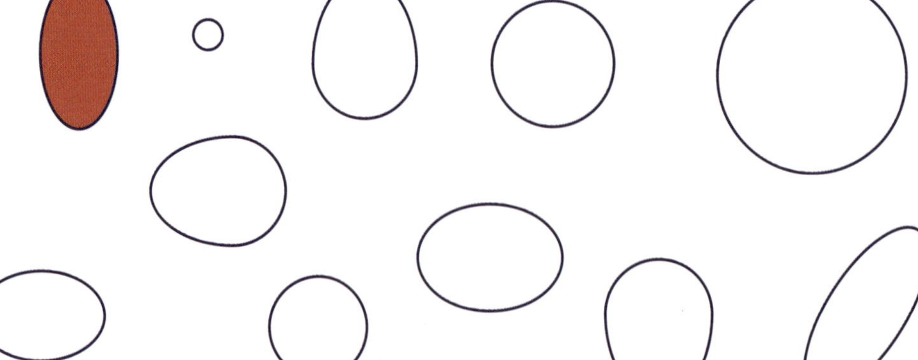

SELF CHECK Tick how you feel		
Got it! ☐	Need help... ☐	I don't get it ☐

Check your answers
How many did you get correct? ☐

  ISBN: 9781925726176

PRACTICE

1 Draw a red dot at the centre of each circle.

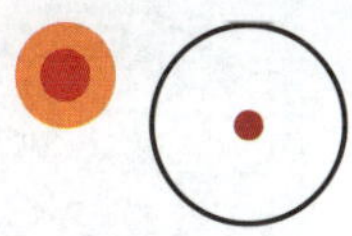

a 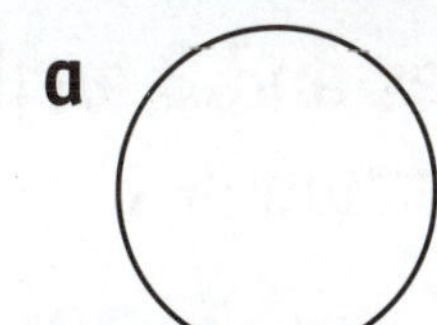

b 

c 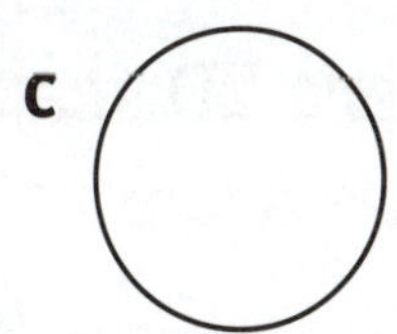

d 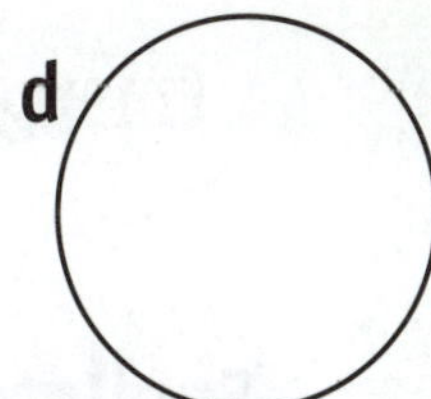

2 Use the labelled circle to complete the definitions.

- C e n t r e: the middle of the circle.
- **a** D _ _ _ _ _ _ _ _: a line from one side of the circle to another through the centre.
- **b** R _ _ _ _ _: a line from the centre of the circle to the circumference; the plural is radii.
- **c** C _ _ _ _: a straight line from one side of the circle to the other that does not pass through the centre.
- **d** C _ _ _ _ _ _ _ _ _ _ _ _ _: the name given to the curved edge of a circle.
- **e** S _ _ _ _ _: a part of a circle made from two radii and an arc.
- **f** A _ _: part of the circumference of a circle.

3 Write the letters on the diagram that mark each part of the circle.

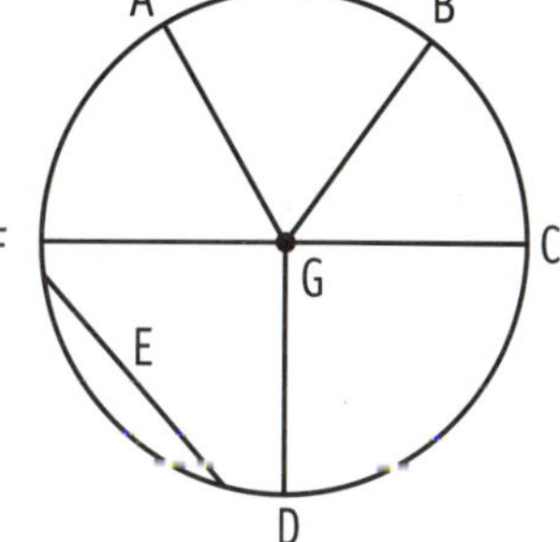

- A radius GD
- **a** The centre __
- **b** A diameter __
- **c** A chord __

4 Use the diagram at the right to complete these tasks.

- **a** Use purple to trace the circumference.
- **b** Use orange to trace an arc.
- **c** Colour a sector red.

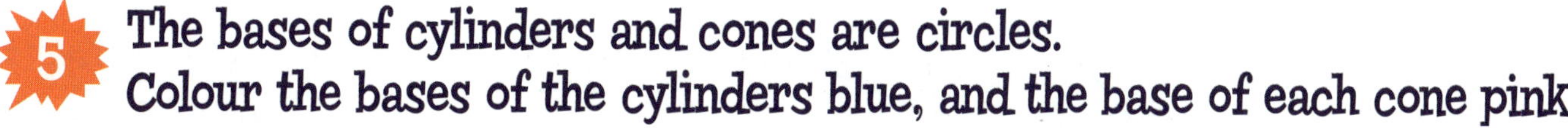

5 The bases of cylinders and cones are circles.
Colour the bases of the cylinders blue, and the base of each cone pink.

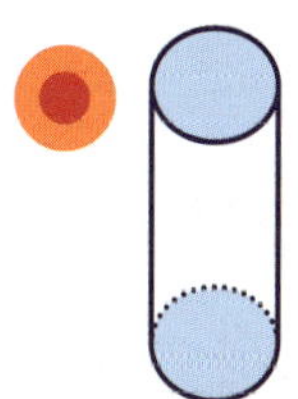

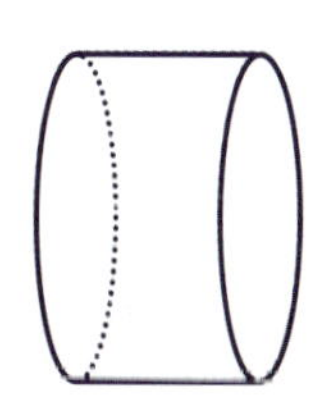
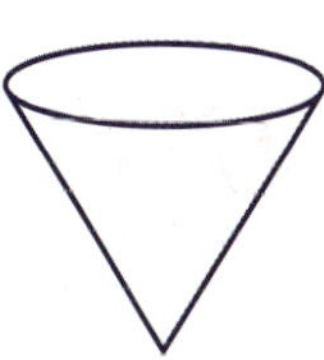
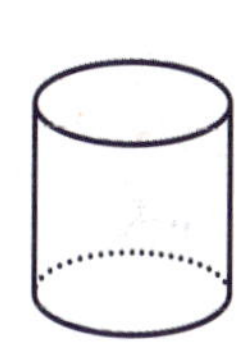
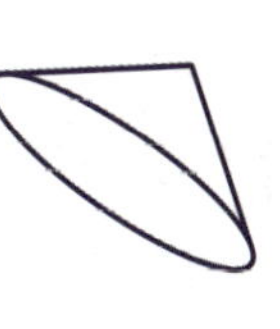
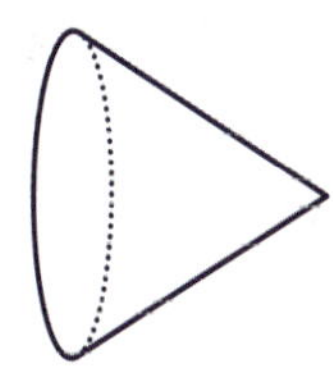

TRIANGLES

Triangles are 2D shapes that have 3 sides and 3 angles. There are four different types of triangles

Equilateral
- All sides the same length
- All angles the same size
- Regular triangle

Isosceles
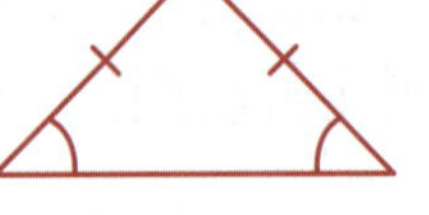

- Two sides the same length
- Two angles the same size
- Irregular triangle

Right angled
- One right angle
- Irregular triangle

Scalene

- All sides different lengths
- All angles different sizes
- Irregular triangle

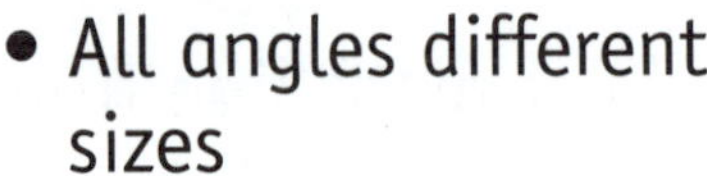

Examples: Trace the regular triangles in red.

a 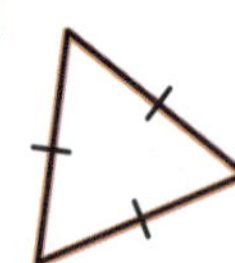
b 
c 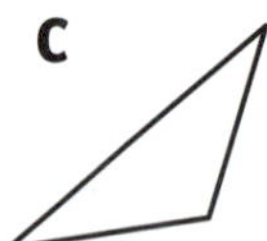
d 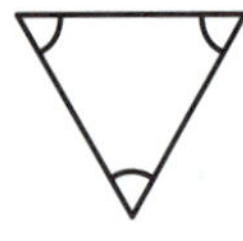
e 

Check the answers on the video!

Your turn

Trace the equilateral triangles in red, isosceles triangles in blue, scalene triangles in yellow and right-angled triangles in green.

 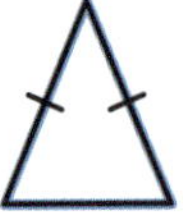
b 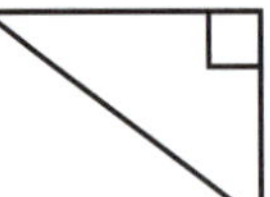
d 
f 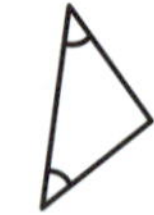

a 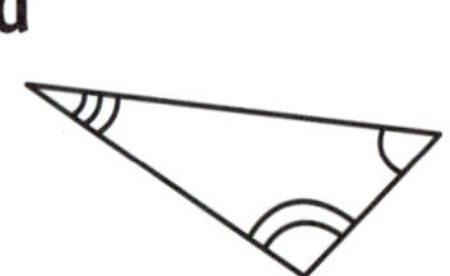
c 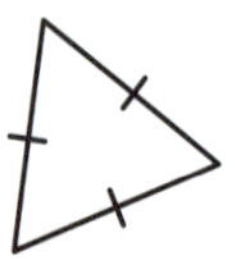
e 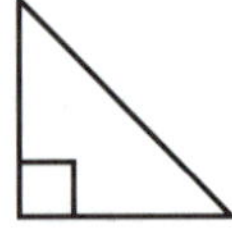
g 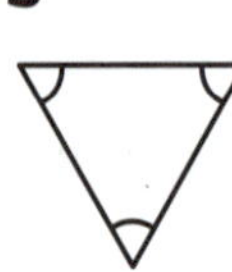

SELF CHECK Tick how you feel		
Got it!	Need help...	I don't get it

Check your answers
How many did you get correct?

CATCH UP MATHS YEAR 5 BOOK B © PASCAL PRESS ISBN: 9781925726176

PRACTICE

1 Measure the sides and angles of each triangle and then label it.

● isosceles ______ d ______

a  ______ e ______

b 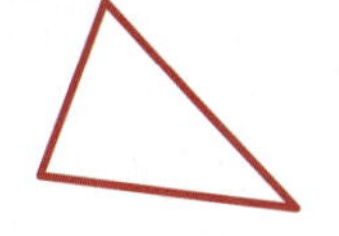 ______ f  ______

c 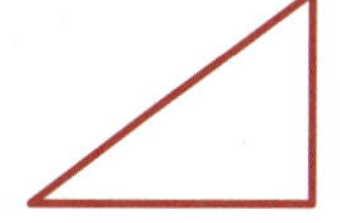 ______ g  ______

2 Draw three of each type of triangle.

Isosceles triangles	Equilateral triangles
Scalene triangles	**Right-angled triangles**

3 Write a definition for each triangle.

a isosceles triangle ______________________

____

b equilateral triangle ______________________

____

c scalene triangle ______________________

____

d right-angled triangle ______________________

____

QUADRILATERALS

Quadrilaterals have four straight sides and four corners.
They are two-dimensional shapes.

All angles the same size.

All sides the same length.

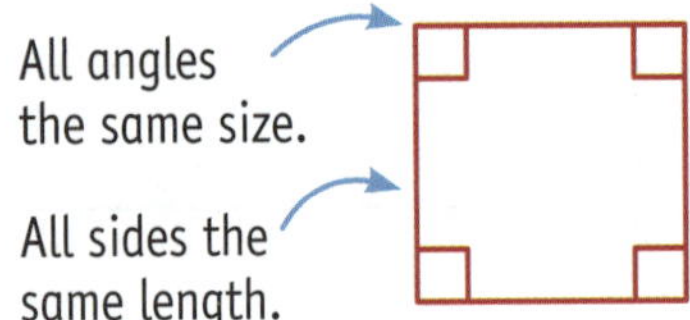

Irregular quadrilaterals

Rectangle (oblong)
- Opposite sides equal in length
- All angles 90°

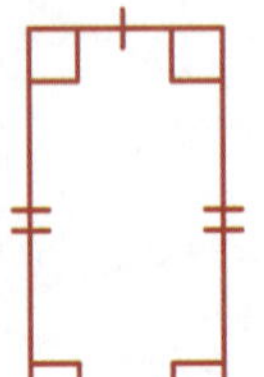

Rhombus
- All sides equal in length
- Two pairs of parallel sides
- Opposite angles equal in size

Parallelogram
- Opposite sides equal in length
- Two pairs of parallel sides
- Opposite angles equal in size

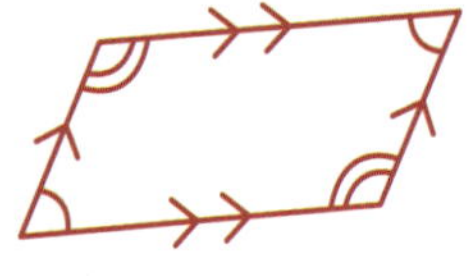

Kite
- Two pairs of sides the same length
- No parallel sides

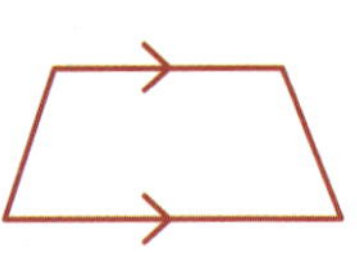

Trapezium
- One pair of parallel sides

Examples:
Label each quadrilateral as regular or irregular.

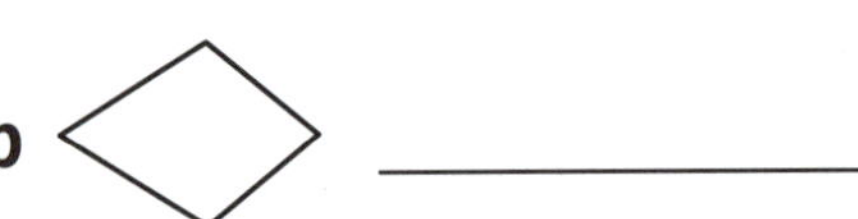

a irregular

c ______

b ______

d ______

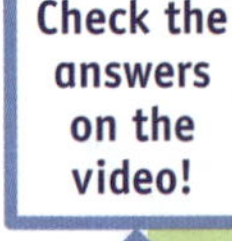

Name each quadrilateral in the Examples above.

a trapezium

c ______

b ______

d ______

Check your answers
How many did you get correct?

CATCH UP MATHS YEAR 5 BOOK B © PASCAL PRESS ISBN: 9781925726176

PRACTICE

Trace the squares with red, the rectangles with blue, rhombuses with green, parallelograms with yellow, trapeziums with orange, and kites with brown.

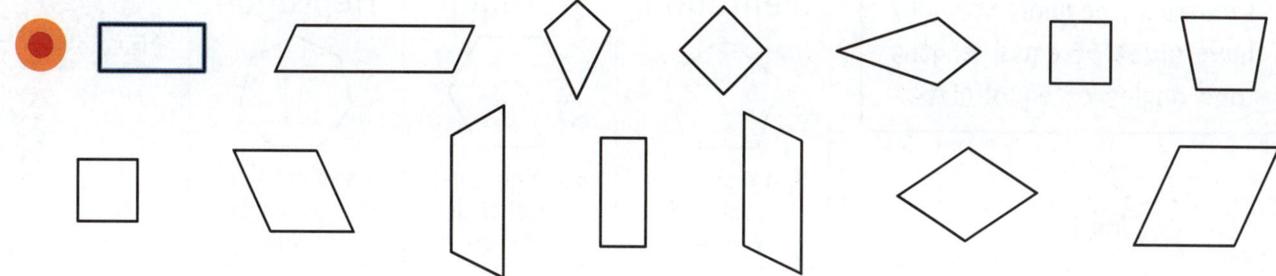

Write a definition for each quadrilateral and then draw an example.

	Definition	Diagram
a	square	
b	rectangle	
c	kite	

	Definition	Diagram
d	rhombus	
e	parallelogram	
f	trapezium	

Trace the parallel lines with purple and the perpendicular lines with red.

a

c

e

g

POLYGONS

Polygons are named for the number of sides they have.

Remember, regular polygons have sides of equal lengths and angles of equal sizes.

Irregular polygons have the same number of sides and angles, but the lengths and sizes are not equal.

Pentagon 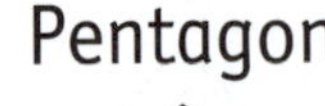	Hexagon	Heptagon
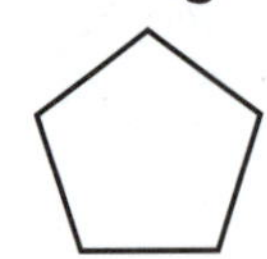		
5 straight sides 5 equal angles 5 corners	6 straight sides 6 equal angles 6 corners	7 straight sides 7 equal angles 7 corners

Octagon 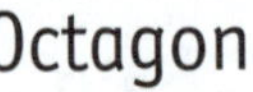	Nonagon	Decagon	Dodecagon
	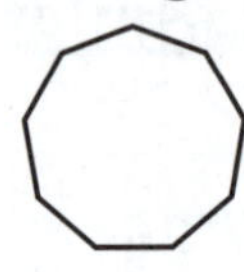	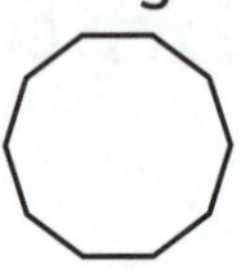	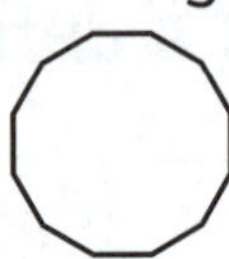
8 straight sides 8 equal angles 8 corners	9 straight sides 9 equal angles 9 corners	10 straight sides 10 equal angles 10 corners	12 straight sides 12 equal angles 12 corners

Examples: Trace the dotted lines to complete the shapes and then label them.

a  regular ______ octagon ______

c  ______ ______

b 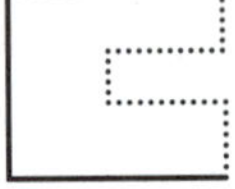 ______ ______

d 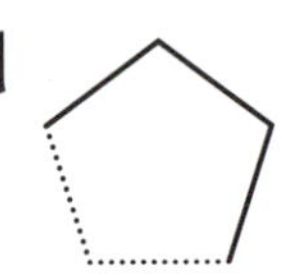 ______ ______

Your turn

Trace over the regular polygons in red and the irregular polygons in blue. Then write the number of sides.

Check the answers on the video!

● 8 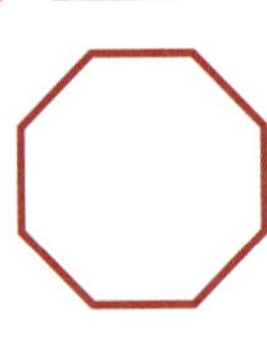

a ___ 

b ___ 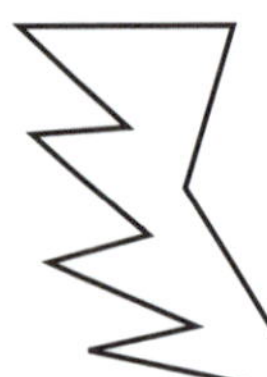

c ___ 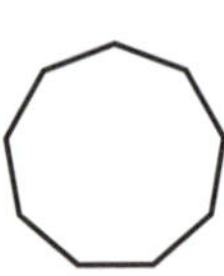

d ___ 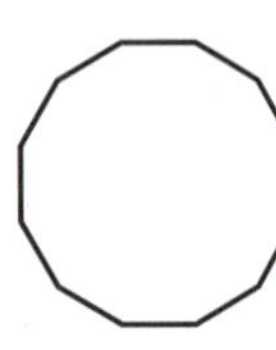

SELF CHECK Tick how you feel

Got it!	Need help...	I don't get it
 ☐	  ☐	 ☐

Check your answers
How many did you get correct? ☐

CATCH UP MATHS YEAR 5 BOOK B © PASCAL PRESS ISBN: 9781925726176

PRACTICE

1 Draw three polygons that match each description.

Irregular decagons	Irregular octagons
Irregular pentagons	Irregular hexagons
Irregular dodecagons	Irregular nonagons
Irregular heptagons	Irregular quadrilaterals

2 Colour the regular pentagons purple, the regular hexagons blue, put an X on the irregular pentagons and tick the irregular hexagons.

a 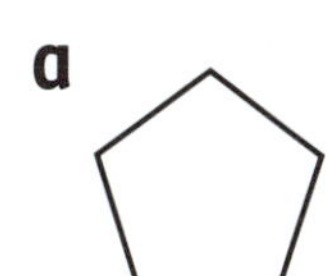 b 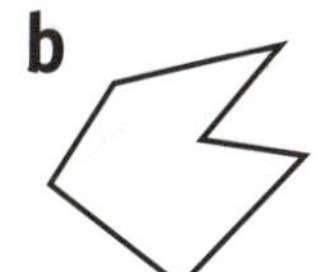 c 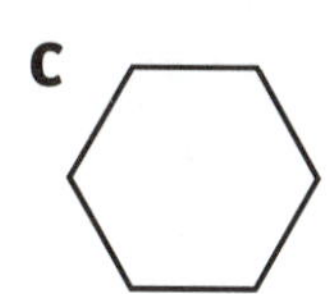 d 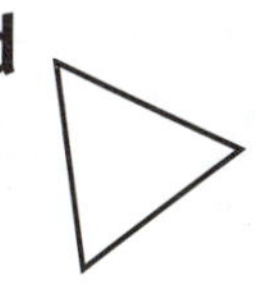 e 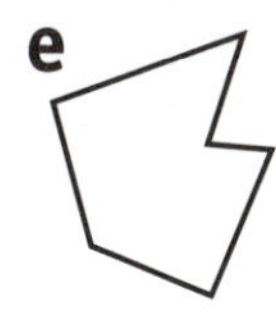 f 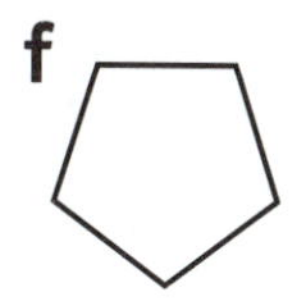 g 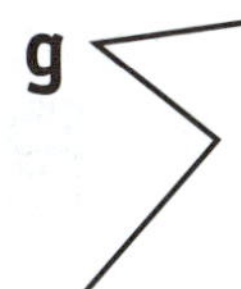

3 Colour the regular heptagons orange, the regular octagons yellow, and put an X on the irregular heptagons and tick the irregular octagons.

a  b  c d 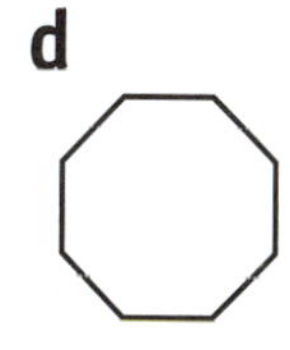 e 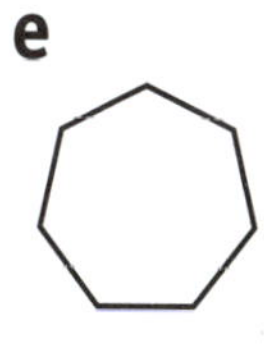 f 

4 Colour the regular nonagons red, the regular decagons green, the regular dodecagons black, and put an X on the irregular nonagons, tick the irregular decagons and circle the irregular dodecagons.

a 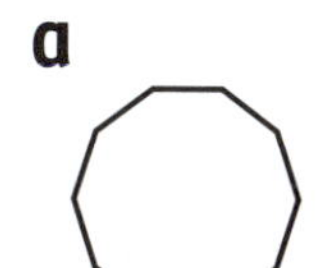 b 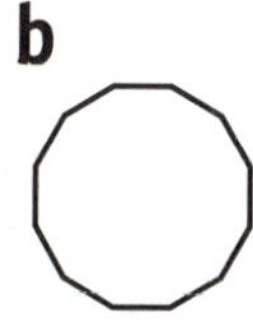 c 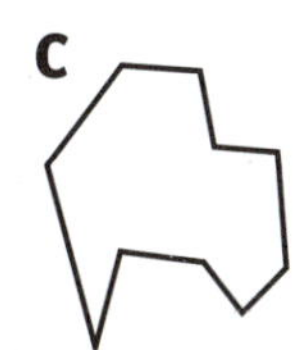 d 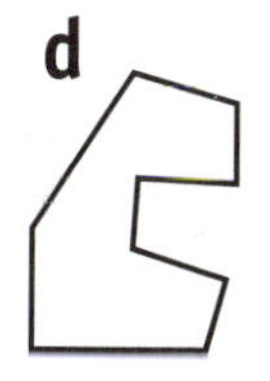 e 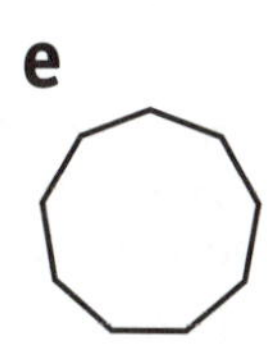 f 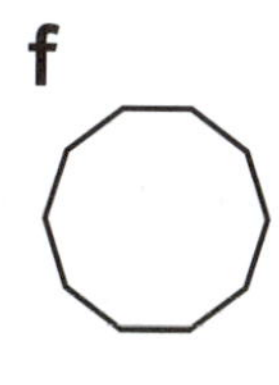 g 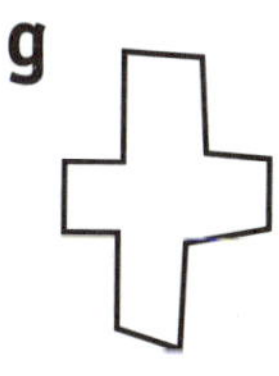

  ISBN: 9781925726176

TRANSFORMATIONS

A transformation is when a shape is moved to a different position.

Flip (reflection) 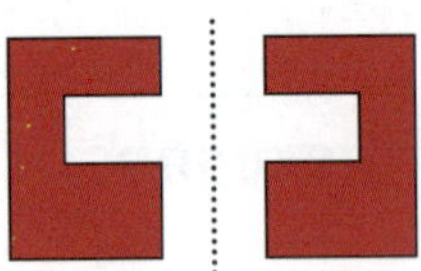 When an object is flipped, it is a mirror image.

Slide (translation)  When an object slides, it is not lifted or turned. The direction doesn't change.

Turn (rotation) 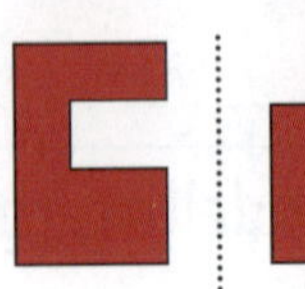  The object is turned or rotated around a point.

Examples:

Trace the reflections.	Trace the translations.	Trace the rotations.
a Q | Q	a 4 | 4	a
b B | B	b 3 | 3	b
c D | D	c 8 | 8	c
d E | E	d 5 | 5	d 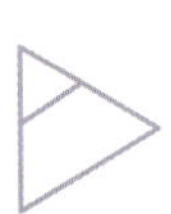

Check the answers on the video!

Your turn

Label each transformation as reflection, translation or rotation.

● translation　　a ______________　　b ______________

SELF CHECK　Tick how you feel

Got it!	Need help...	I don't get it
☐	☐	☐

Check your answers
How many did you get correct? ☐

CATCH UP MATHS YEAR 5 BOOK B © PASCAL PRESS ISBN: 9781925726176

PRACTICE

1 Label each transformation as a reflection, rotation or translation.

translation

a  ______

b     ______

c     ______

2 Flip (reflect) the shapes to continue the pattern.

a ______ ______ ______ ______

b ______ ______ ______ ______

c ______ ______ ______ ______

3 Slide (translate) the shapes to continue the pattern.

a ______ ______ ______ ______ ______

b ______ ______ ______ ______ ______

c ______ ______ ______ ______ ______

4 Turn (rotate) the shapes to continue the pattern.

a  ______ ______ ______ ______ ______

b  ______ ______ ______ ______ ______

c  ______ ______ ______ ______ ______

TESSELLATIONS

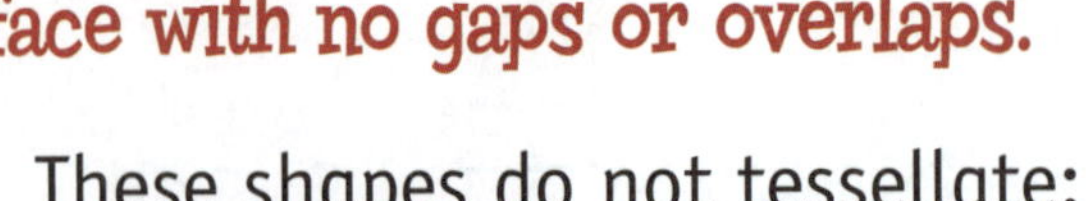

A tessellation is a pattern of 2D shapes.
The shapes cover a flat surface with no gaps or overlaps.

SCAN to watch video

These rectangles tessellate:

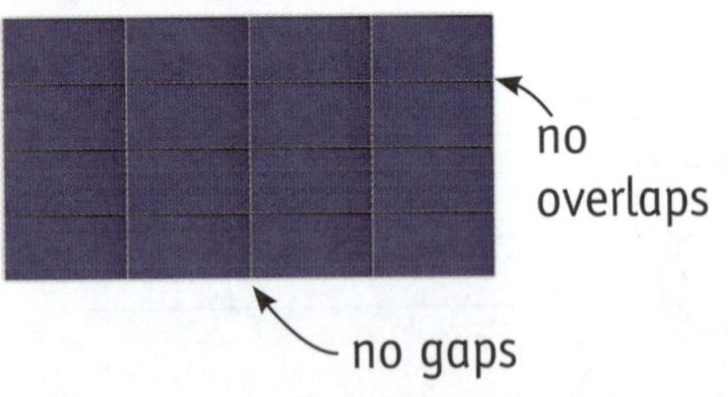

These shapes do not tessellate:

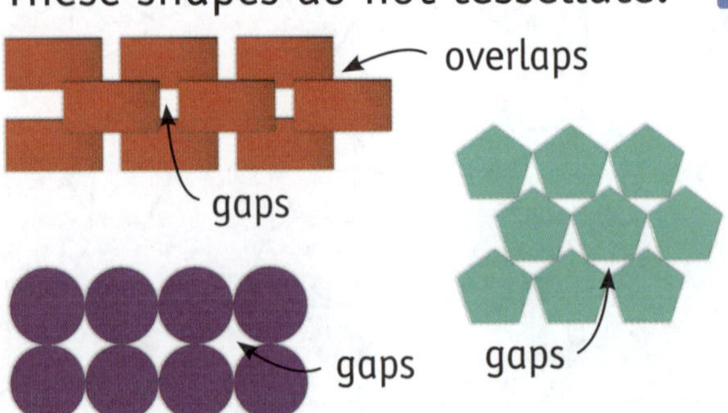

You can use just one shape to make a tessellation. Here is a hexagon tessellation:

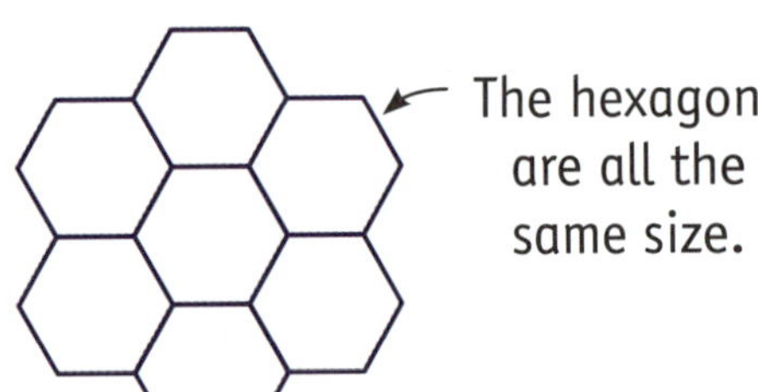

You can also use a combination of shapes to make a tessellation. Here, hexagons and triangles make a tessellating pattern:

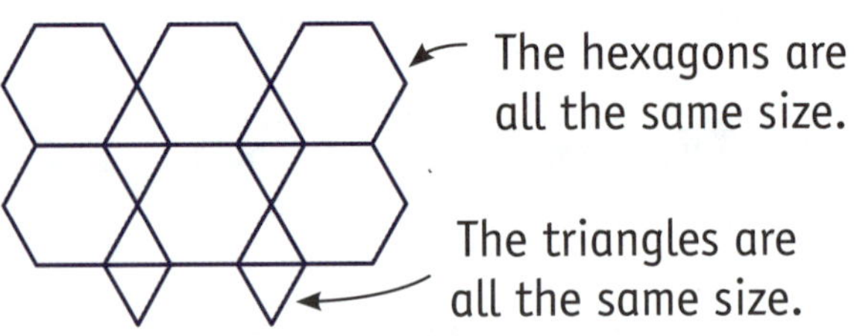

Examples: Trace the shapes to continue the tessellations.

a Chess board

b Tiles on a floor

c A beehive

Check the answers on the video!

Your turn

Tick the shapes that will tessellate on their own.
Cross the shapes that will not tessellate.

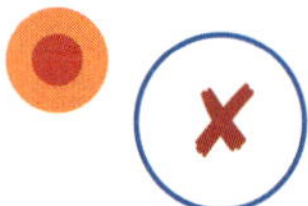

a

b

c

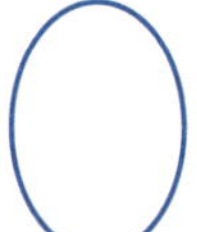

d

SELF CHECK Tick how you feel

Got it!	Need help...	I don't get it

Check your answers

How many did you get correct?

CATCH UP MATHS YEAR 5 BOOK B © PASCAL PRESS ISBN: 9781925726176

PRACTICE

1 Continue the tessellations and then colour them to show a pattern.

a

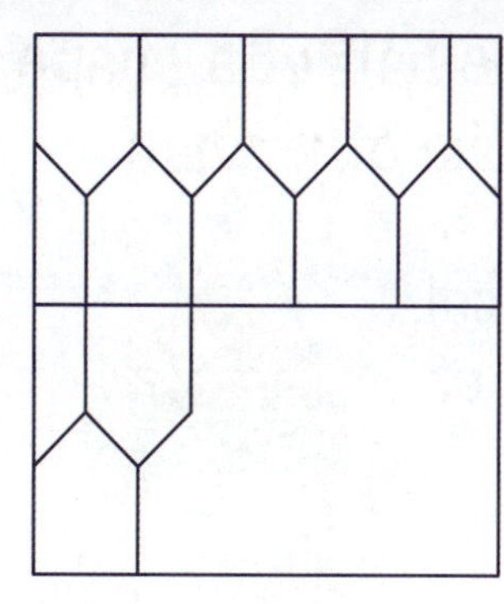

b

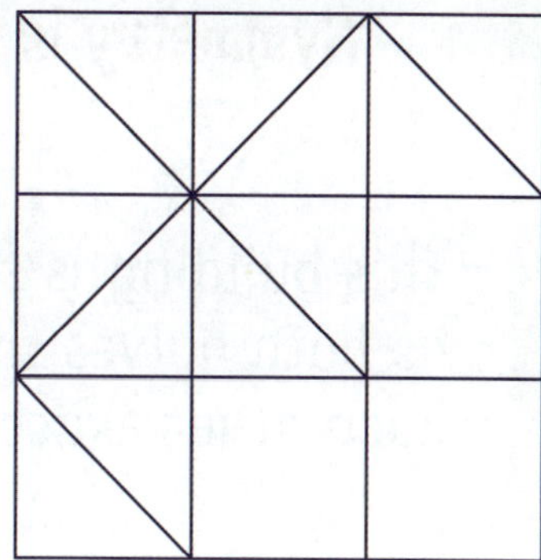

2

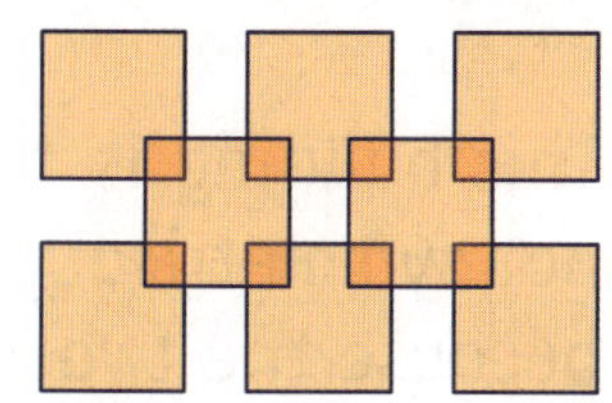

Is this a tessellation? ______

Explain your answer.

____

3 Shapes were reflected to make these tessellations.
Continue the tessellations.

a

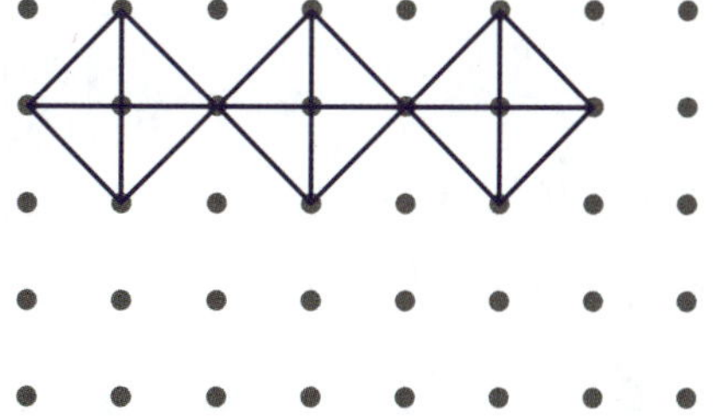

b

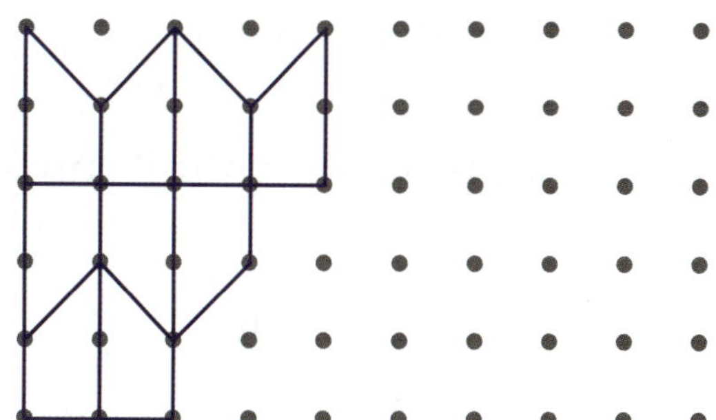

4 Make your own tessellations with these shapes.
You can rotate, reflect or slide the shapes.

a

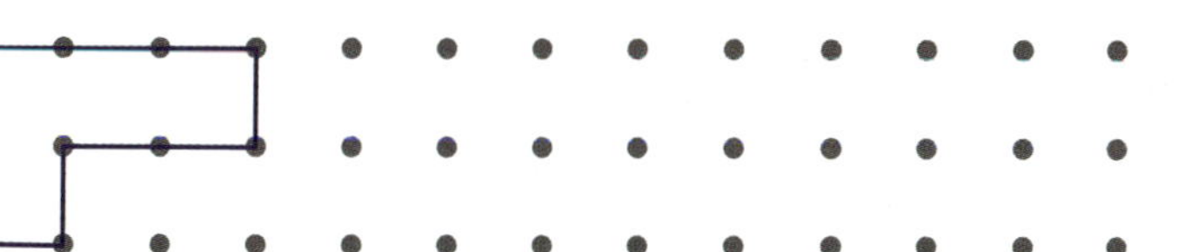

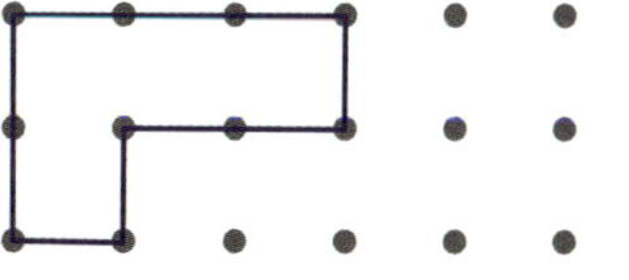

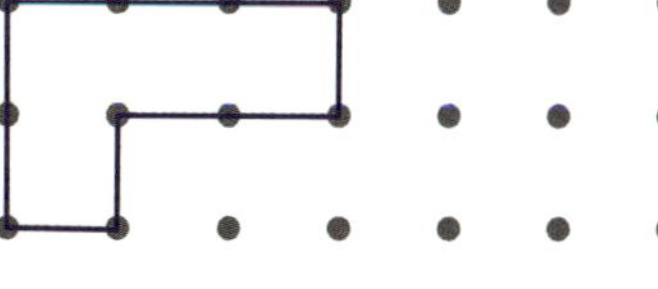

b

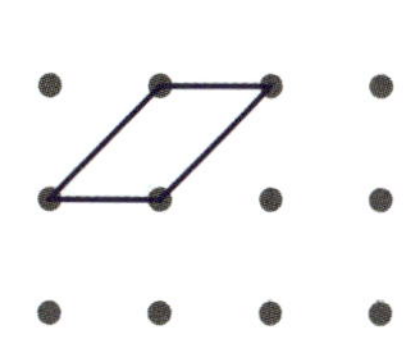

SYMMETRY

Symmetry is when one half of a shape is a reflection of the other half.

This building is symmetrical as both halves fit on top of each other exactly.

A line of symmetry is always dotted.

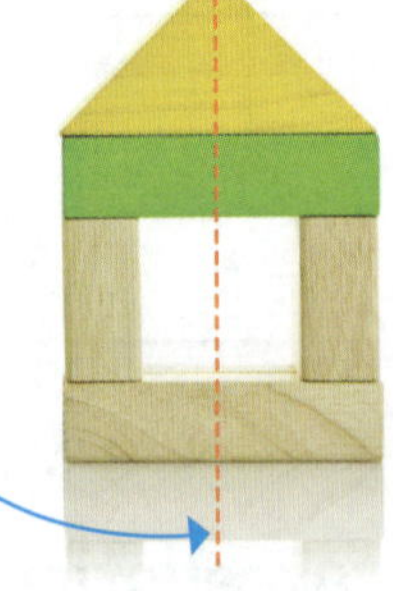

This building is not symmetrical because the two sides are not the same size.

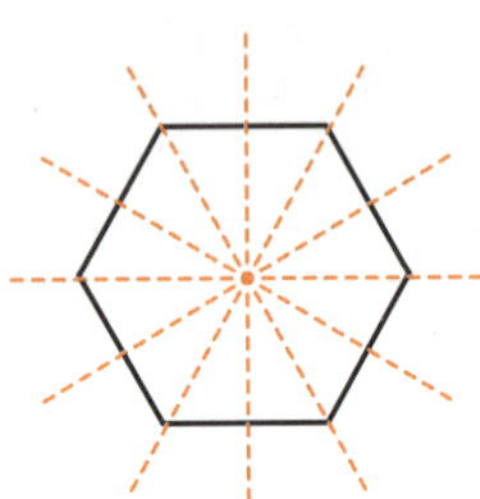

This hexagon has 6 lines of symmetry.

Examples: Circle the shapes that are symmetrical and cross the shapes that are not symmetrical.

a

b

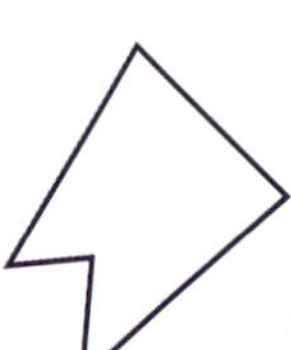

c

d

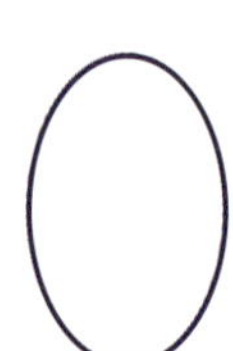

e

Your turn

Draw the lines of symmetry.

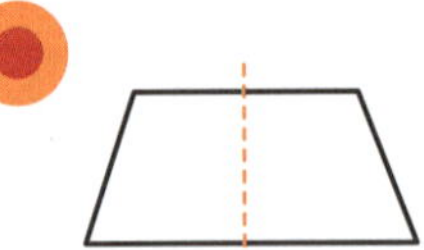

a

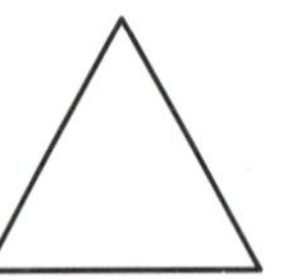

b

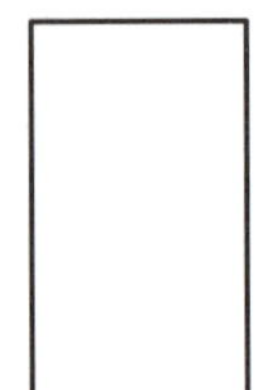

c

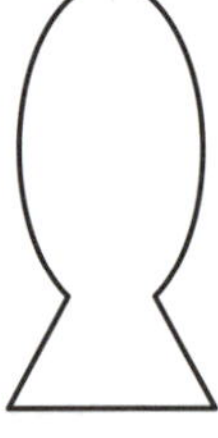

SELF CHECK Tick how you feel		
Got it!	Need help...	I don't get it

Check your answers

How many did you get correct?

CATCH UP MATHS YEAR 5 BOOK B © PASCAL PRESS ISBN: 9781925726176

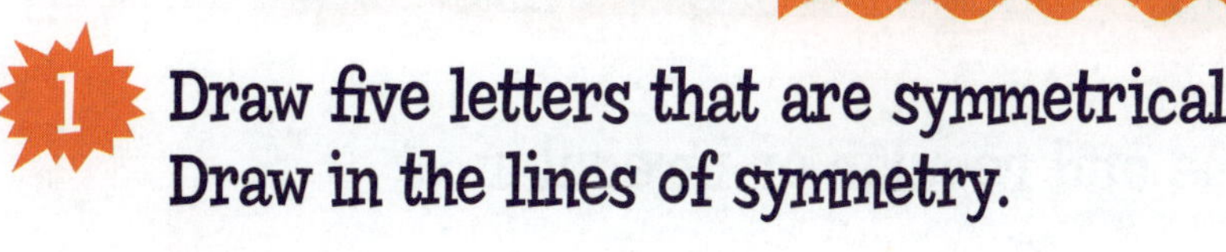

1 Draw five letters that are symmetrical.
Draw in the lines of symmetry.

    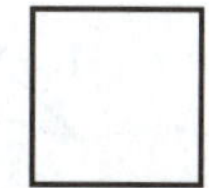

2 Write your name in block letters.
Draw in any lines of symmetry on the letters.

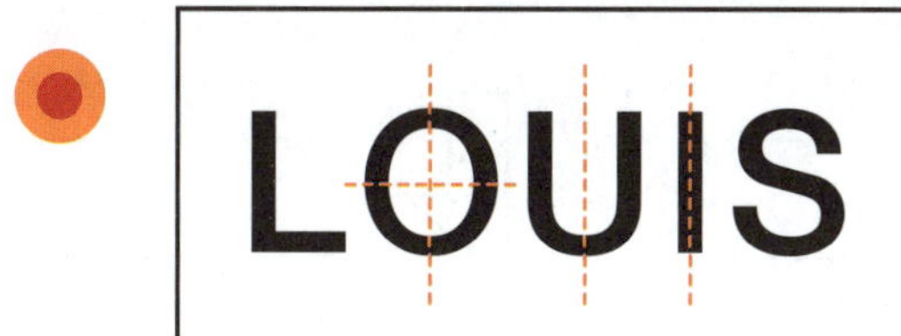

3 Use these shapes to answer the following questions.

a Draw any lines of symmetry on the shapes above.

b Tick the shapes that are symmetrical.

c Cross the shapes that are not symmetrical.

d Circle the shapes that have more than one line of symmetry.

4 Complete the other half of the picture along the axis of symmetry.

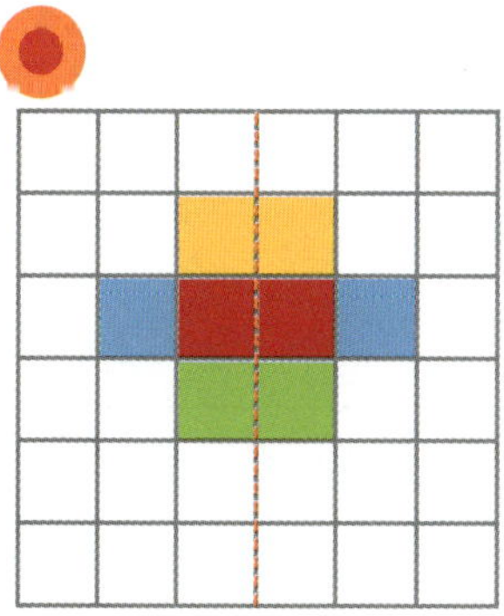

a

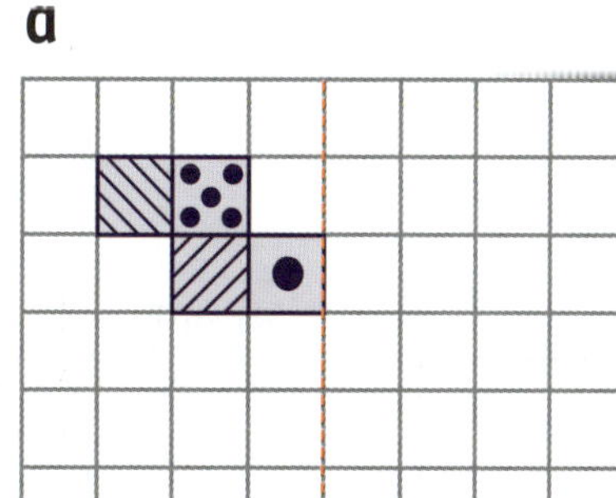

b

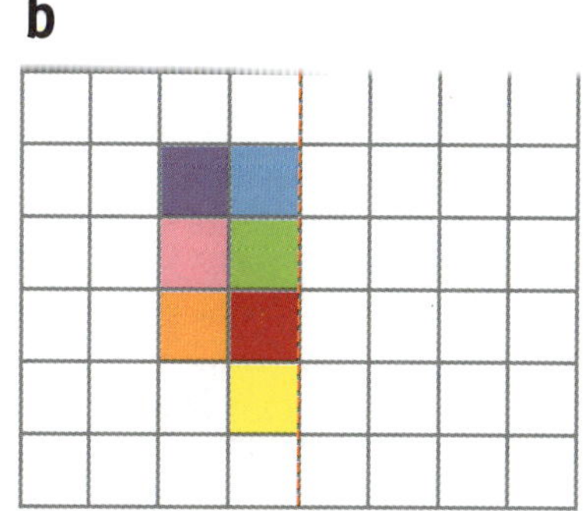

2D SHAPES REVIEW

1 Label each shape with the name and regular or irregular.

a ______ ______

b ______ ______

c ______ ______

d ______ ______

e ______ ______

f ______ ______

g ______ ______

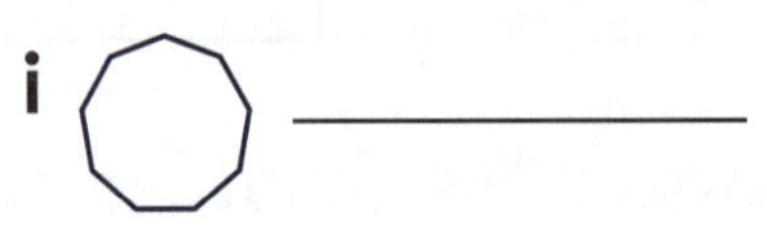

h ______ ______

i ______ ______

j ______ ______

k ______ ______

l ______ ______

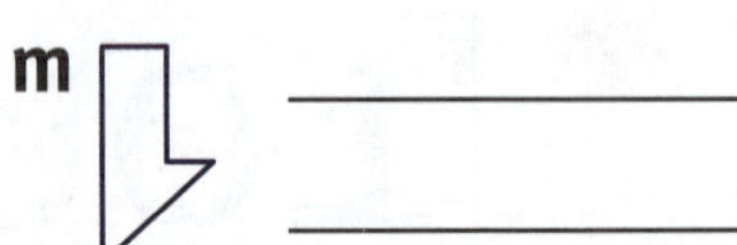

m ______ ______

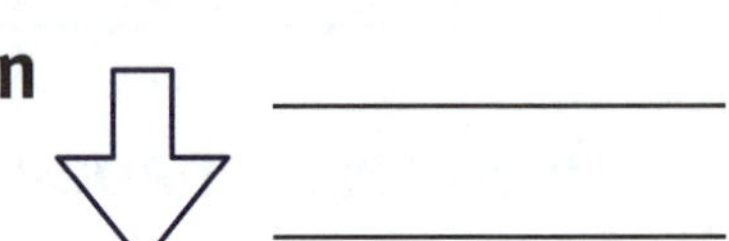

n ______ ______

o ______ ______

2 Label the lines as vertical, horizontal, parallel or perpendicular.

a ______

b ______

c ______

d ______

e ______

f ______

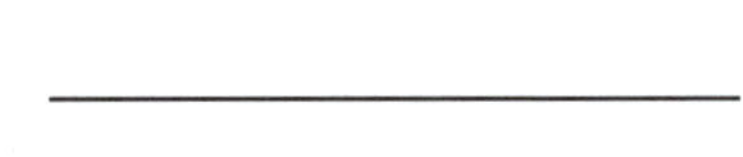

g ______

h ______

i ______

j ______

ISBN: 9781925726176

3 Draw three examples of each type of line.

Vertical	Perpendicular	Parallel	Horizontal

4 Are these regular circles or irregular circles?

a ____________

b ____________

c ____________

d ____________

5 Use the diagram to complete the sentences.

This is a ____________.

It is half of a ____________.

6 Complete the instructions on the circle diagram.

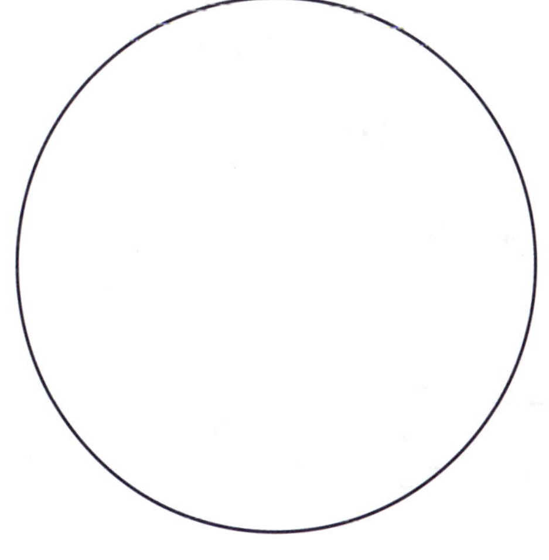

a Draw a • at the centre of the circle.

b Draw a radius in blue.

c Trace around the circumference in purple.

d Draw a diameter in yellow.

e Draw a chord in green.

f Trace an arc in orange.

g Colour a sector in red.

REVIEW

7 Colour the circular bases of the cylinders in red and the base on each cone in blue.

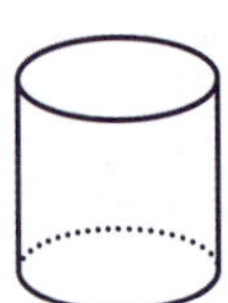 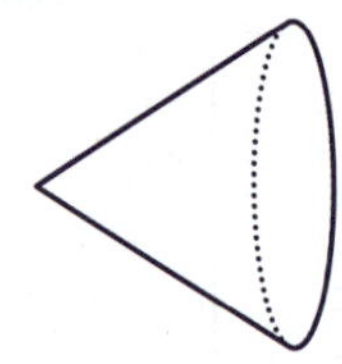 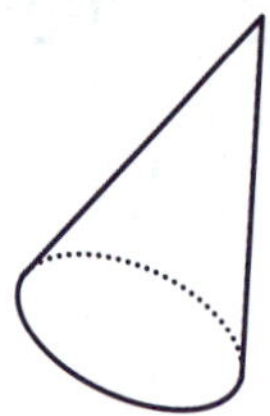  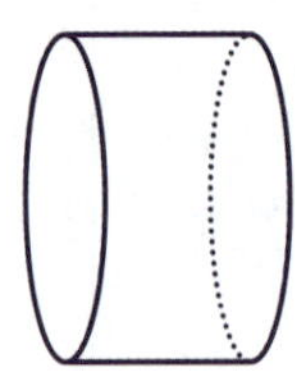 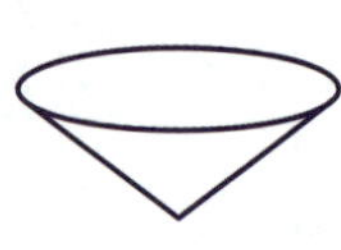

8 Label the triangles.

a 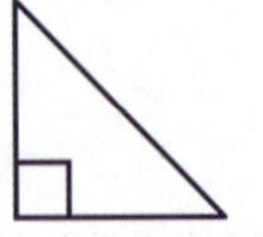 ______________________

c  ______________________

b  ______________________

d 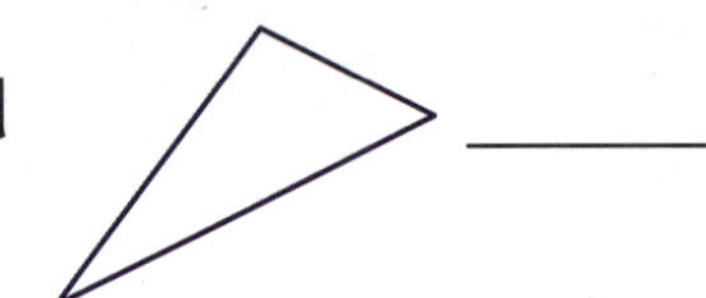 ______________________

9 Match the descriptions to the triangle types.

Descriptions	Triangle types	Descriptions
Two sides the same length	Equilateral triangle	One right angle
All sides different lengths	Isosceles triangle	All sides the same length
Two angles the same size	Scalene triangle	All angles different sizes
	Right-angled triangle	All angles the same size

10 Label the quadrilaterals.

a  ______________________

d 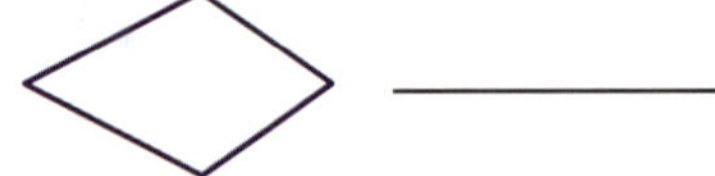 ______________________

b  ______________________

e 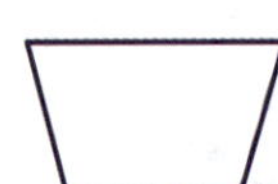 ______________________

c  ______________________

f  ______________________

CATCH UP MATHS YEAR 5 BOOK B © PASCAL PRESS ISBN: 9781925726176

11 Use the quadrilaterals in Question 10 to answer the questions.

a Which quadrilaterals have parallel lines? ________

b Which quadrilaterals have perpendicular lines? ________

12 Write the number of sides each polygon has.

a octagon __

b nonagon __

c irregular heptagon __

d dodecagon __

e irregular decagon __

f quadrilateral __

g pentagon __

h irregular hexagon __

i heptagon __

j triangle __

13 Label each transformation as flip, turn or slide.

a 9 9 9 9 9 ______________

b ______________

c      ______________

14 Reflect the shapes to continue the pattern.

a     _____ _____ _____ _____

b  _____ _____ _____ _____

c     _____ _____ _____ _____

REVIEW

15 Translate the shapes to continue the pattern.

a M M M M ____ ____ ____ ____

b 4 4 4 4 ____ ____ ____ ____

c    ____ ____ ____ ____

16 Rotate the shapes to continue the pattern.

a ____ ____ ____ ____

b ____ ____ ____ ____

c    ____ ____ ____ ____

17 Make your own patterns using these transformations.

a Rotation

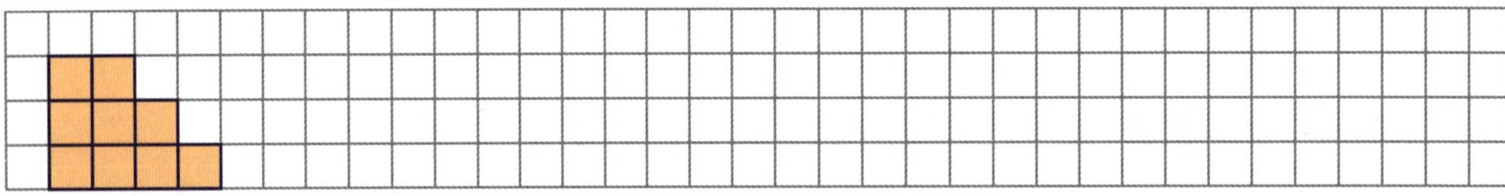

b Translation

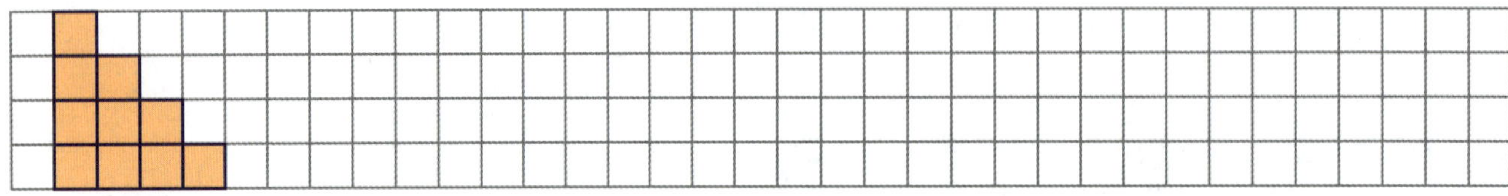

c Reflection

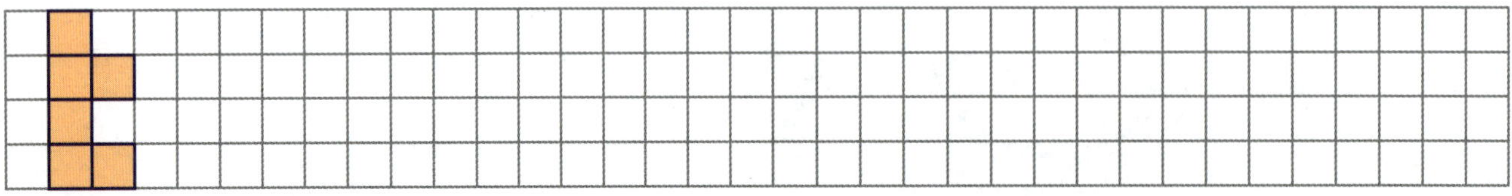

CATCH UP MATHS YEAR 5 BOOK B © PASCAL PRESS ISBN: 9781925726176

18 Name three shapes that tessellate on their own.

____

19 Name three shapes that do not tessellate on their own.

____

20 Make your own tessellations with these shapes.

a 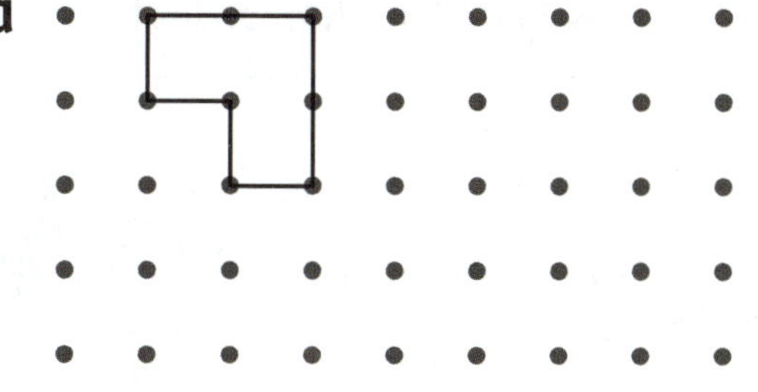

b 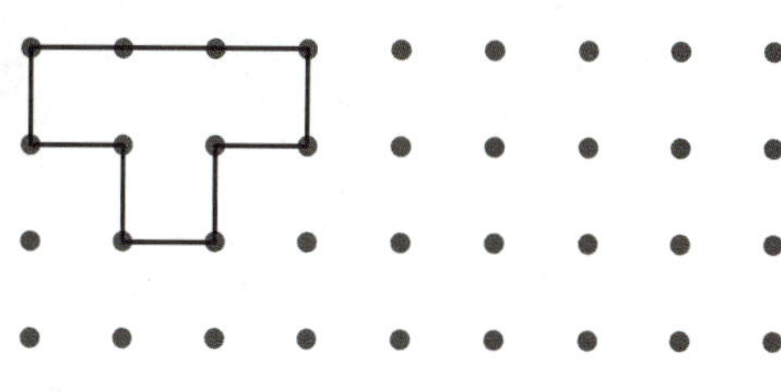

21 Complete the sentence.

Symmetry is when one ______ of a shape is a ______________ of the other half.

22 Write three numbers that are symmetrical.

a ___ b ___ c ___

23 Draw in the lines of symmetry.

a 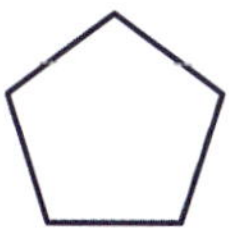

b 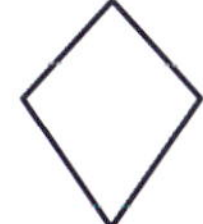

c 

d 

e 

24 Complete the other half of the picture along the axis of symmetry.

a 

b 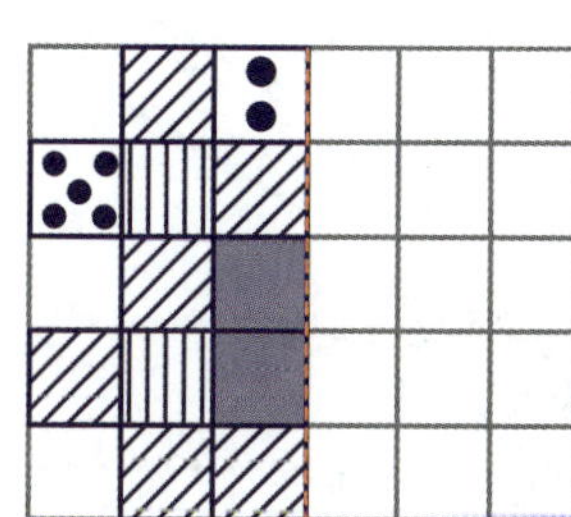

c 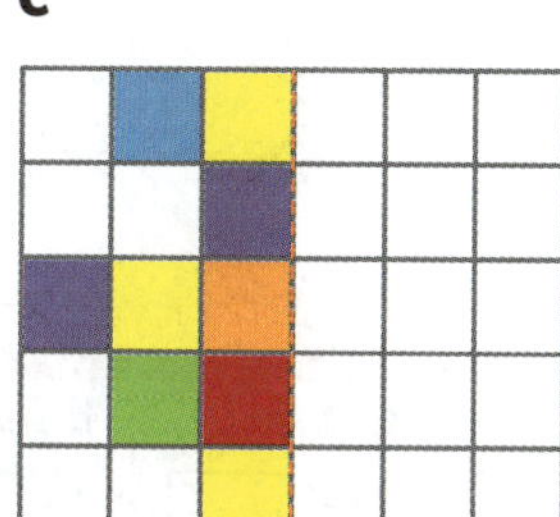

PRISMS

Prisms are 3D objects. They have flat faces (planes).
They have one pair of parallel bases.
The shape of the bases names the prism.
Rectangles join the bases.

SCAN to watch video

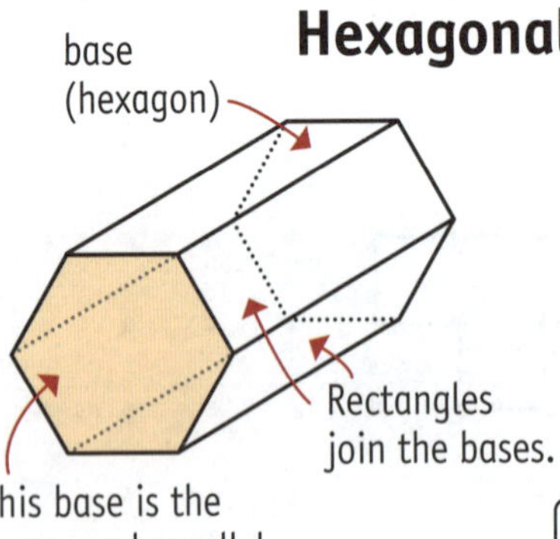

Three-dimensional objects have three dimensions: height, length and depth.

A rectangle is a two-dimensional shape.

Examples: Use blue to colour the bases of each prism.

a 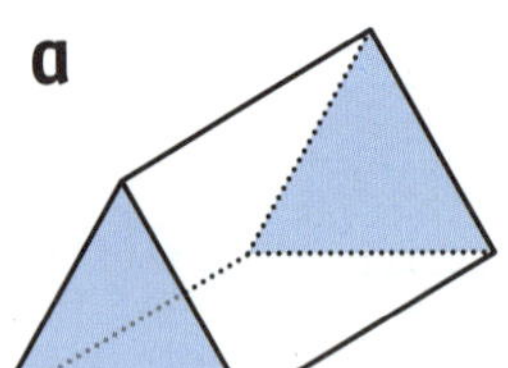

b 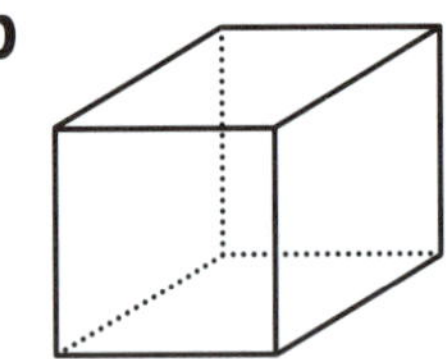

c 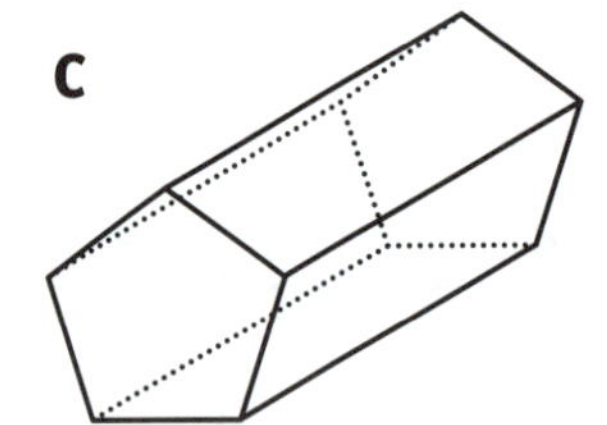

d 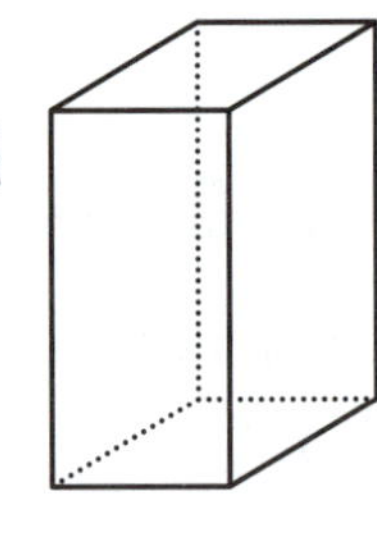

Check the answers on the video!

Remember, rectangular sides join the bases.

Your turn

Use blue to colour the bases of each prism and use red to colour the sides.

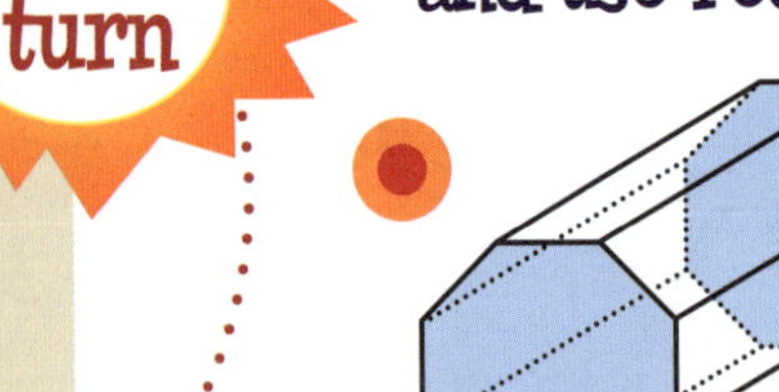

b 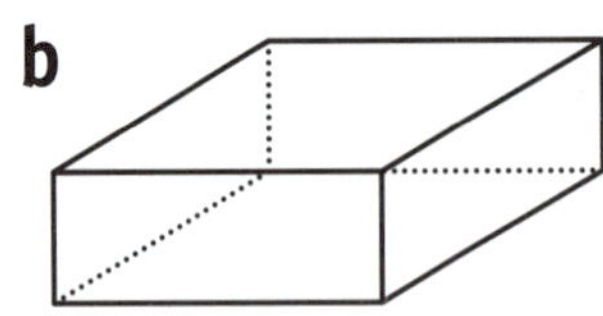

d 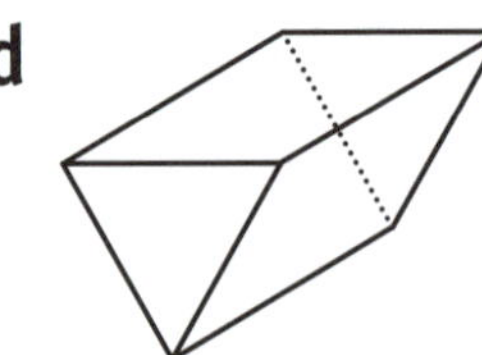

a 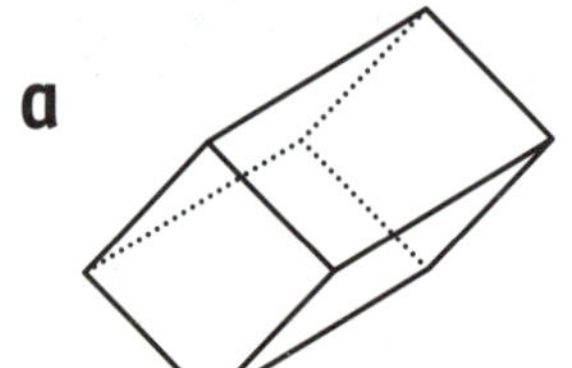

c 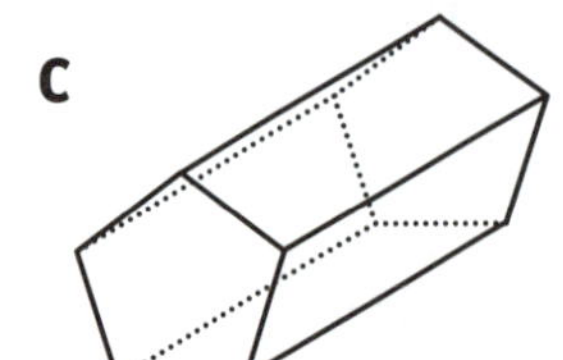

e 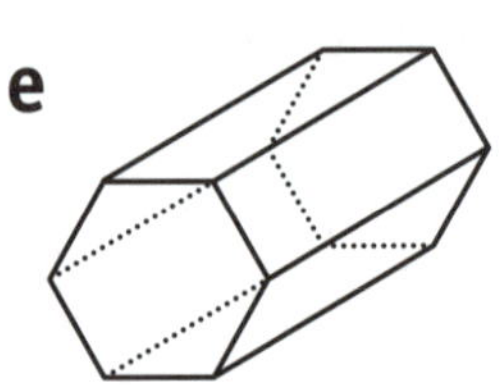

SELF CHECK Tick how you feel		
Got it!	Need help...	I don't get it

Check your answers
How many did you get correct?

CATCH UP MATHS YEAR 5 BOOK B © PASCAL PRESS ISBN: 9781925726176

PRACTICE

1 Fill in the table.

	Name of prism	Picture	Number of vertices	Number of edges	Number of faces
●	cube		8	12	6
a	rectangular prism				
b	pentagonal prism				
c	hexagonal prism				
d	octagonal prism				
e	triangular prism				

2 Which prism am I?

● I have six faces that are all squares.

____

a I have 8 faces and 12 vertices. Two faces are hexagons.

____

b I have 3 rectangular faces and 2 triangular faces.

____

c I have 6 rectangular faces.

____

3 Describe the prism.

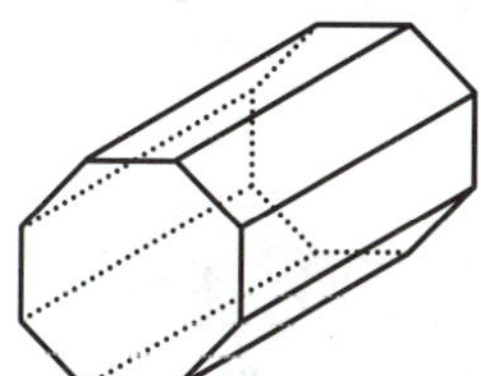

____

____

PYRAMIDS

Pyramids are 3D objects. They have flat faces (planes). They have one base. The shape of the base names the pyramid. The other faces are triangles and they meet at a point called the apex.

Square pyramid

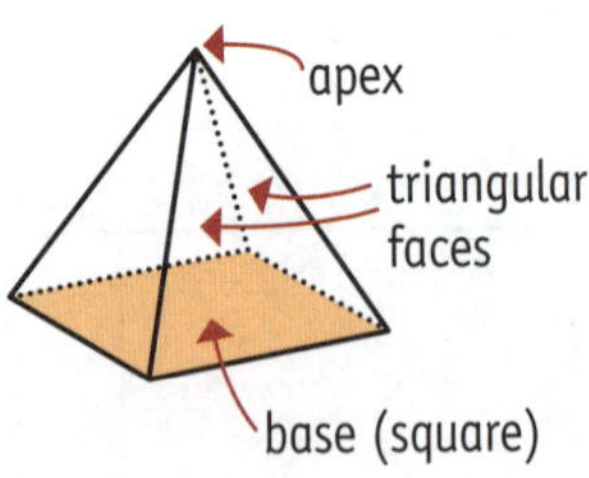

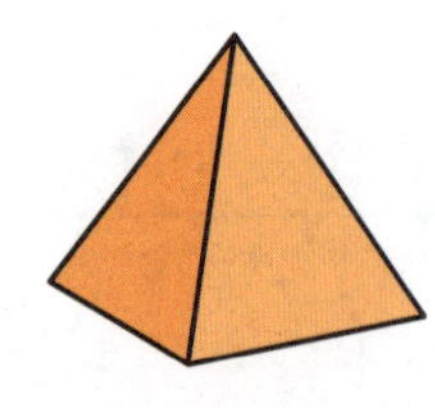

Examples: Colour the base of each pyramid blue.

a 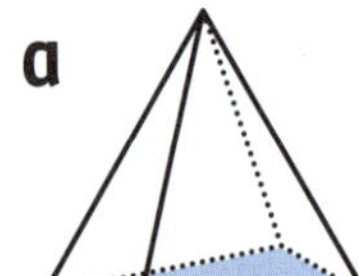

b 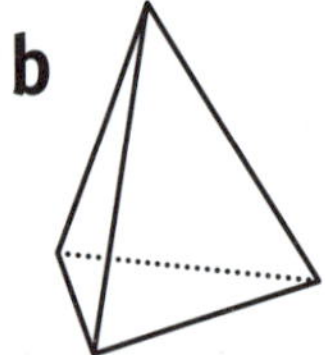

c 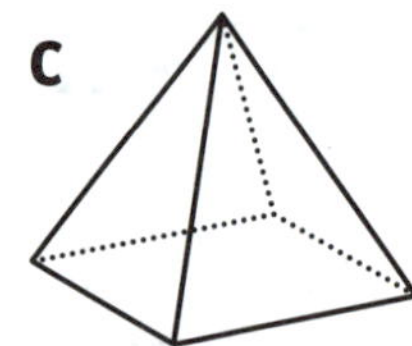

d 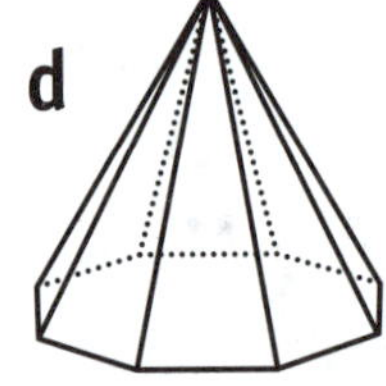

Your turn

1 Colour the base of each pyramid blue and the triangular sides green.

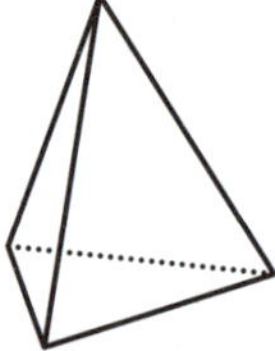

a 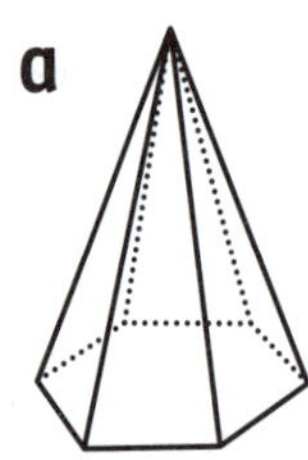

b 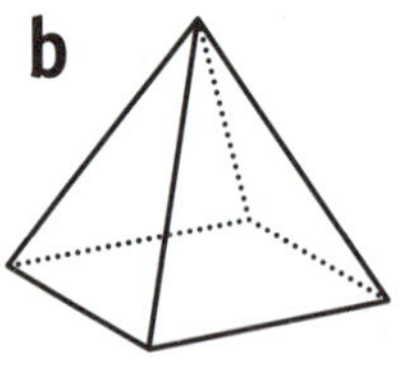

c 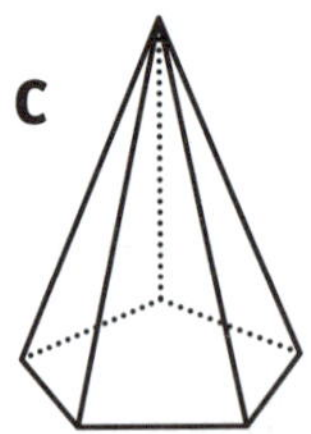

2 Name each pyramid in Question 1.

● triangular pyramid

a ______________________

b ______________________

c ______________________

SELF CHECK Tick how you feel

Got it!	Need help...	I don't get it
☐	☐	☐

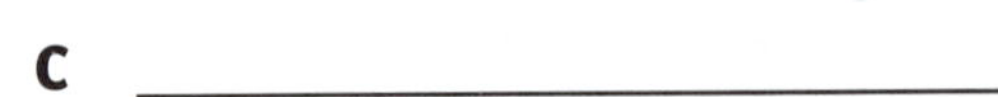

Check your answers

How many did you get correct? ☐

CATCH UP MATHS YEAR 5 BOOK B © PASCAL PRESS ISBN: 9781925726176

PRACTICE

1 Dotted lines are often used to mark hidden edges. Draw the missing dotted lines on the pyramids.

- rectangular pyramid

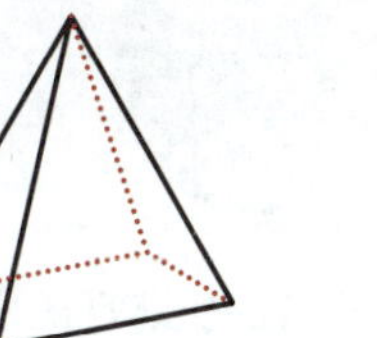

a pentagonal pyramid

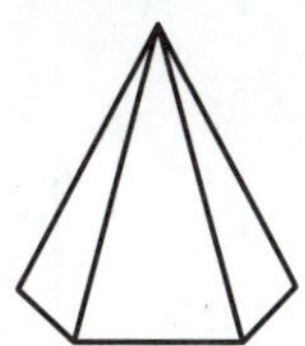

b triangular pyramid

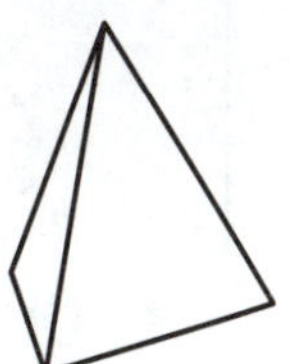

c hexagonal pyramid

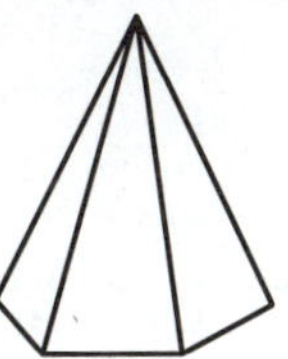

2 Draw a diagram and write the missing numbers.

- square pyramid
  bases: 1
  faces: 4
  apexes: 1

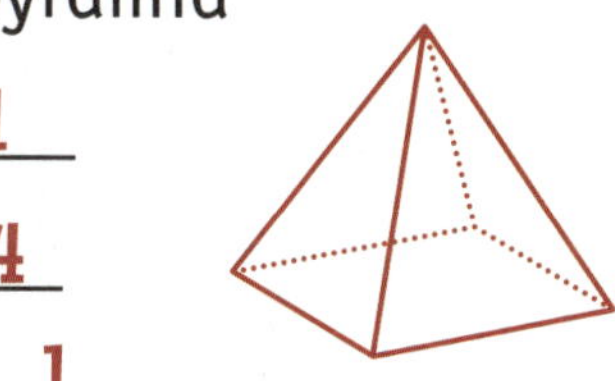

a rectangular pyramid
bases: ___
faces: ___
apexes: ___

b triangular pyramid
bases: ___
faces: ___
apexes: ___

c pentagonal pyramid
bases: ___
faces: ___
apexes: ___

d hexagonal pyramid
bases: ___
faces: ___
apexes: ___

e octagonal pyramid
bases: ___
faces: ___
apexes: ___

3 Match the top view of each pyramid with the correct name.

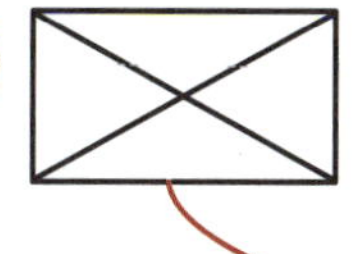

a

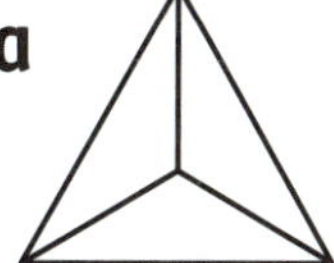

b

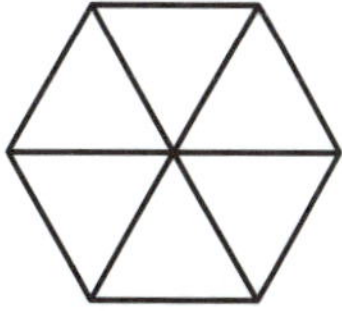

c

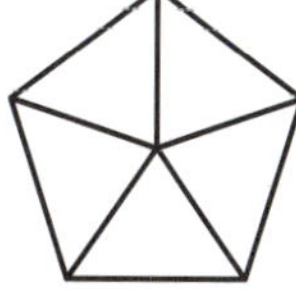

d

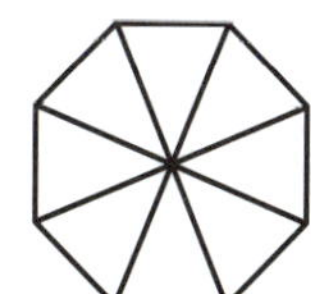

octagonal pyramid	hexagonal pyramid	rectangular pyramid	triangular pyramid	pentagonal pyramid

CURVED SURFACES

Cylinders, cones and spheres have curved surfaces.

Cylinder

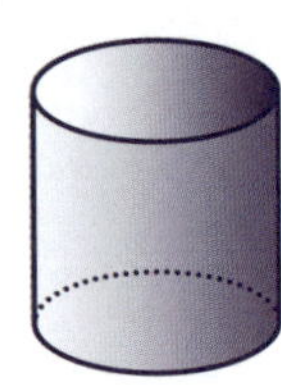

- Two circles for bases (flat surfaces)
- One curved surface
- No faces, edges or corners

Cone

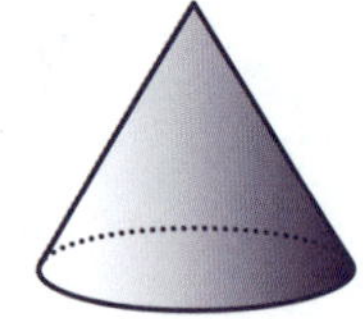

- One circle for the base (flat surface)
- One curved surface
- One apex
- No faces or edges

Sphere

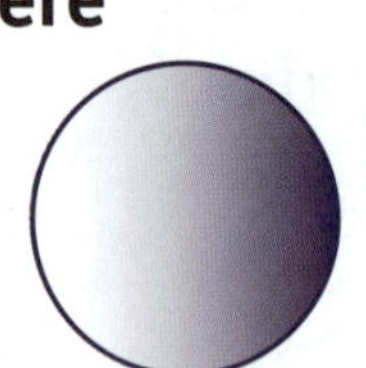

- No flat surfaces (faces)
- No corners
- No straight edges
- One curved surface
- Round

Examples: Shade the objects to show the curves.

a 

b 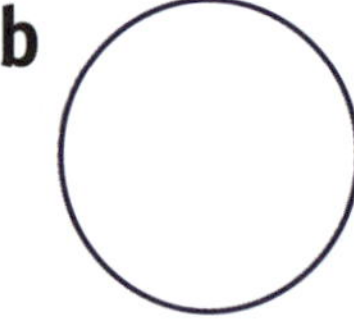

c 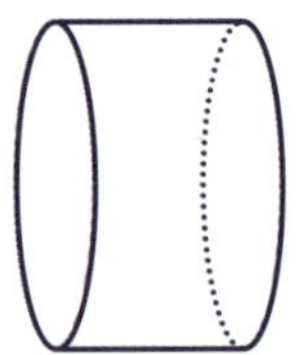

d 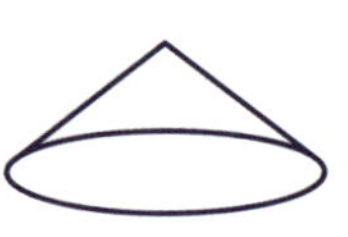

Check the answers on the video!

Your turn

1 Complete and label the 3D objects.
Remember to shade the objects to show the curves.

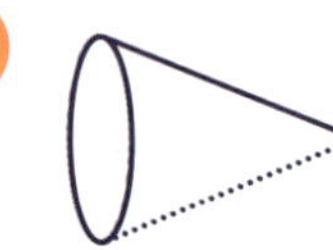
cone

b 
____

d 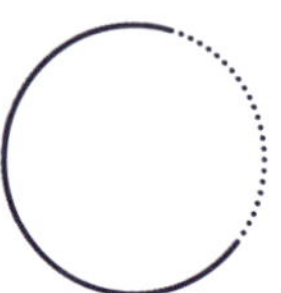
____

f 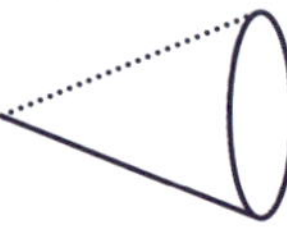
____

a 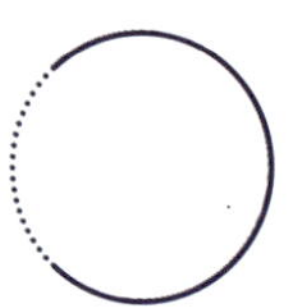
____

c 
____

e 
____

g 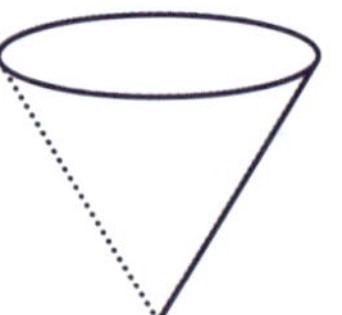
____

2 Colour the 3D objects in Question 1. Use red for cylinders, green for cones and yellow for spheres.

SELF CHECK Tick how you feel		
Got it! ☐	Need help... ☐	I don't get it ☐

Check your answers
How many did you get correct? ☐

CATCH UP MATHS YEAR 5 BOOK B © PASCAL PRESS ISBN: 9781925726176

PRACTICE

Write the letters of the pictures that are shaped like each of these 3D objects.

a sphere ____________ b cone ____________ c cylinder ____________

A

C

E

G

I

B

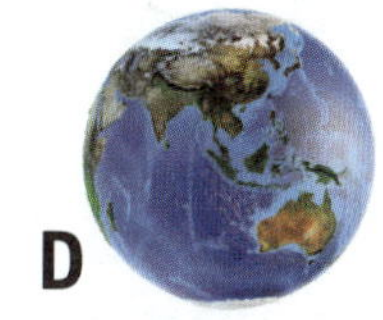
D

F

H

J

Complete the table.

	Solid	Top view	Front view	Side view	Cross-section
a					
b					
c					

Are the nets for a cone or a cylinder?

a cylinder

b ____________

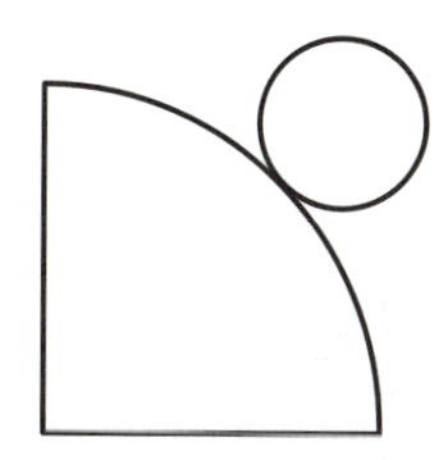

c ____________

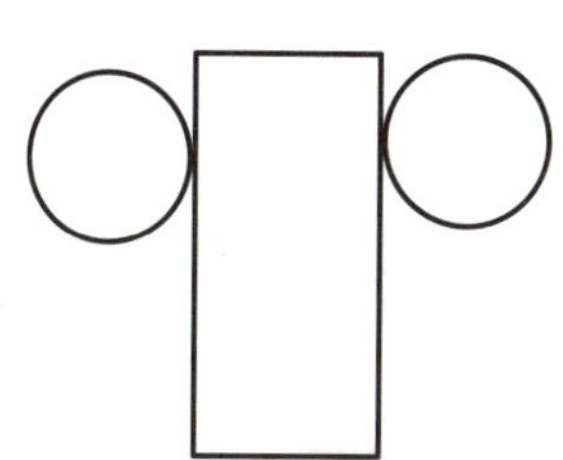

CROSS-SECTIONS

A cross-section is the shape you see when you cut through a 3D object.

SCAN to watch video

In a pyramid, the cross-section is the same shape as the base. It gets smaller the closer it is cut to the apex.

In a prism, the cross-section is the same shape and size as the base.

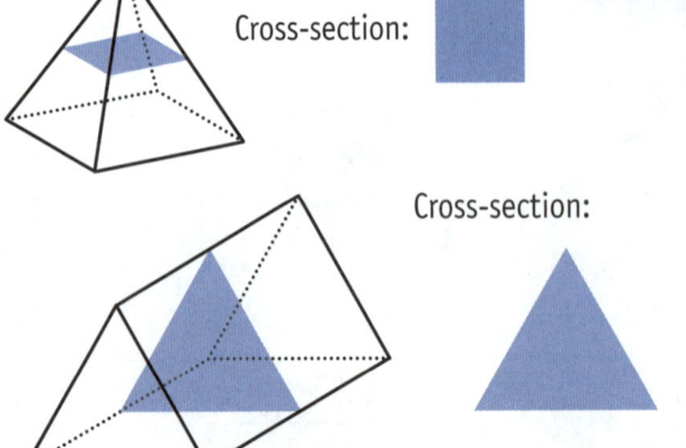

Examples: Circle the cross-section you would see if you cut the 3D objects along the blue line.

a

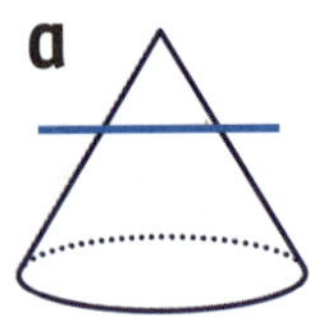

b

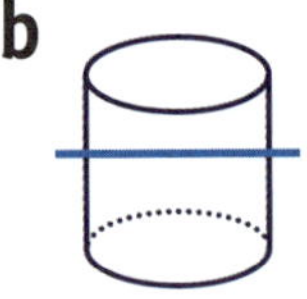

Draw a line on the 3D object to show where the cross-section was made.

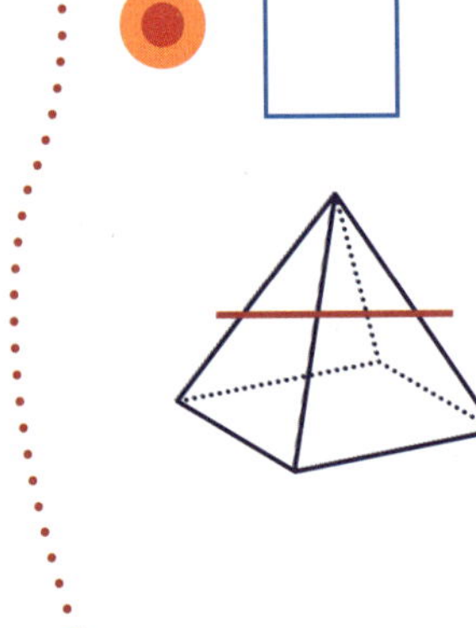

a

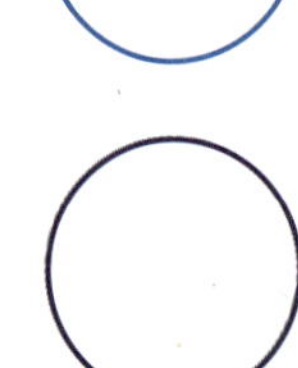

b

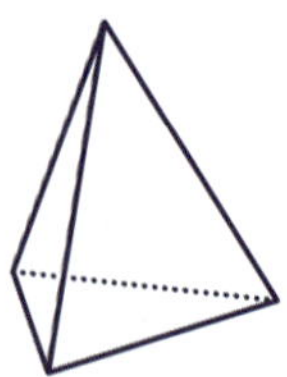

SELF CHECK Tick how you feel

Got it!	Need help...	I don't get it
☐	☐	☐

Check your answers

How many did you get correct? ☐

CATCH UP MATHS YEAR 5 BOOK B © PASCAL PRESS ISBN: 9781925726176

PRACTICE

Draw the cross-section made by the orange line.

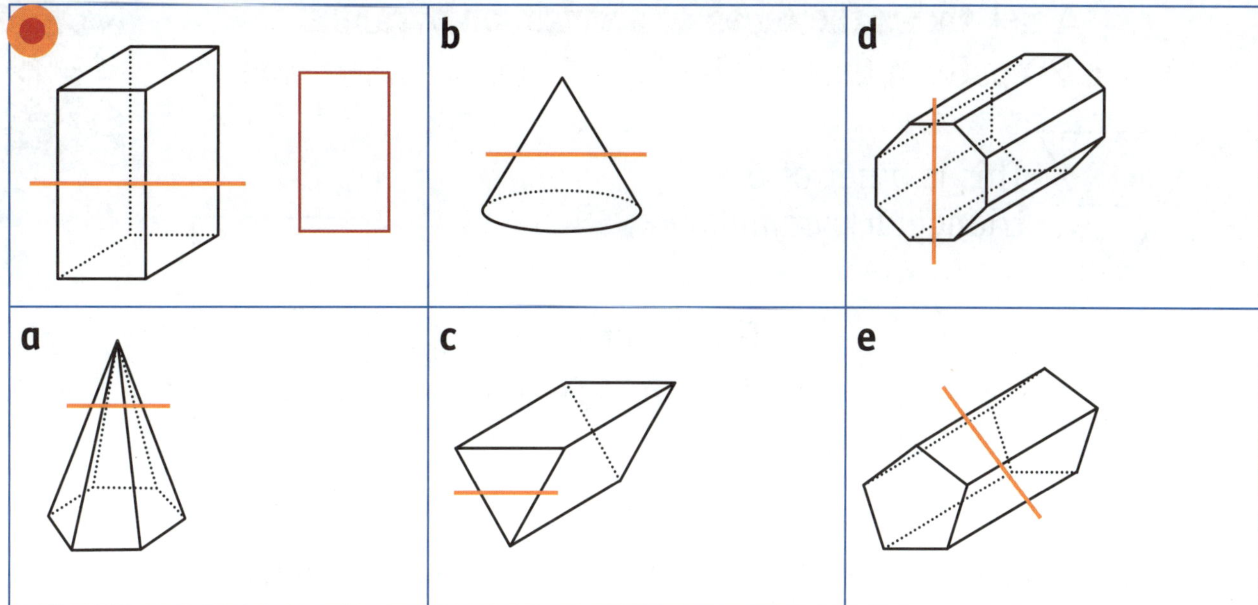

Draw the cross-sections made by the red lines and the blue lines.

	Object	Cross-section: red line	Cross-section: blue line
a			
b			
c			
d			

  ISBN: 9781925726176

NETS OF PRISMS AND PYRAMIDS

A net shows the faces of a prism or pyramid.
It is a 2D pattern that can be folded to make a 3D object.

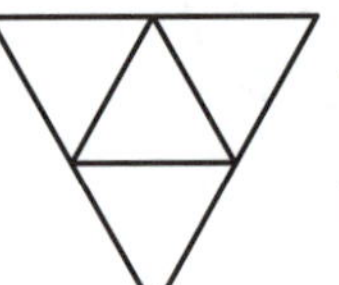

This is a net of a triangular pyramid.

This is a net of a triangular prism.

Examples: Match the net with the name of the 3D object.

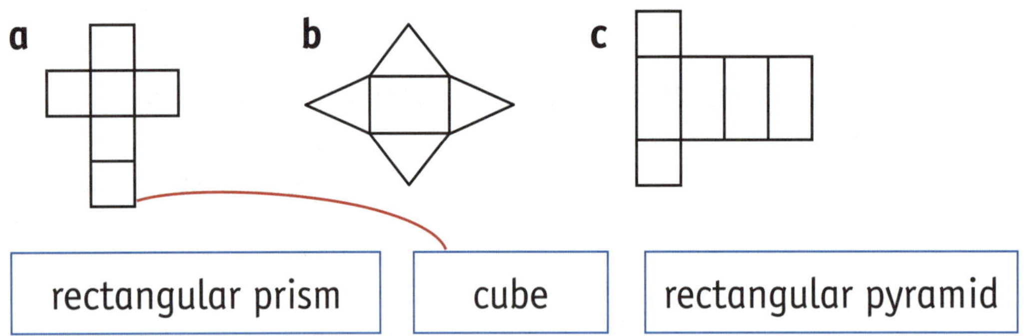

Check the answers on the video!

Draw the 3D object you would make from each net in the Examples above.

a

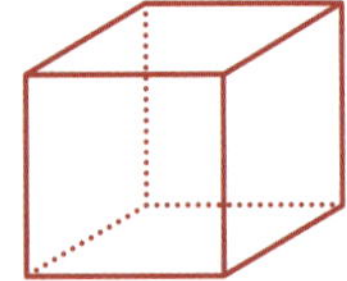

b

c

SELF CHECK Tick how you feel

Got it!	Need help...	I don't get it
☐	☐	☐

Check your answers
How many did you get correct? ☐

CATCH UP MATHS YEAR 5 BOOK B © PASCAL PRESS ISBN: 9781925726176

PRACTICE

1 Name the 3D object each net will make.

b 

d 

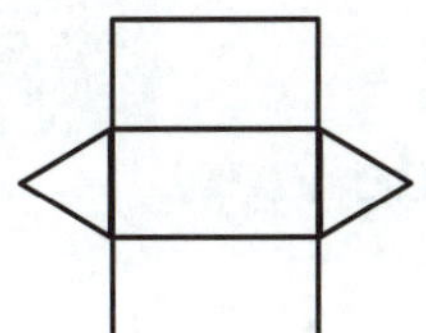

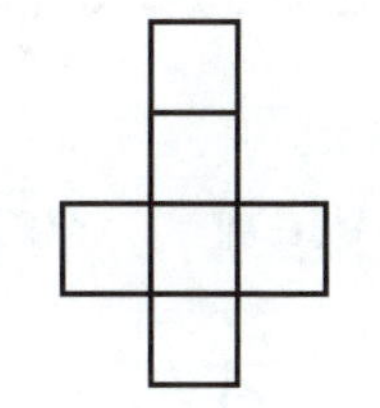

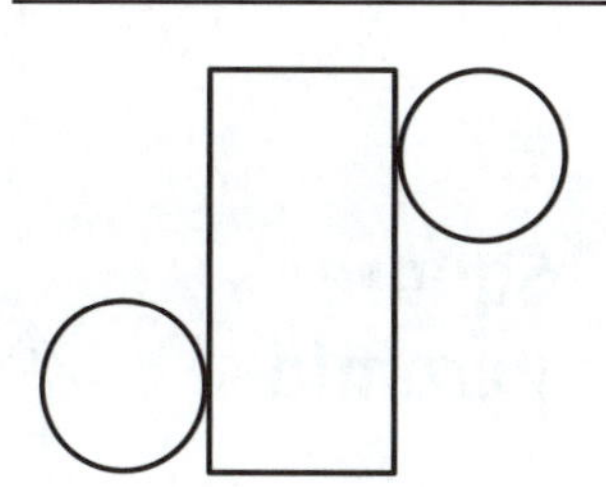

a ____________ c ____________ e ____________

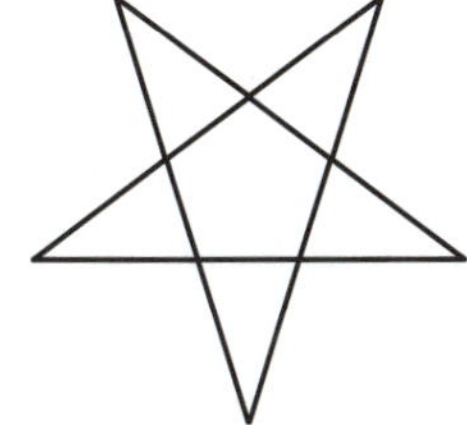

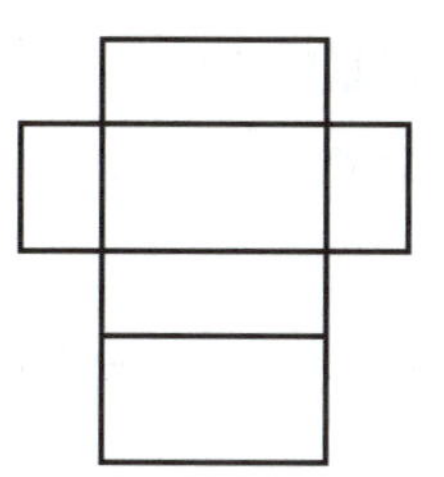

2 Draw a net for each 3D object.

b octagonal prism

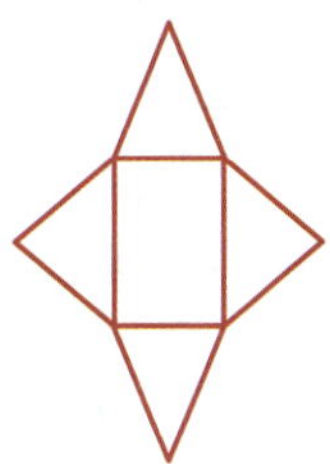

3 Join the nets to the correct label.

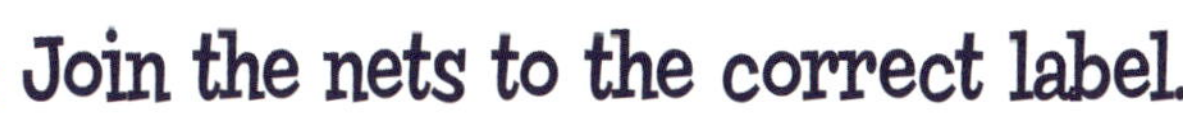

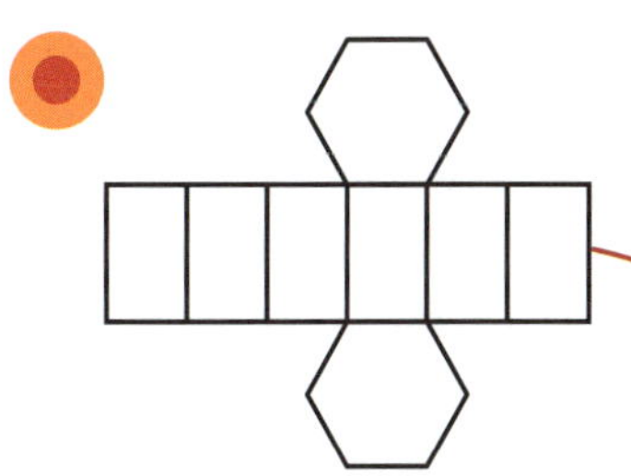

b 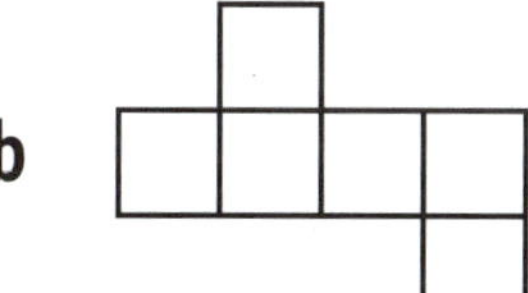

pentagonal pyramid

cube

hexagonal prism

triangular prism

octagonal pyramid

c

a 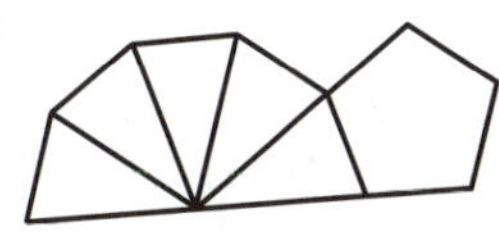

d 

VIEWS

Three-dimensional (3D) objects look different when you look at them from different views.

SCAN to watch video

	Top view	Front view	Side view
Square pyramid 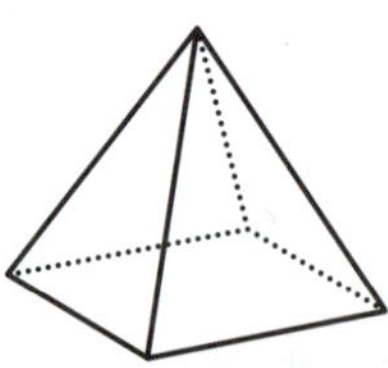			
Rectangular prism 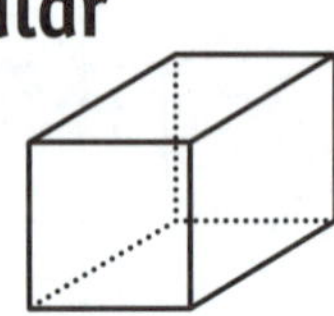			

Examples: Label the views as top, front or side.

a  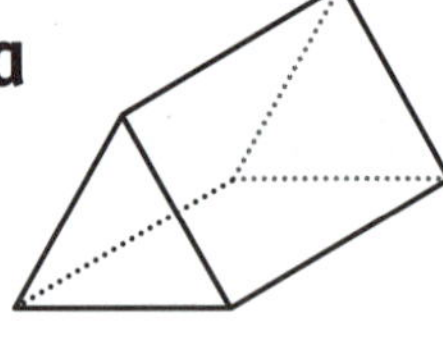  

Front Side Top

b 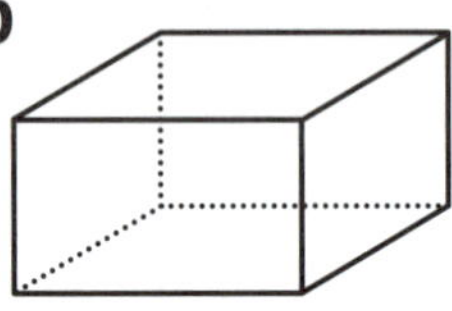   

______ ______ ______

Check the answers on the video!

Your turn

Draw the front view of each 3D object.

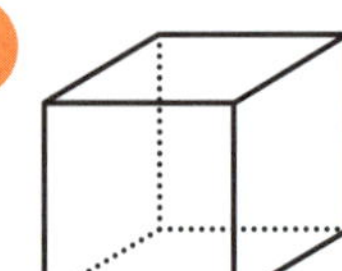 

a 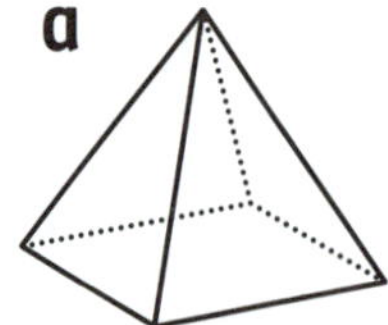

b 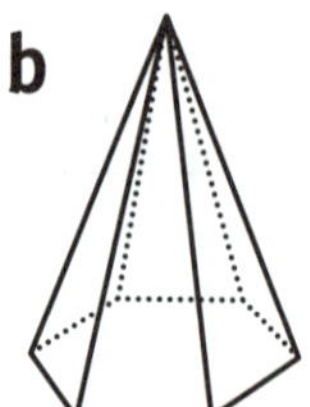

SELF CHECK Tick how you feel

Got it!	Need help...	I don't get it

Check your answers

How many did you get correct?

CATCH UP MATHS YEAR 5 BOOK B © PASCAL PRESS ISBN: 9781925726176

PRACTICE

1 Draw the view of each 3D object you would see from the arrow.

b

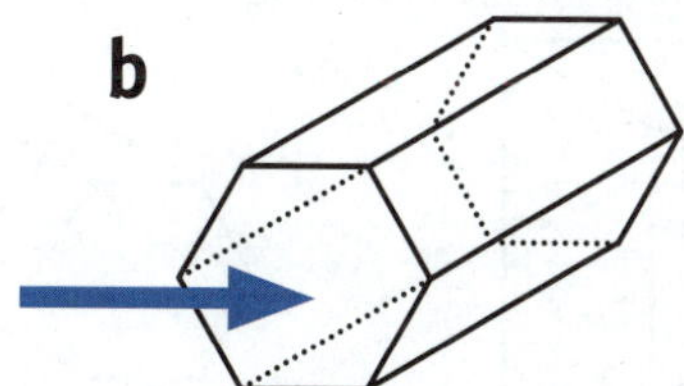

d

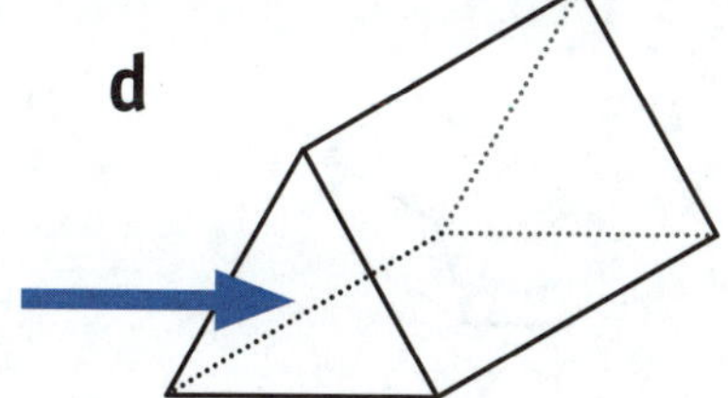

a

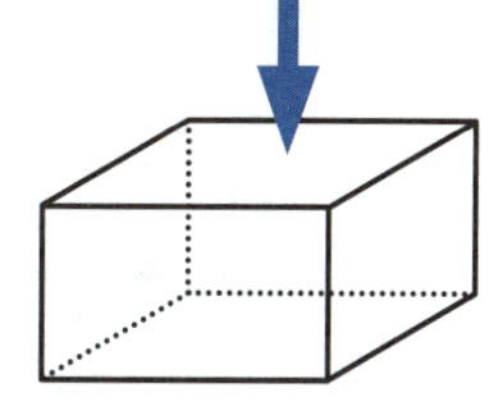

c

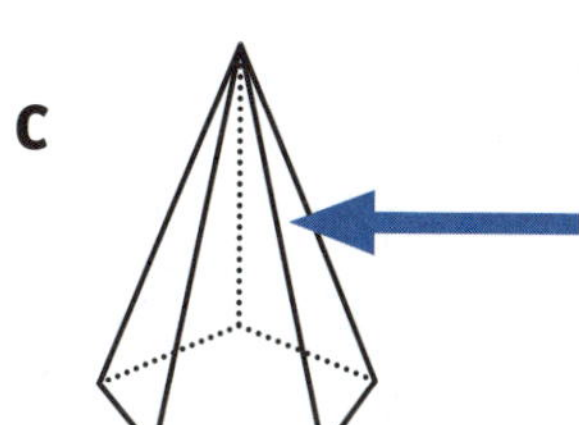

e

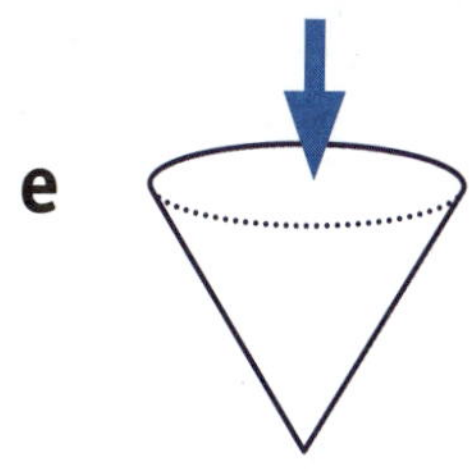

2 Draw the top, front and side view of each 3D object.

	Object	Top view	Front view	Side view
a				
b				
c				

	Object	Top view	Front view	Side view
d				
e				
f				
g				

3D OBJECTS REVIEW

Colour the bases of each prism blue.

a 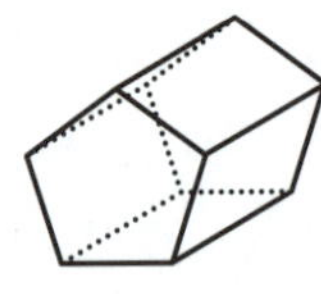

b 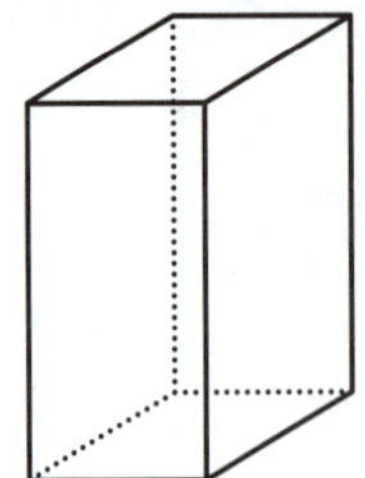

c 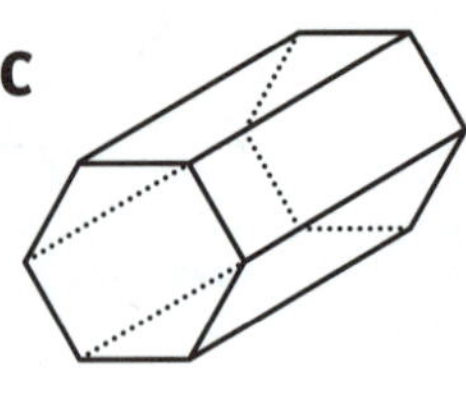

d 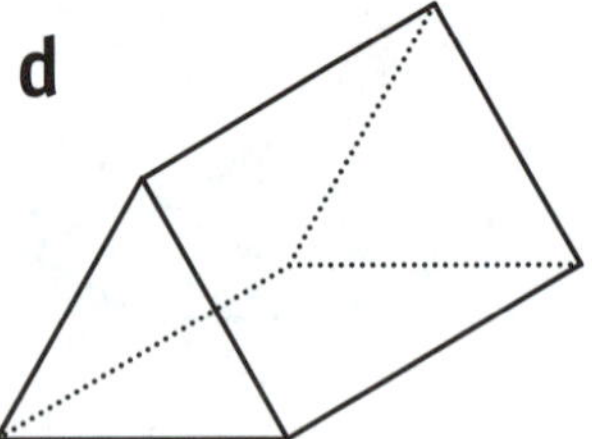

Colour the rectangular faces of each prism red.

a 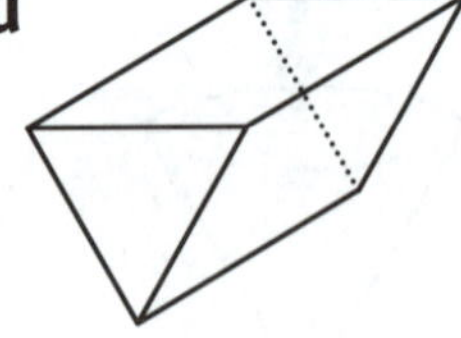

b 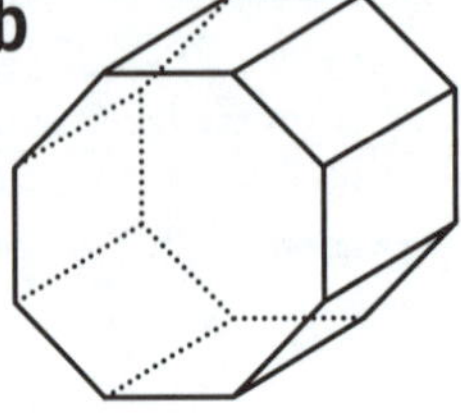

c 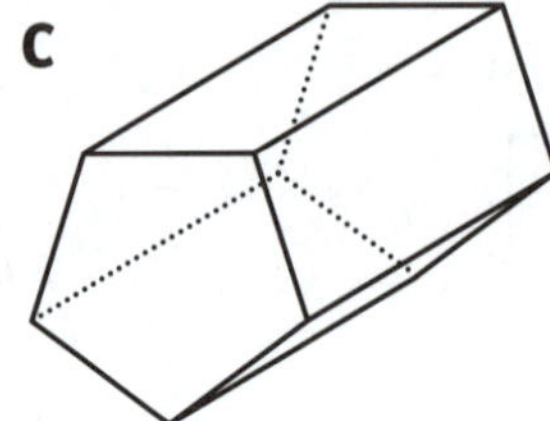

d 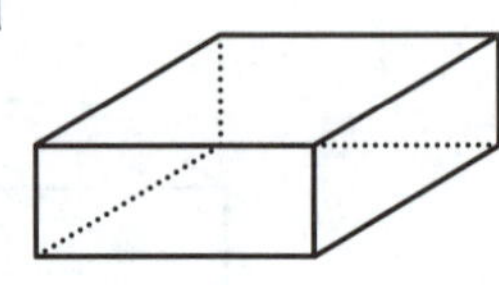

Colour the base of each pyramid blue.

a 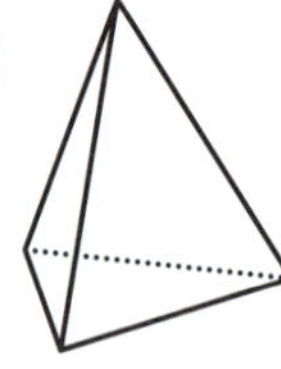

b 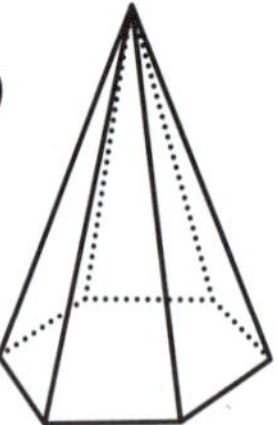

c 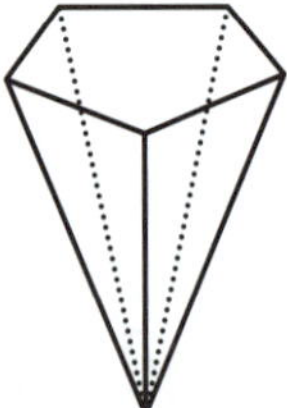

d 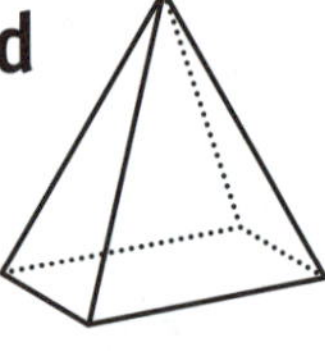

Colour the triangular faces of each pyramid orange.

a 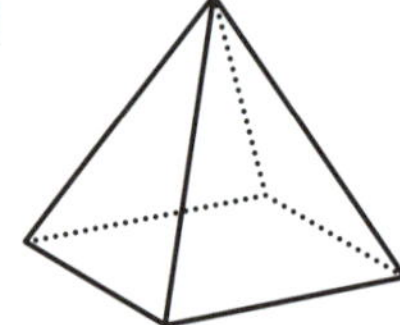

b 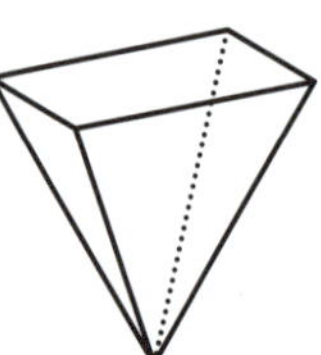

c 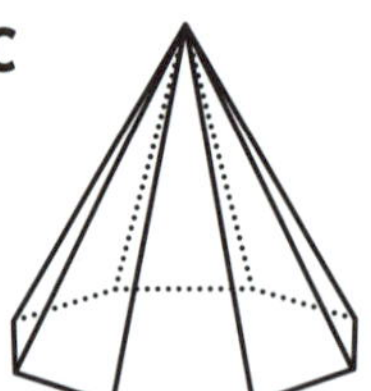

d 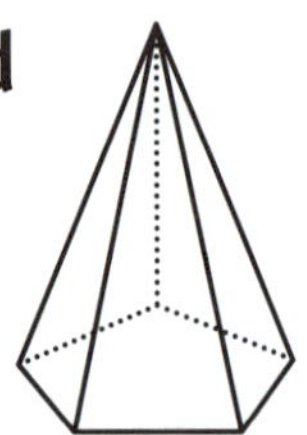

Draw a diagram of each 3D object.

a cube	b triangular prism	c rectangular prism

  ISBN: 9781925726176

d rectangular pyramid	f pentagonal pyramid	h triangular pyramid
e octagonal prism	g hexagonal prism	i octagonal pyramid

6 Complete the table.

	Name	Diagram	Number of vertices	Number of edges	Number of faces
a	triangular prism				
b	rectangular prism				
c	octagonal prism				
d	pentagonal prism				
e	hexagonal prism				
f	cube				

REVIEW

7 Complete the table.

	Name	Diagram	Number of vertices	Number of edges	Number of faces
a	pentagonal pyramid				
b	triangular pyramid				
c	square pyramid				
d	octagonal pyramid				

8 Write the name of each 3D object.

a ____________________ **b** ____________________ **c** ____________________

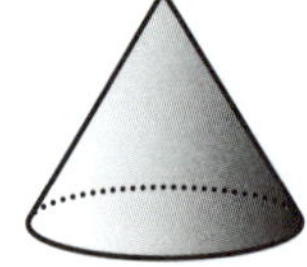

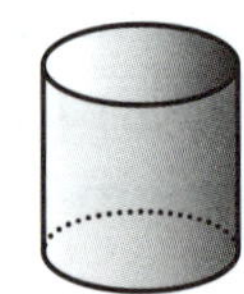

9 Complete the table.

	3D object	Top view	Front view	Side view
a				
b				
c				

CATCH UP MATHS YEAR 5 BOOK B © PASCAL PRESS ISBN: 9781925726176

List four things shaped like each 3D object.

a cone ______________________

b cylinder ______________________

c sphere ______________________

Name the 3D object each net will make.

a ______________________

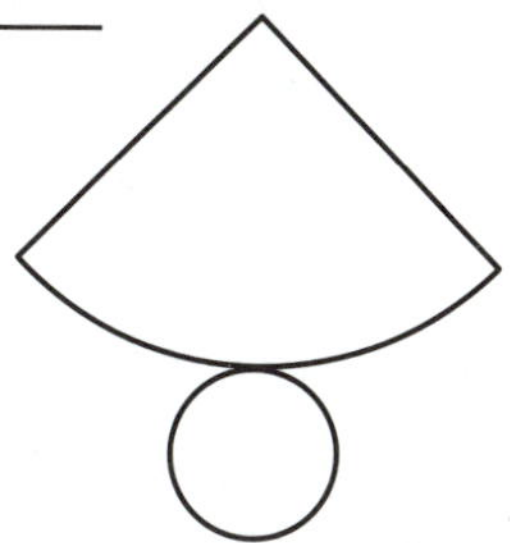

b ______________________

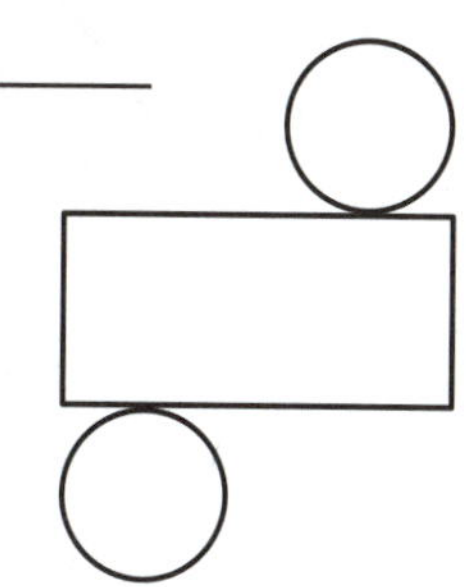

Draw the cross-section made by the orange line.

a 	d 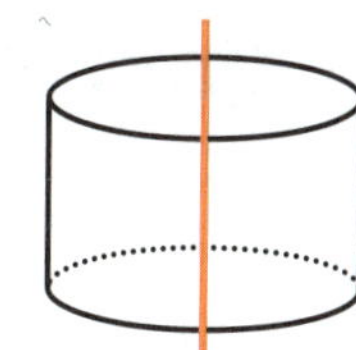
b 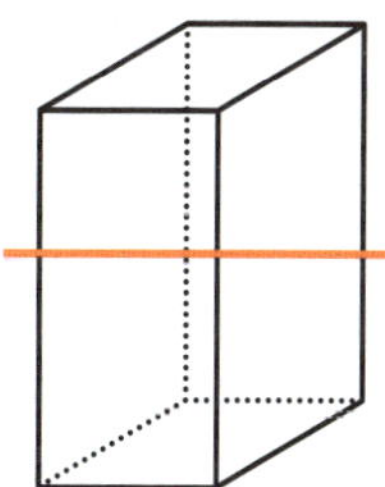	e 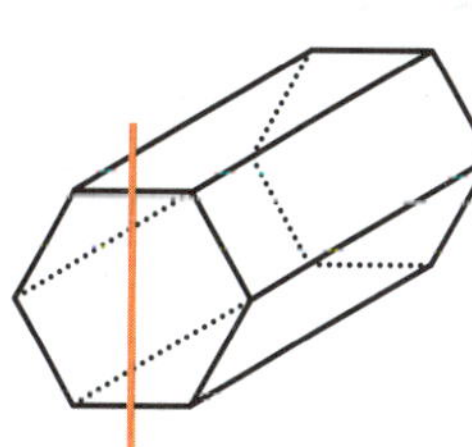
c 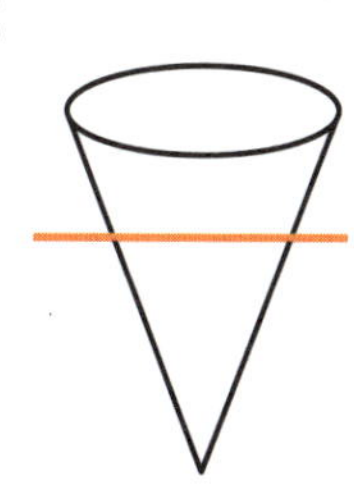	f 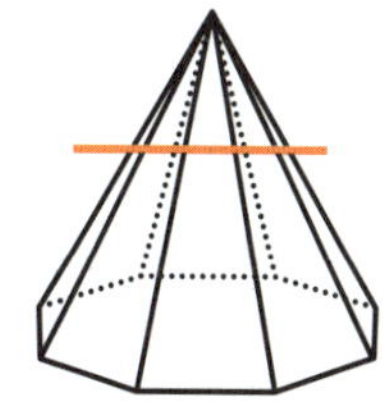

13 Draw a line on each 3D object to match the cross-section shown.

a	d
b	e
c	f

14 Join the nets to the correct labels.

a

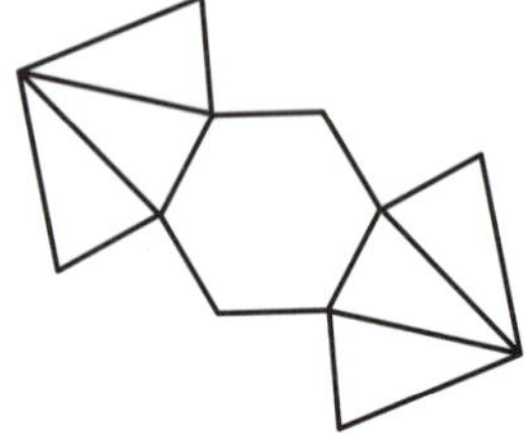

cone

hexagonal pyramid

pentagonal pyramid

cube

cylinder

triangular prism

rectangular prism

d

e

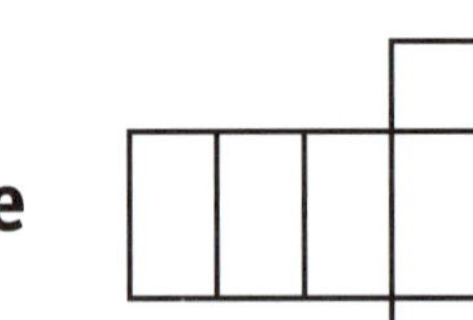

b

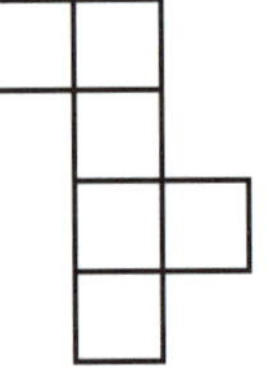

f

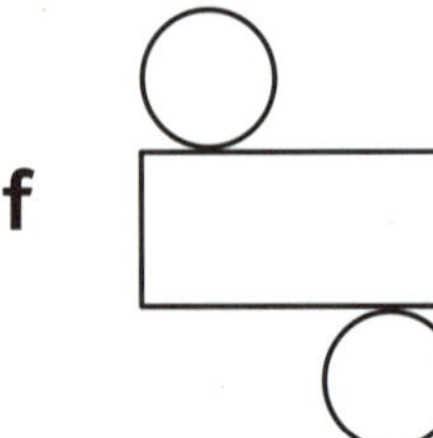

c

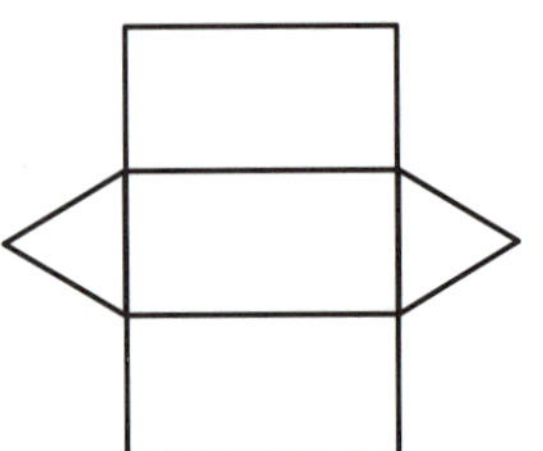

g

CATCH UP MATHS YEAR 5 BOOK B © PASCAL PRESS ISBN: 9781925726176

15 Complete the tables.

	Object	Top view	Front view	Side view
a				
b				
c				
d				

	Object	Top view	Front view	Side view
e				
f				
g				
h				

16 Circle the diagram that matches the view.

a		Top view			
b		Side view			
c		Front view			

THE SQUARE CENTIMETRE

When measuring small areas, we use a unit of measurement called the square centimetre (cm^2).

one square centimetre

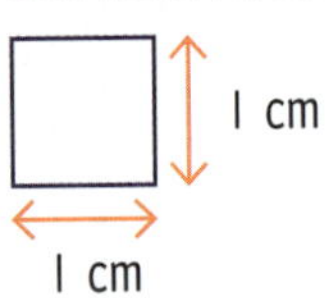

There are 10 square centimetres in this shape.

The area of this shape is 10 cm^2.

Examples: Find the area of each shape by counting the square centimetres.

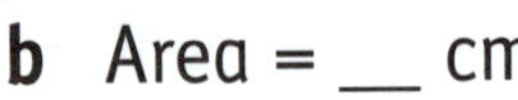

a Area = 6 cm^2 **b** Area = __ cm^2 **c** Area = __ cm^2

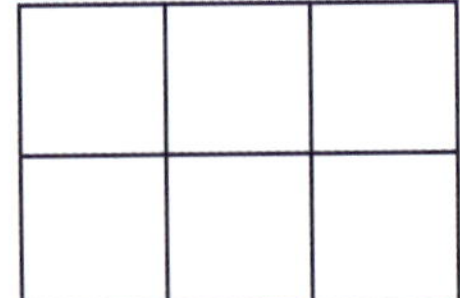

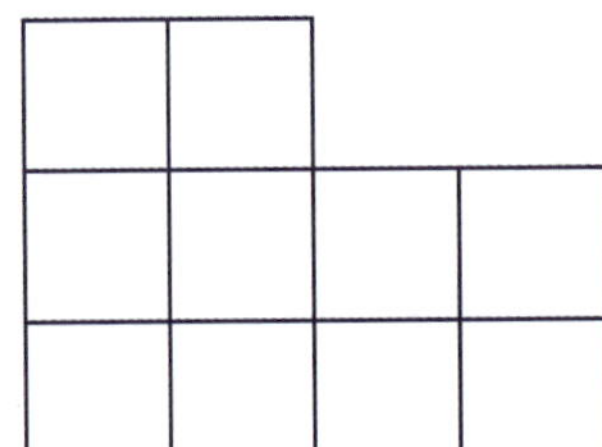

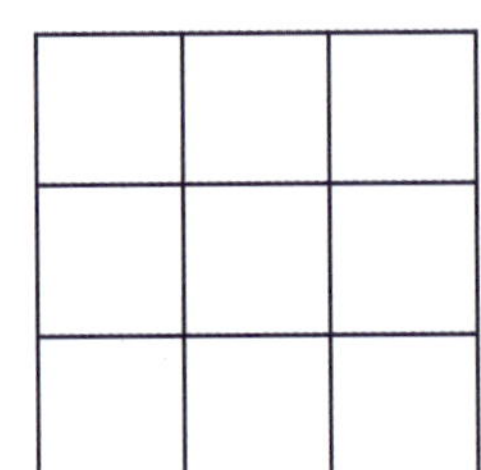

Check the answers on the video!

Your turn

Calculate the area of each shape and record your answer.

☐ = 1 cm^2

● A = 24 cm^2 **a** A = ______ **b** A = ______

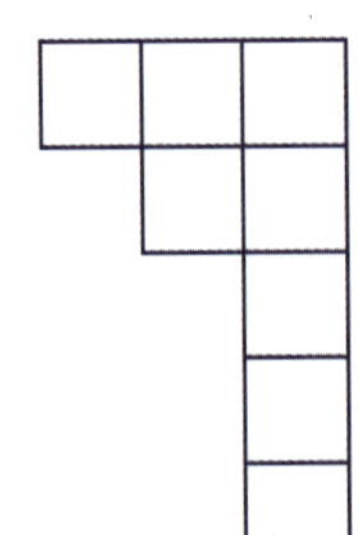

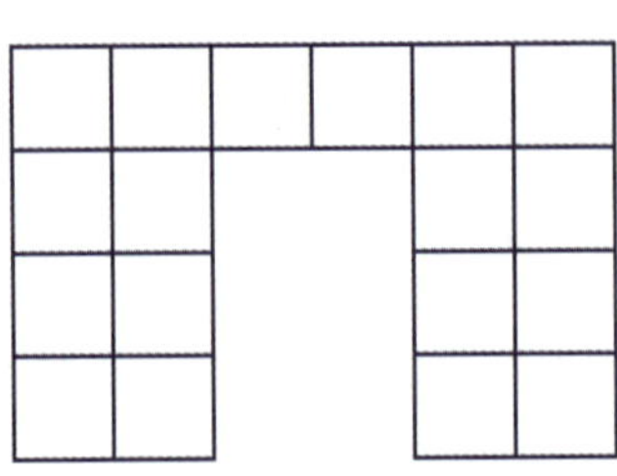

Check your answers

How many did you get correct?

  ISBN: 9781925726176

PRACTICE

1 What is the area of each coloured shape? □ = 1 cm^2 ◩ = $\frac{1}{2}$ cm^2

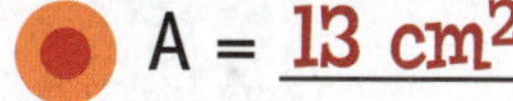

● A = 13 cm^2 **b** A = ______   **d** A = ______   **f** A = ______

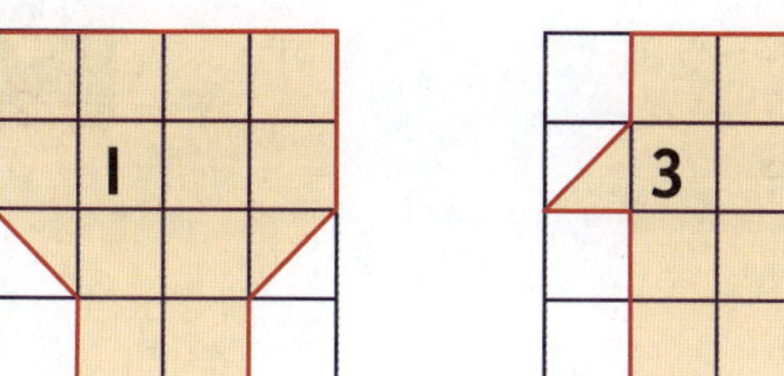

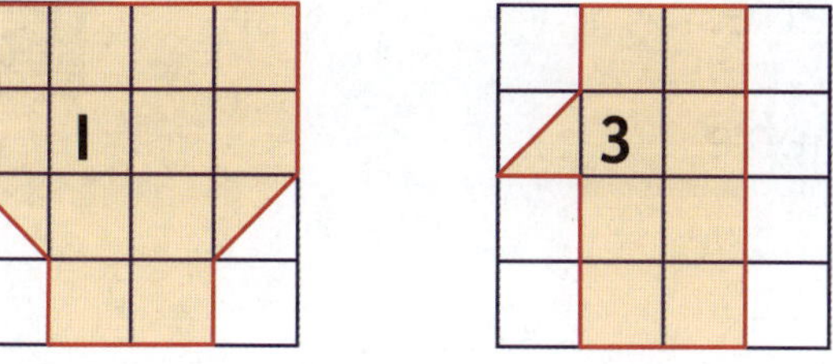

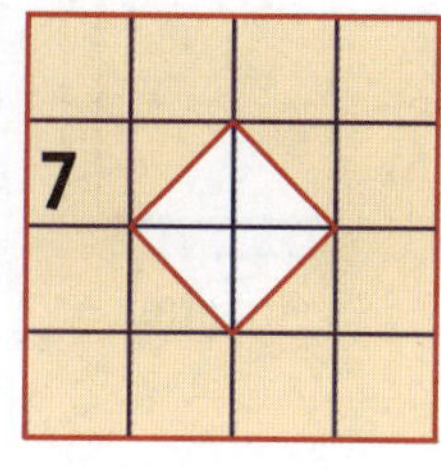

a A = ______   **c** A = ______   **e** A = ______   **g** A = ______

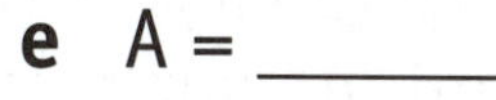

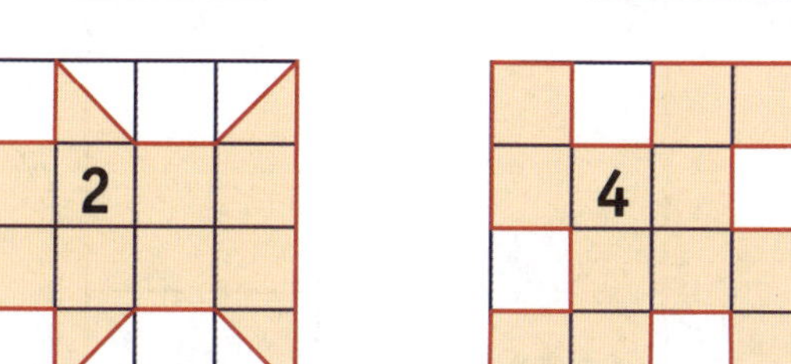

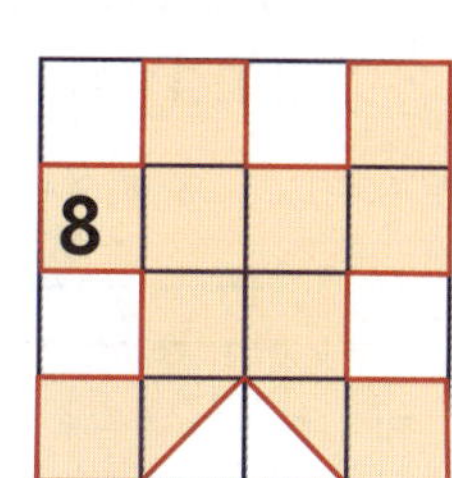

2 Use the shapes in Question 1 to answer the following questions.

● Which shape has the smallest area? 3

a Which shape has the largest area? __

b Which two shapes have the same area? __ and __

c What is the difference in area between Shape 6 and Shape 3? ______

d Shape __ is 3 cm^2 larger in area than Shape 8.

e Shape __ is $\frac{1}{2}$ cm^2 smaller in area than Shape 2.

f Shape 1 is ___ cm^2 larger in area than Shape 3.

3 Draw five different shapes that have an area of 8 cm^2. □ = 1 cm^2

  ISBN: 9781925726176

THE SQUARE METRE

Square metres (m^2) are used to measure larger areas.

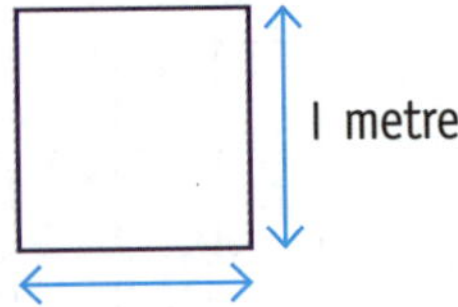

This is one square metre.

Area = 1 metre × 1 metre

= 1 m^2

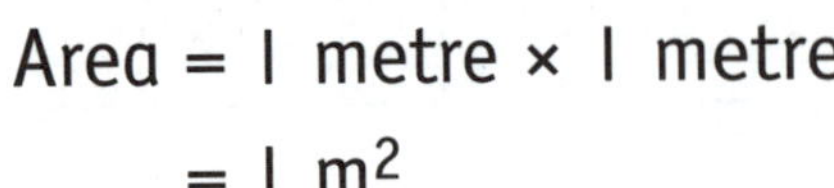

You would use square metres to measure the board in your classroom and your bedroom floor.

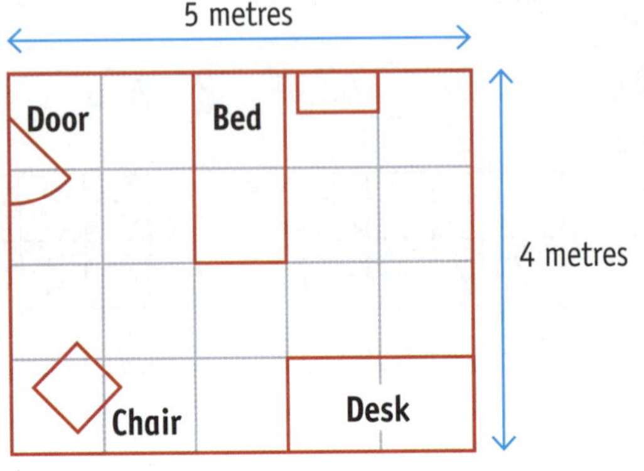

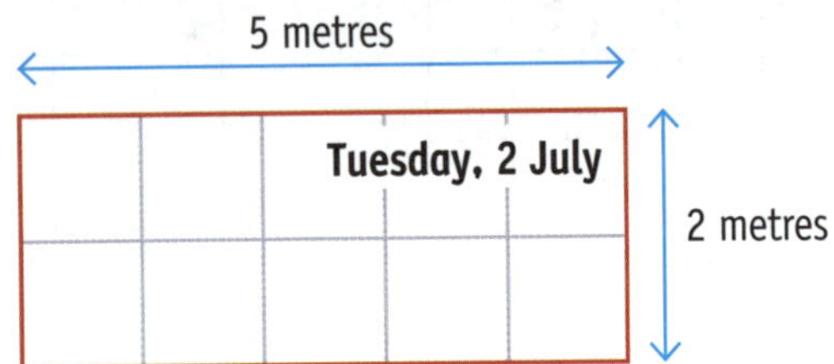

The area of the board is 10 m^2.

The area of the bedroom floor is 20 m^2.

Examples: Calculate the area. ☐ = 1 m^2

a Area = 8 m^2

4 m

2 m

b Area = ___ m^2

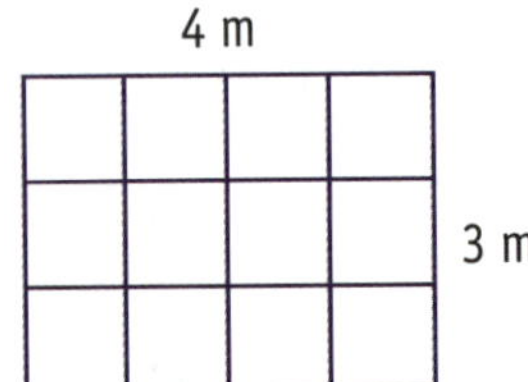

c Area = ___ m^2

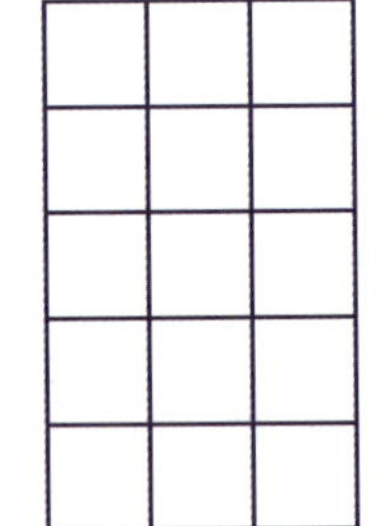

Check the answers on the video!

Calculate the area.

A = 5 m^2

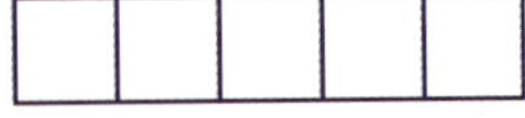

a A = ______

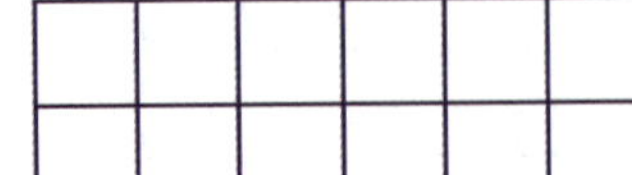

b A = ______

SELF CHECK Tick how you feel

Got it!	Need help...	I don't get it
 ☐	 ☐	 ☐

Check your answers

How many did you get correct? ☐

CATCH UP MATHS YEAR 5 BOOK B © PASCAL PRESS ISBN: 9781925726176

PRACTICE

1 Circle the items that would be measured using m^2.

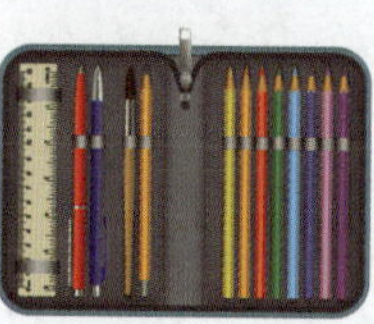

2 Rewrite using numbers and m^2.

- twelve square metres $\underline{12\ m^2}$
- **a** four square metres ______
- **b** eighteen square metres ______
- **c** one hundred and two square metres ______
- **d** fifty-seven square metres ______
- **e** eighty-nine square metres ______

3 Jay wants to lay flooring in each room. How much of each product will she need to buy?

Garage
Family
Kitchen
Laundry
Entry
Hallway
Outdoor Eating
Study
Dining
Sitting

- Carpet for the study $\underline{6\ m^2}$
- **a** Carpet for the family room ____
- **b** Tiles for the kitchen, laundry and sitting area ____
- **c** Tiles for the outdoor eating area ____
- **d** Vinyl for the garage ____
- **e** Tiles for the entry ____

4 Jay buys 7 m^2 of tiles for the dining room. Will she have enough? ____

5 Write the correct unit of measurement (cm^2 or m^2) for these areas.

- basketball court 495 $\underline{m^2}$
- **a** notepad 150 ____
- **b** flag 4 ____
- **c** novel cover 400 ____
- **d** postal stamp 4 ____
- **e** hall floor 800 ____

SQUARE KILOMETRES

A square kilometre (km^2) is 1 000 000 m^2.
Very large areas are measured in square kilometres.

Our national parks are measured using square kilometres.

Sydney Olympic Park is 6.63 square kilometres.

Examples:
Tick the places you would measure in square kilometres.

a ☑ Australia
b ☐ Daintree Rainforest
c ☐ a football field
d ☐ the Great Barrier Reef
e ☐ a cricket ground
f ☐ a basketball court

Name five things you would measure in square kilometres.

A suburb in a city
____
____
____
____

SELF CHECK Tick how you feel

Got it!	Need help...	I don't get it
☐	☐	☐

Check your answers
How many did you get correct? ☐

CATCH UP MATHS YEAR 5 BOOK B © PASCAL PRESS ISBN: 9781925726176

PRACTICE

If ☐ = 1 km^2, find the area (A) of each shape.

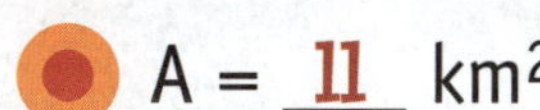

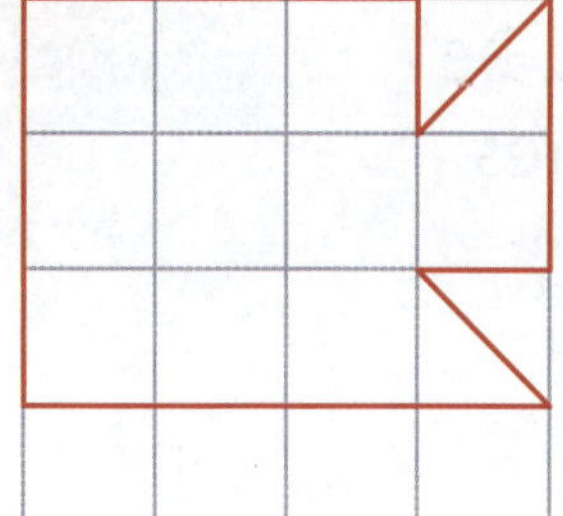

b A = ___ km^2

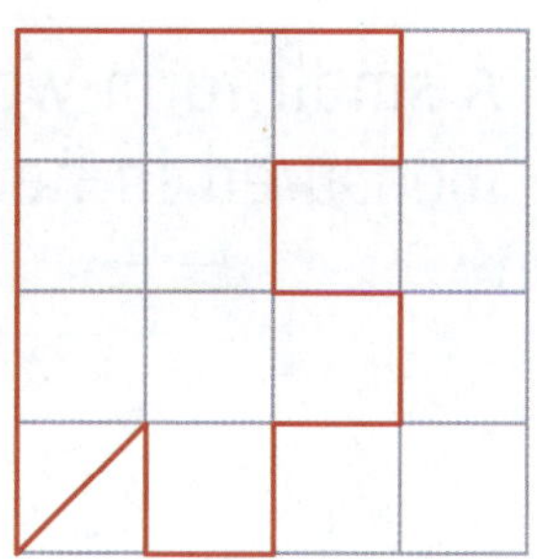

d A = ___ km^2

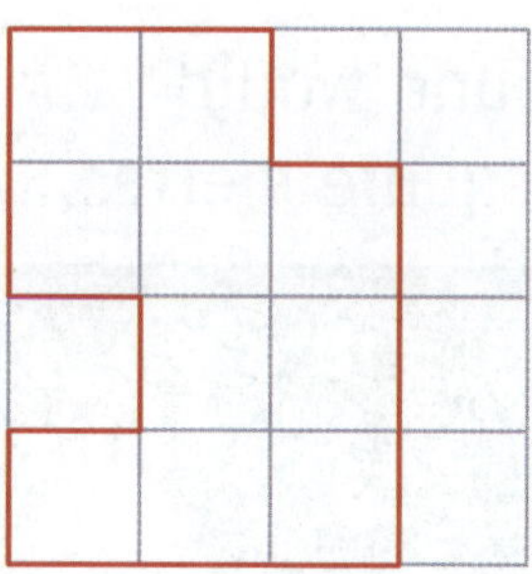

f A = ___ km^2

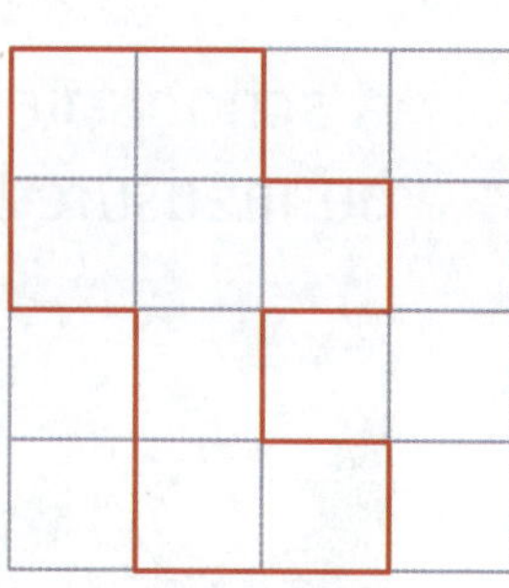

a A = ___ km^2

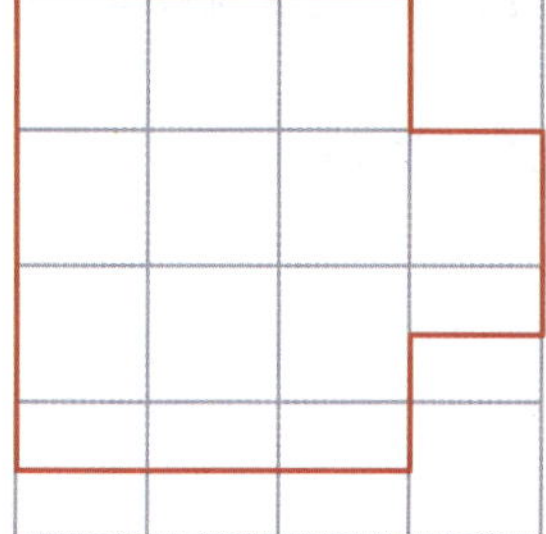

c A = ___ km^2

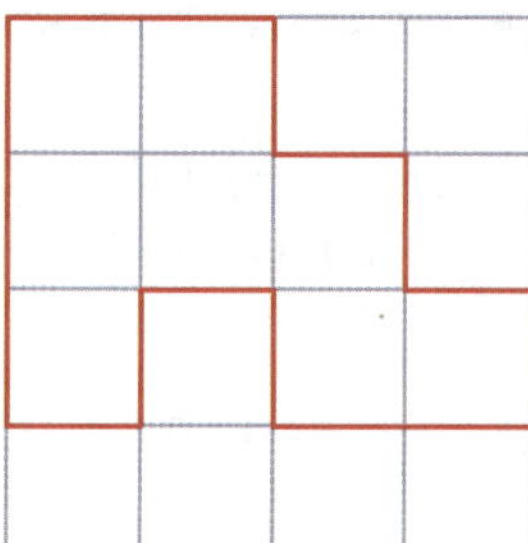

e A = ___ km^2

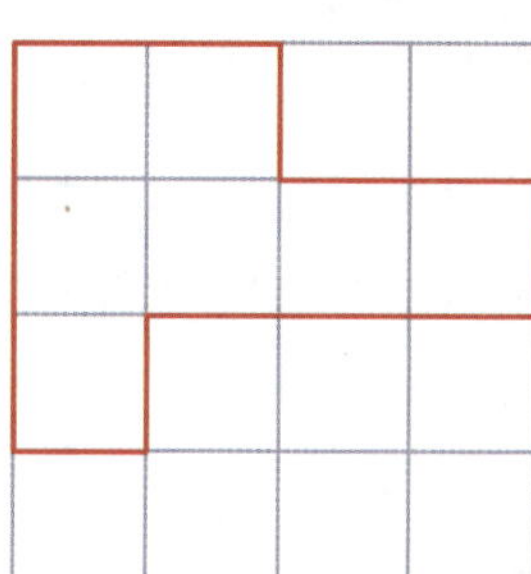

g A = ___ km^2

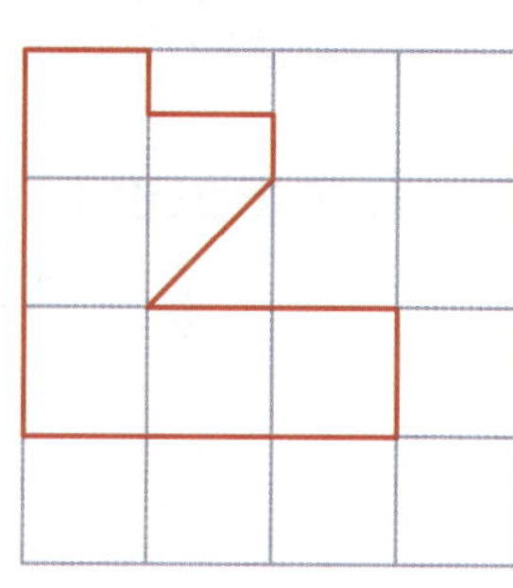

2 Use the map to answer the following questions.

a Which state or territory has the smallest area?

____

b Which state or territory has the largest area?

____

c Which state or territory is larger in area than Queensland?

____

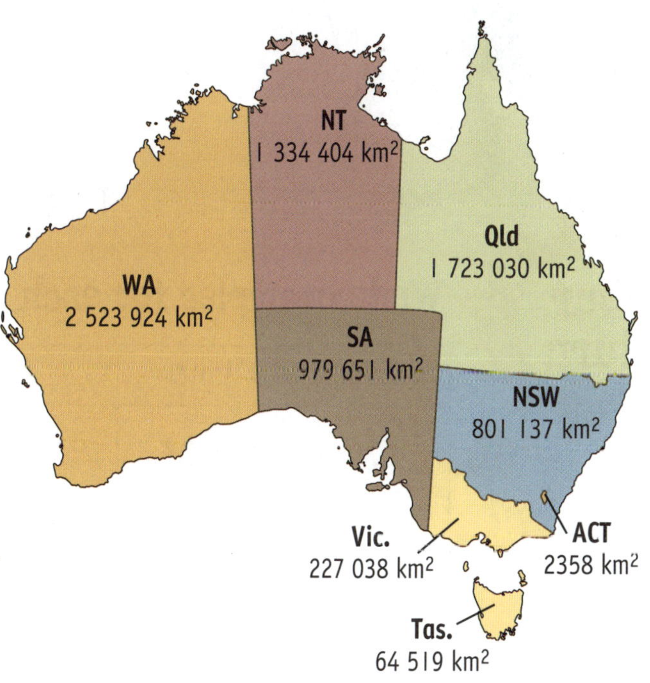

d List the states and territories in ascending order of area. Use the abbreviations in your list.

______ ______ ______ ______ ______ ______ ______ ______

HECTARES

A hectare (ha) is 100 m × 100 m. That's 10 000 m².
Hectares are used to measure large areas.

A school playground would be measured in square metres.

A small farm would be measured in hectares.

Examples: Would you use square metres (m^2) or hectares (ha) to measure the areas of these things?

Why do we need different units for measuring area?

a your classroom m^2

b a baseball field ___

c a large local park ___

d a large shopping centre ___

e a swimming pool ___

f an airport ___

Check the answers on the video!

Your turn Write examples for each area.

Less than 1 ha	About 1 ha	More than 1 ha
A soccer field	A local park	A national park

SELF CHECK Tick how you feel

Got it!	Need help...	I don't get it
☐	☐	☐

Check your answers
How many did you get correct? ☐

CATCH UP MATHS YEAR 5 BOOK B © PASCAL PRESS ISBN: 9781925726176

PRACTICE

1 **Use the abbreviation ha to write these areas.**

- 6 hectares 6 ha
- a 17 hectares ______
- b 28 hectares ______
- c 304 hectares ______
- d 100 hectares ______
- e 1 hectare ______
- f 349 hectares ______
- g 757 hectares ______

2 **Write the long form of these areas.**

- 8 ha 8 hectares
- a 16 ha ______
- b 22 ha ______
- c 41 ha ______
- d 109 ha ______
- e 324 ha ______
- f 603 ha ______
- g 890 ha ______

3

Great Barrier Reef Marine Park 34.44 million ha	Booderee National Park 6312 ha	Kakadu National Park 1.98 million ha
Uluru–Kata Tjuta National Park 132 600 ha	Murramarang National Park 2200 ha	

a Order the national parks above from largest to smallest in area.

______

______

b How much bigger than Murramarang National Park is Booderee National Park? ______

c How much smaller than the Great Barrier Reef Marine Park is Kakadu National Park? ______

AREA – USING MULTIPLICATION

You can use multiplication to calculate the area of squares and rectangles.

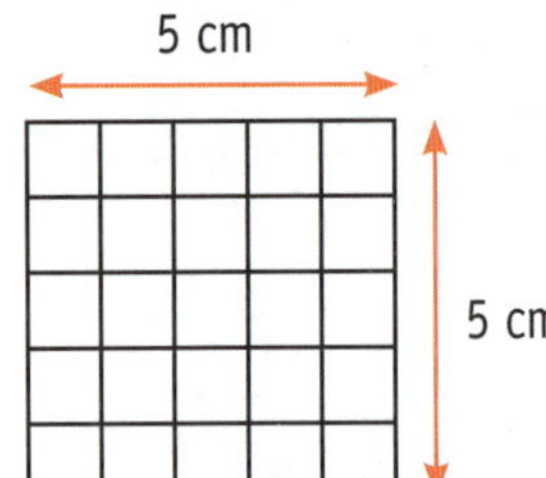

Area = length × width
= 5 cm × 5 cm
= 25 cm^2

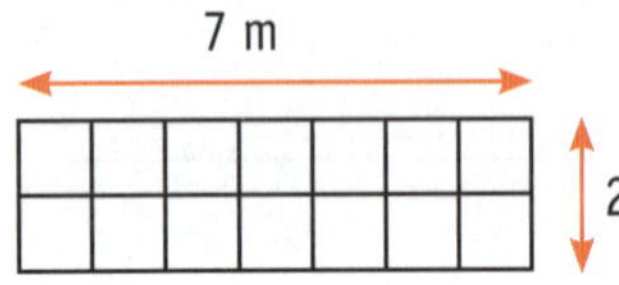

Area = length × width
= 2 m × 7 m
= 14 m^2

Examples: Fill in the gaps.

a Area = length × <u>width</u>
= 2 cm × 4 cm
= <u>8</u> cm^2

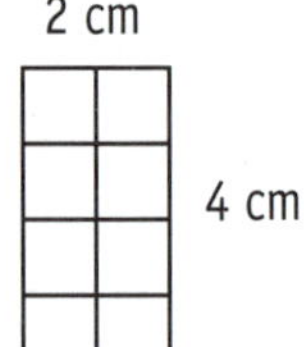

b Area = ________ × width
= 3 cm × ______
= ___ cm^2

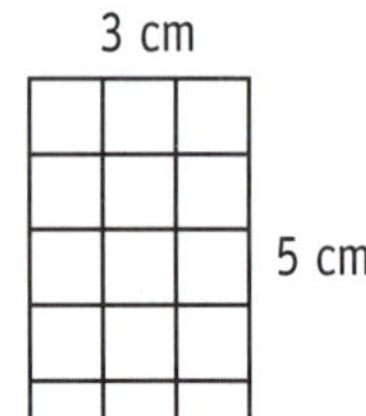

Your turn

Write the missing information.

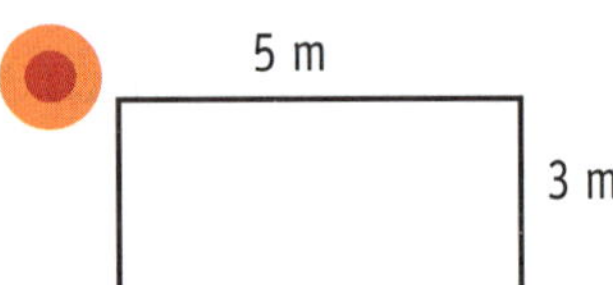

Area = <u>length</u> × <u>width</u>
= <u>5 m</u> × <u>3 m</u>
= <u>15</u> m^2

a

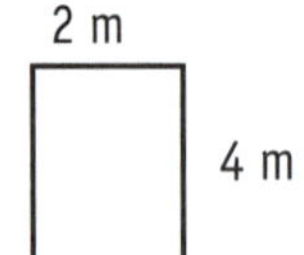

Area = ________ × ________
= _____ × _____
= ___ m^2

Got it!	Need help...	I don't get it

Check your answers
How many did you get correct?

CATCH UP MATHS YEAR 5 BOOK B © PASCAL PRESS ISBN: 9781925726176

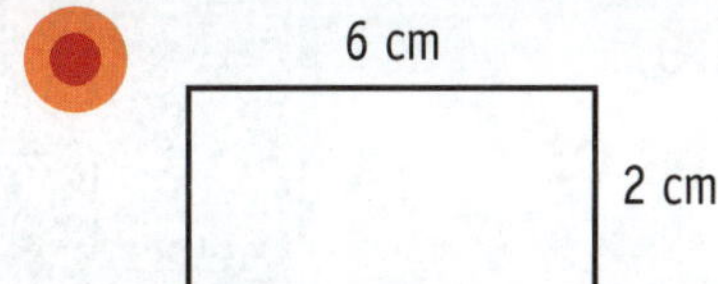

PRACTICE

1 Calculate the area using multiplication.

Example: rectangle 6 cm × 2 cm

Area = length × width
= 6 cm × 2 cm
= 12 cm^2

a rectangle 5 m × 4 m

Area = ______ × ______
= _____ × _____
= ____ m^2

b rectangle 6 m × 1 m

Area = ______ × ______
= _____ × _____
= ____ m^2

c rectangle 4 cm × 3 cm

Area = ______ × ______
= _____ × _____
= ____ cm^2

2 Calculate the area using multiplication.

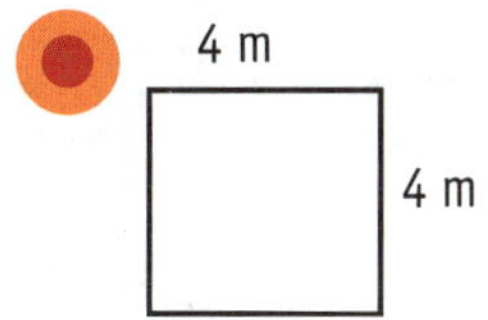

Example: square 4 m × 4 m

A = l × w
= 4 m × 4 m
= 16 m^2

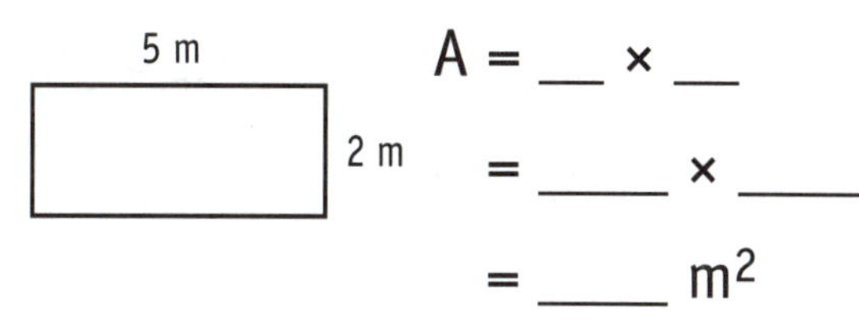

c rectangle 5 m × 2 m

A = __ × __
= _____ × _____
= ____ m^2

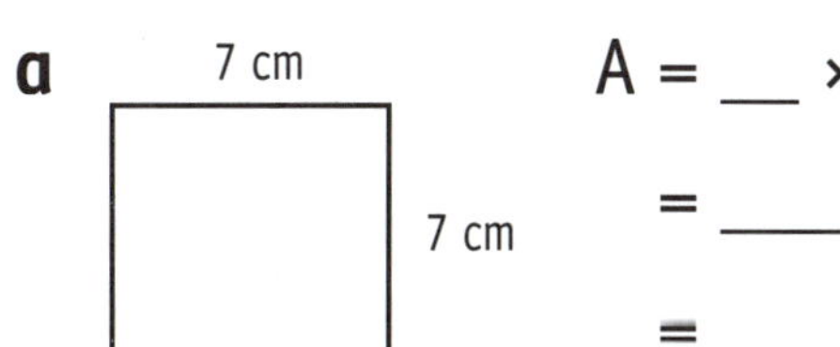

a square 7 cm × 7 cm

A = __ × __
= _____ × _____
= ____ cm^2

d square 6 m × 6 m

A = __ × __
= _____ × _____
= ____ m^2

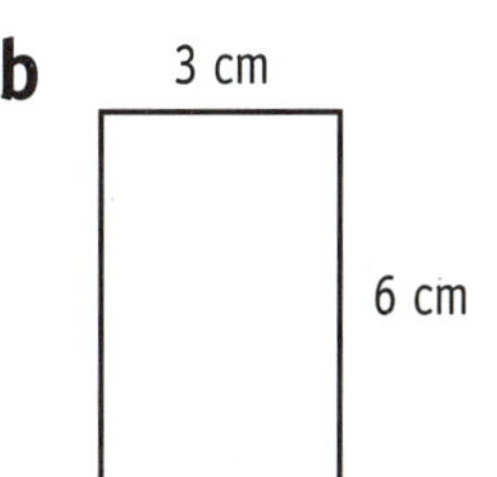

b rectangle 3 cm × 6 cm

A = __ × __
= _____ × _____
= ____ cm^2

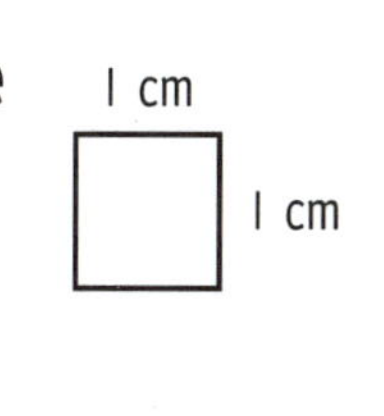

e square 1 cm × 1 cm

A = __ × __
= _____ × _____
= ____ cm^2

PERIMETER

SCAN to watch video

The distance around the outside of a shape is the perimeter (P). To find the perimeter, add the lengths of all the sides.

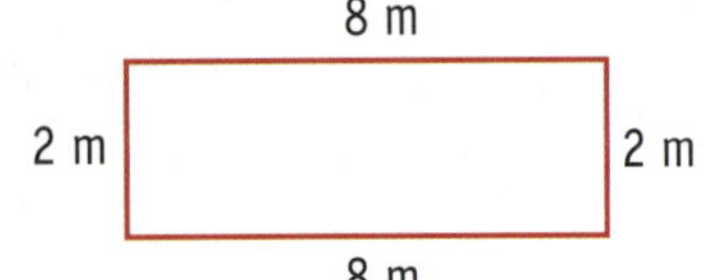

Perimeter (P) = 2 m + 8 m + 2 m + 8 m
= 20 m

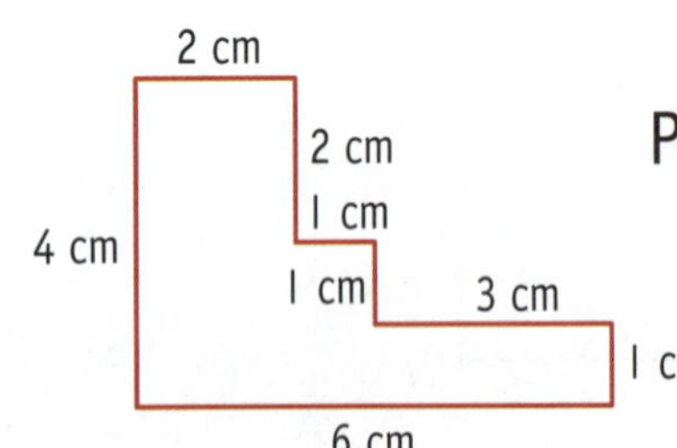

Perimeter (P) = 4 cm + 2 cm + 2 cm + 1 cm
+ 1 cm + 3 cm + 1 cm + 6 cm
= 20 cm

Examples: Calculate the perimeter.

a

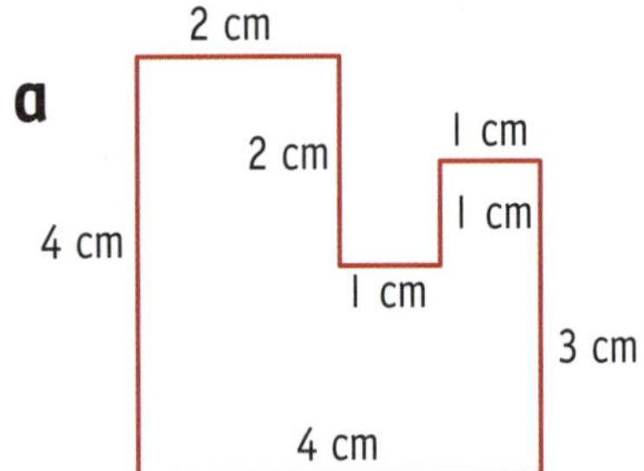

P = 4 cm + 2 cm + 2 cm + 1 cm
+ 1 cm + 1 cm + 3 cm + 4 cm
= 18 cm

b

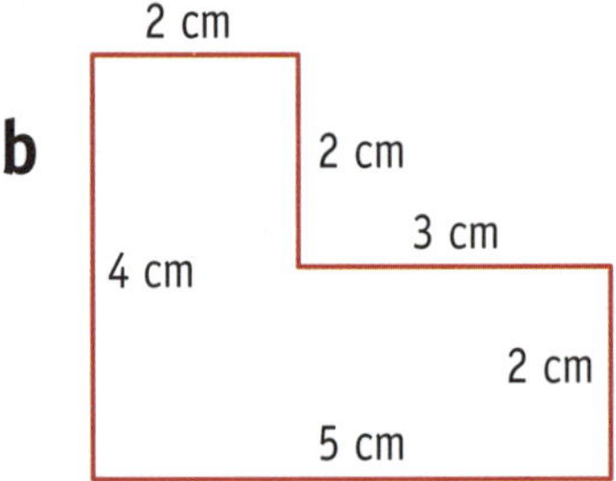

P = 4 cm + 2 cm + 2 cm
+ 3 cm + 2 cm + 5 cm
= ____ cm

Check the answers on the video!

Your turn

Find the perimeter.

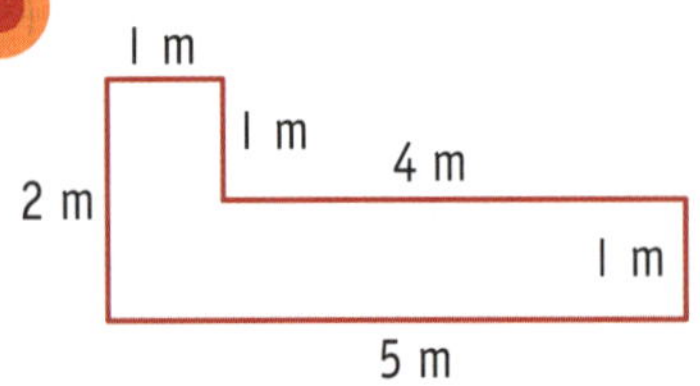

P = 2 + 1 + 1 + 4 + 1 + 5
= 14 m

a

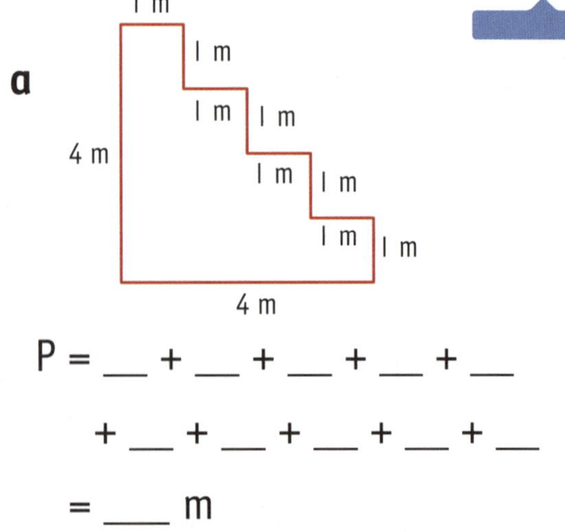

P = __ + __ + __ + __ + __
+ __ + __ + __ + __ + __
= ____ m

SELF CHECK Tick how you feel

Got it!	Need help...	I don't get it
☐	☐	☐

Check your answers

How many did you get correct? ☐

CATCH UP MATHS YEAR 5 BOOK B © PASCAL PRESS ISBN: 9781925726176

1 Calculate the perimeter.

2 m

P = 2 + 2 + 2 + 2

= 8 m

a

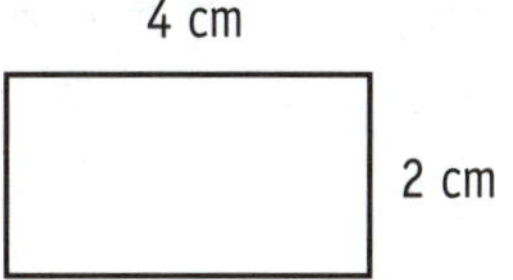

P = __ + __ + __ + __

= ____ cm

b

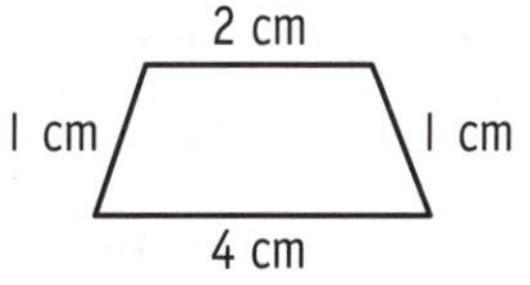

P = __ + __ + __ + __

= ____ cm

c

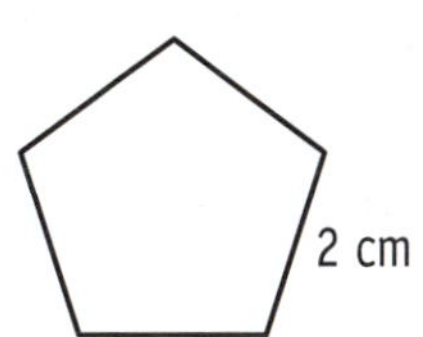

P = __ + __ + __ + __ + __

= ____ cm

d

3 cm

1.5 cm

P = __ + __ + __ + __

= ____ cm

e

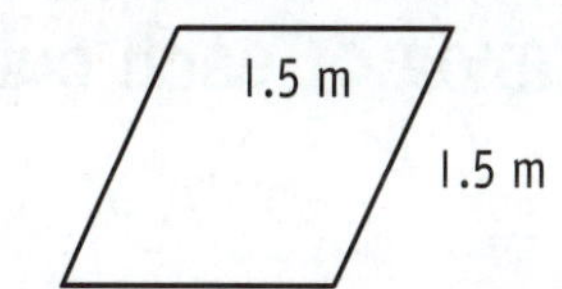

P = __ + __ + __ + __

= ____ m

f

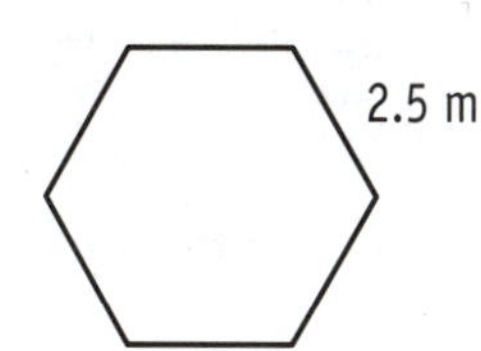

P = __ + __ + __ + __ + __ + __

= ____ m

g

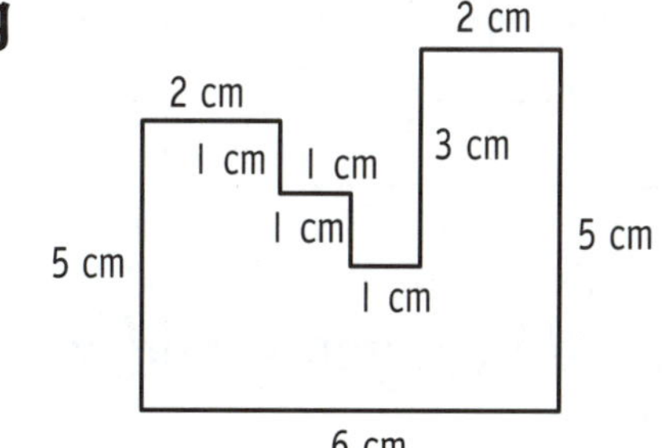

P = __ + __ + __ + __ + __ +

__ + __ + __ + __ + __

= ____ cm

h

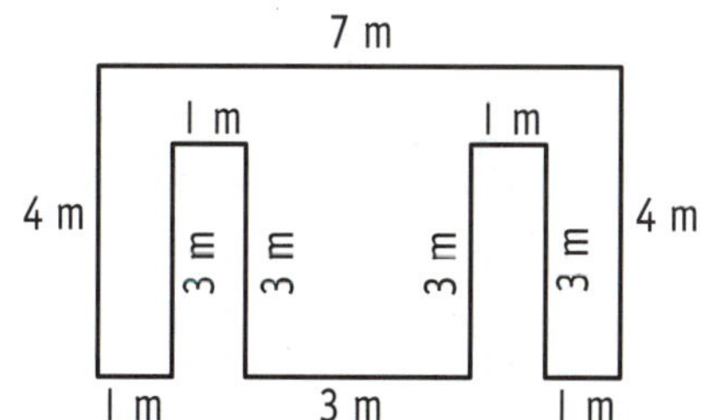

P = __ + __ + __ + __ + __ + __

+ __ + __ + __ + __ + __ + __

= ____ m

AREA REVIEW

1 What is the area of each coloured shape? □ = 1 cm^2

a A = _______ b A = _______ c A = _______ d A = _______

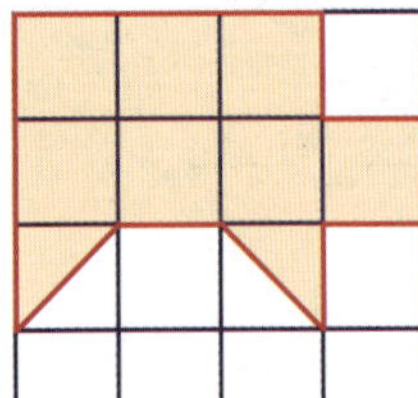

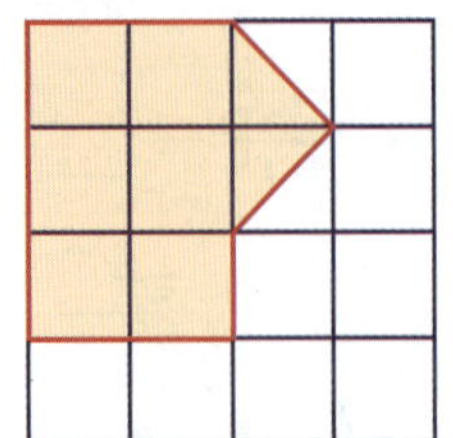

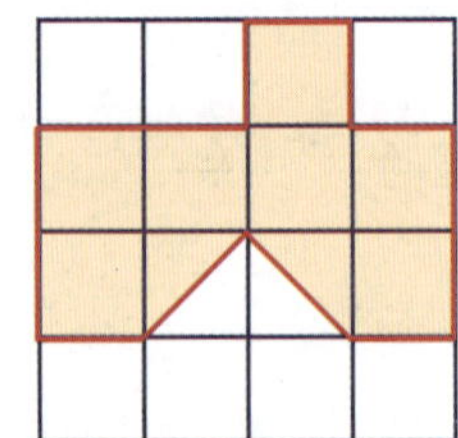

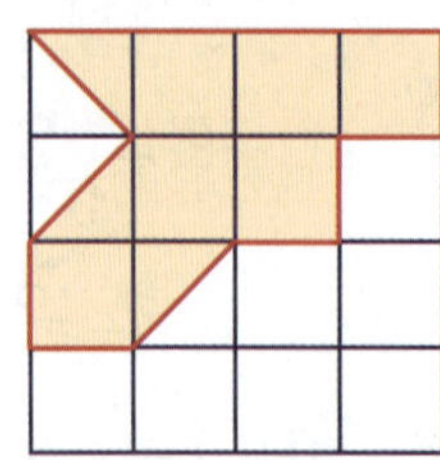

2 Draw 5 different shapes that have an area of 12 cm^2.

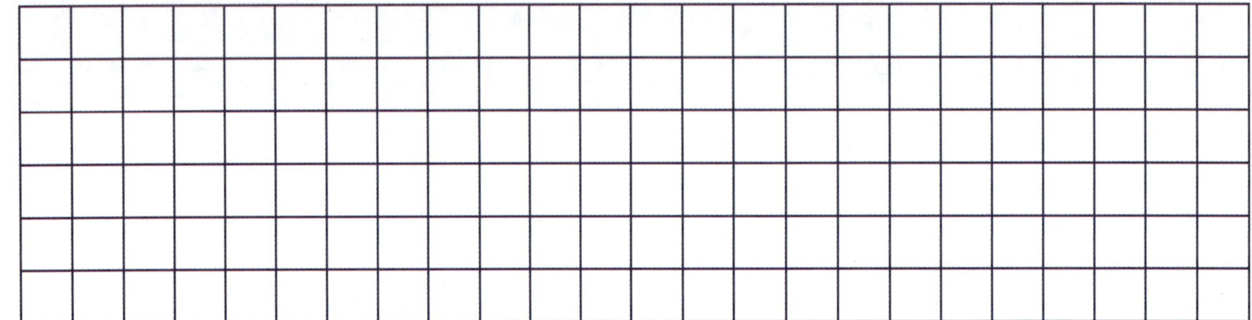

3 Tick the things you would measure using cm^2.

- a □ sticky note
- b □ backyard
- c □ envelope
- d □ book cover
- e □ large desk top
- f □ garage door
- g □ phone case
- h □ watch face

4 If □ = 1 m^2, find the area (A) of each shape.

a A = _____ m^2 b A = _____ m^2 c A = _____ m^2 d A = _____ m^2

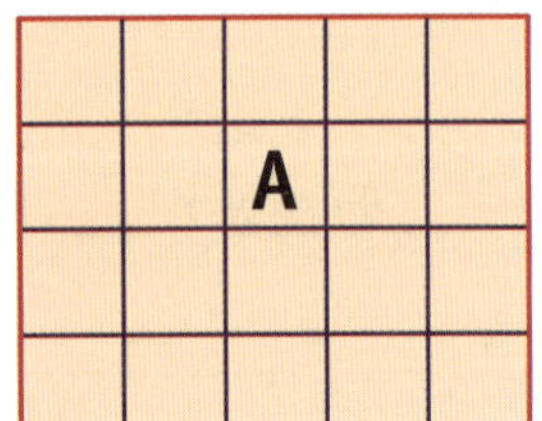

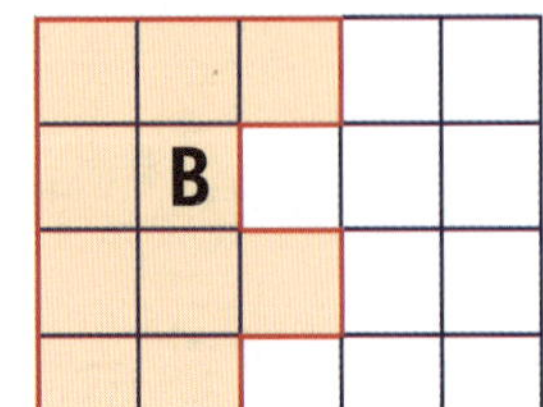

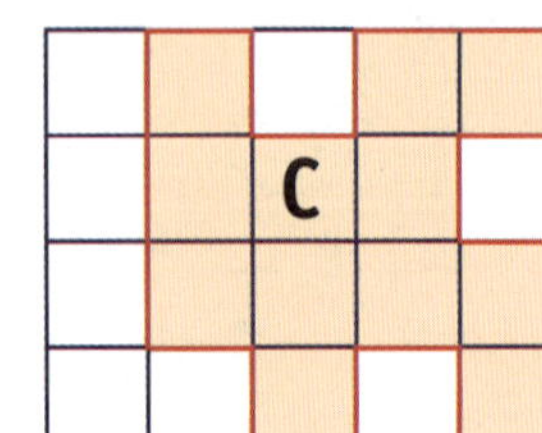

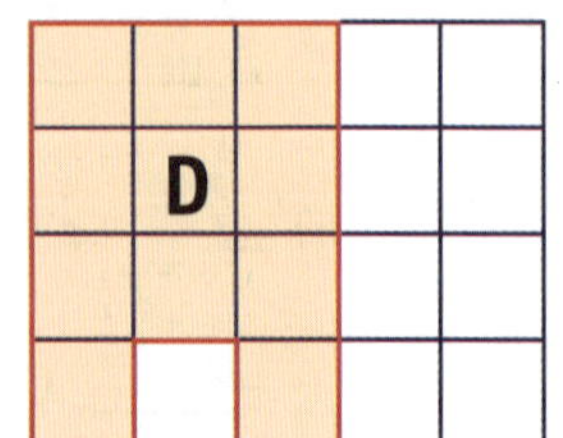

  ISBN: 9781925726176

e A = ____ m^2 f A = ____ m^2 g A = ____ m^2 h A = ____ m^2

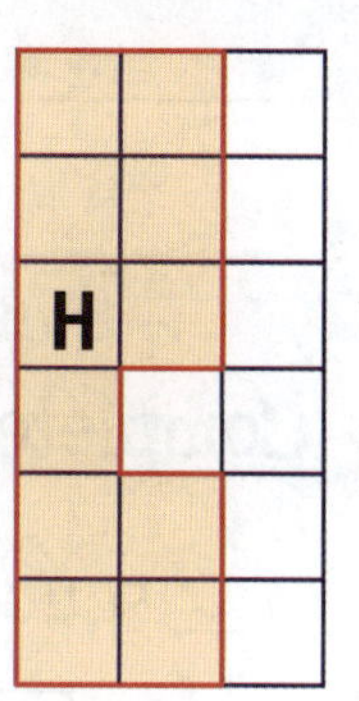

5 Use the shapes in Question 4 to answer the following questions.

a Which shapes have an area of 11 m^2? ________

b Which shapes have an area larger than 12 m^2? ________

c Which shapes have an area of 20 m^2? ________

d Which shape has an area of 18 m^2? ________

e What is the difference in area between Shape E and Shape B? ________

6 How many m^2 of tiles will Christian need for these spaces? □ = 1 m^2

a Pool (floor only) ______

b Kitchen ______

c Barbecue area ______

d Family room ______

e Hallway ______

f Bedroom ______

g Office ______

h Entry ______

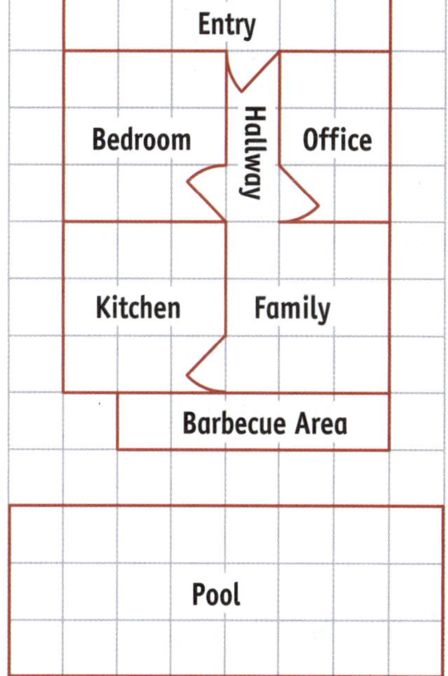

7 How many square metres of tiles altogether does Christian need?

____

  ISBN: 9781925726176

REVIEW

8 Name five things measured using square kilometres.

____

____

9 Complete the labels.

__ square kilometre

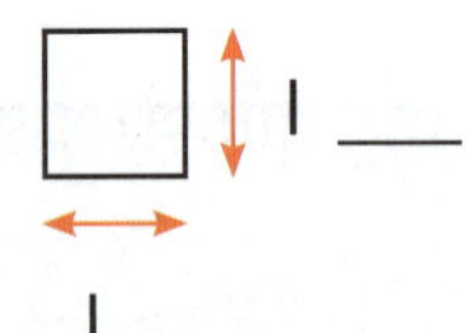

10 What is the difference in km^2?

a Western Australia and Queensland

____

b New South Wales and Victoria

____

c Tasmania and Australian Capital Territory

____

d Northern Territory and South Australia ______________

11 List the states and territories in descending order of area. Use the abbreviations in your list.

_____ _____ _____ _____ _____ _____ _____ _____

12 Fill in the gaps.

A hectare is _____ m by _____ m.

It is __________ m^2.

The abbreviation for hectare is ___.

CATCH UP MATHS YEAR 5 BOOK B © PASCAL PRESS ISBN: 9781925726176

13 Use the abbreviation ha to write these areas.

a 9 hectares ______

b 12 hectares ______

c 31 hectares ______

d 97 hectares ______

e 104 hectares ______

f 212 hectares ______

g 703 hectares ______

h 390 hectares ______

14 Write the long form of each measurement.

a 7 ha ______________

b 136 ha ______________

c 28 ha ______________

d 211 ha ______________

e 630 ha ______________

f 909 ha ______________

g 100 ha ______________

h 687 ha ______________

15 Suburbs in Pleasant Town

Sunshine 624 ha	Beachview 736 ha	Riverside 209 ha
Rainbow Valley 502 ha	Splashville 379 ha	

a List the suburbs in Pleasant Town in ascending order of size.

____

____

b What suburb is smaller in area than Splashville? ______________

c The difference in the areas of Sunshine and Beachview is ________.

d The difference in the areas of Riverside and Rainbow Valley is ________.

e The difference in the areas of Beachview and Riverside is ________.

  ISBN: 9781925726176

16 Calculate the area using multiplication.

a

1 m

1 m

Area = ________ × ________

= ______ × ______

= _____ m^2

b

7 cm

1 cm

Area = ________ × ________

= ______ × ______

= _____ cm^2

c

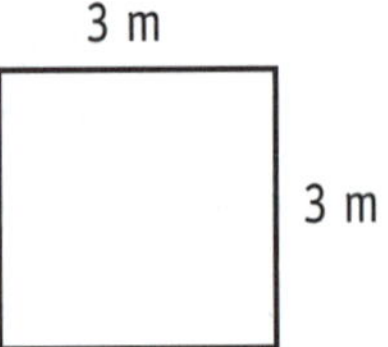

3 m

3 m

Area = ________ × ________

= ______ × ______

= _____ m^2

d

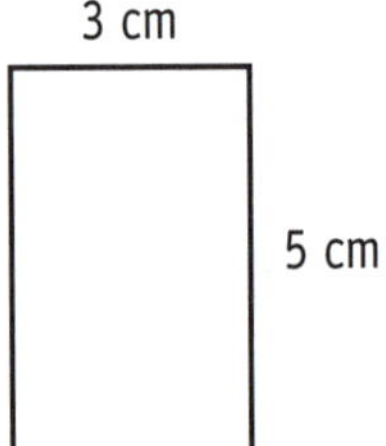

3 cm

5 cm

Area = ________ × ________

= ______ × ______

= _____ cm^2

e

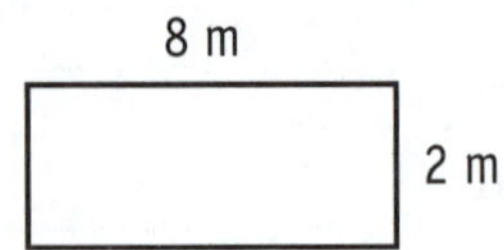

8 m

2 m

A = __ × __

= _____ × _____

= _____ m^2

f

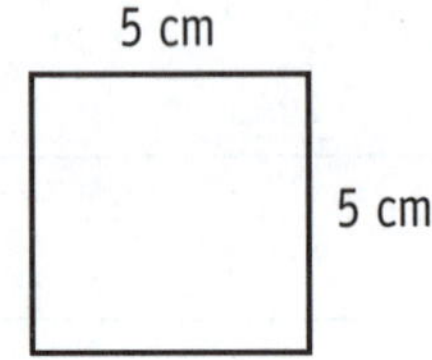

5 cm

5 cm

A = __ × __

= _____ × _____

= _____ cm^2

g

4 cm

3 cm

A = __ × __

= _____ × _____

= _____ cm^2

h

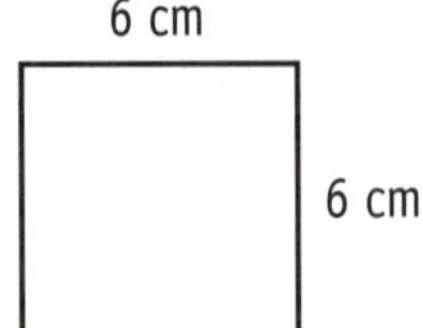

6 cm

6 cm

A = __ × __

= _____ × _____

= _____ cm^2

  ISBN: 9781925726176

17 Calculate the perimeter.

a

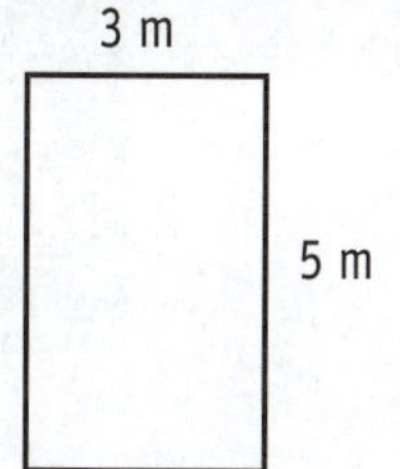

P = ______________________

= ____ m

b

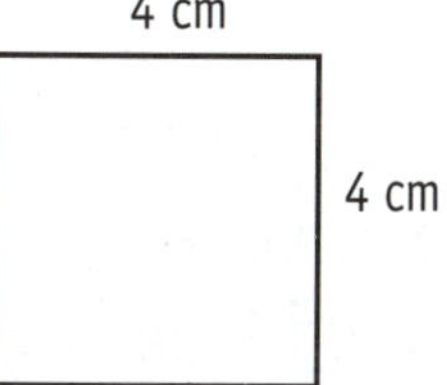

P = ______________________

= ____ cm

c

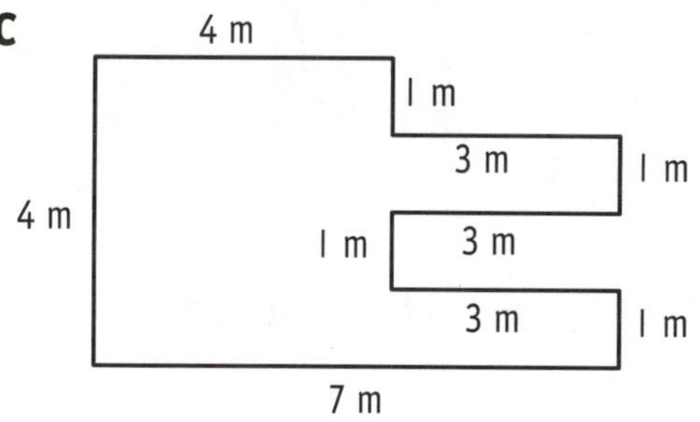

P = ______________________

____

= ____ m

d

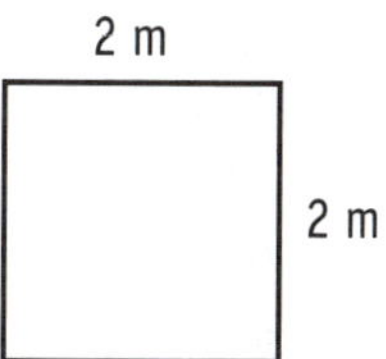

P = ______________________

= ____ m

e

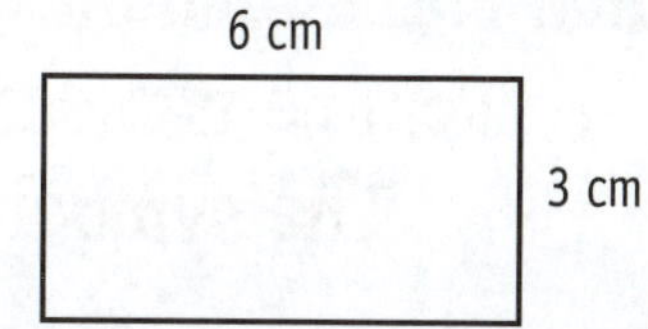

P = ______________________

= ____ cm

f

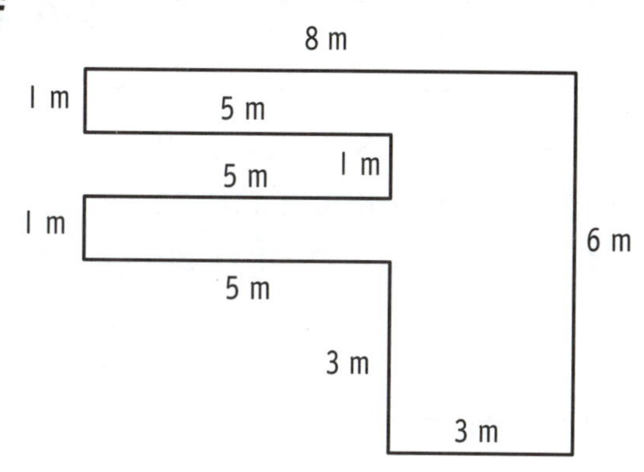

P = ______________________

____

= ____ m

g

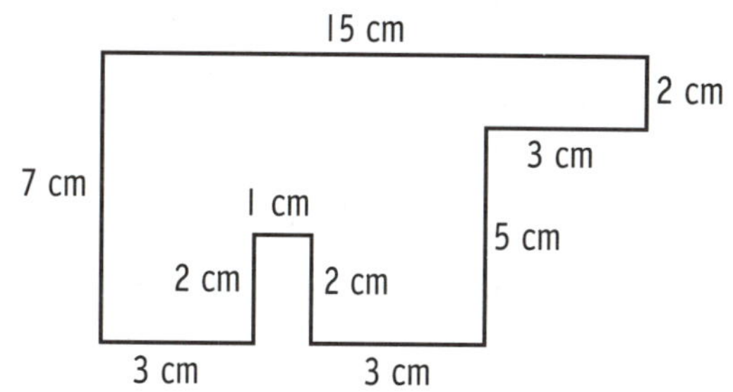

P = ______________________

____

= ____ cm

VOLUME

Volume is the amount of space an object takes up.
Volume is measured in cubic units
The symbol for volume is V.

This is one cubic centimetre, 1 cm^3.

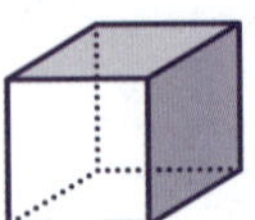

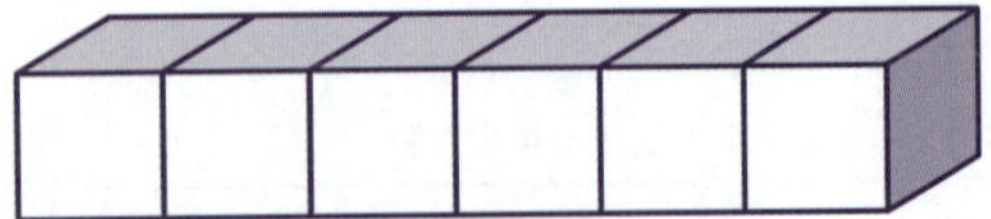

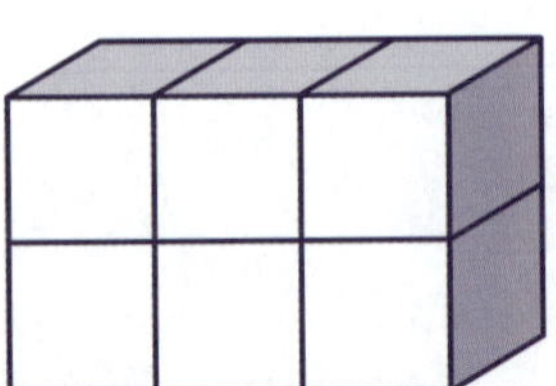

This object is made from 6 one-centimetre cubes. It has a volume of 6 cm^3.

This object is also made from 6 one-centimetre cubes It has a volume of 6 cm^3.

Examples: What is the volume?

a

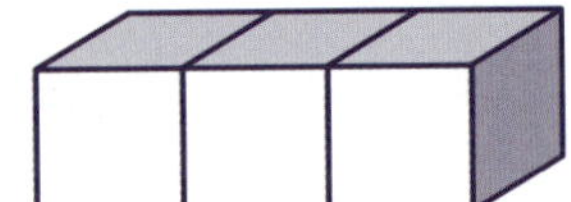

V = __ cubic centimetres
= __ cm^3

b

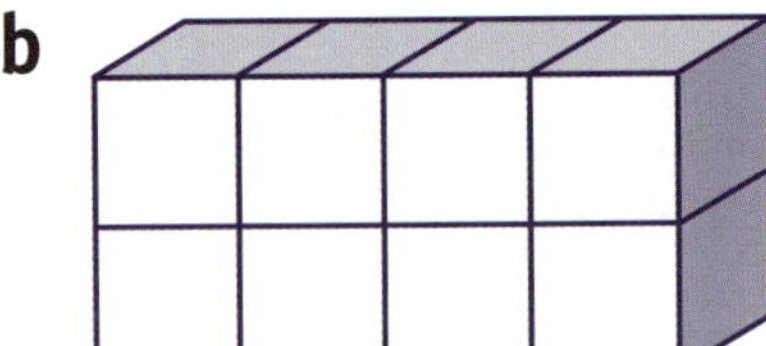

V = __ cubic centimetres
= __ cm^3

Check the answers on the video!

Your turn

How many cubic centimetres were used to make the objects?

● 6 cm^3

a ___ cm^3

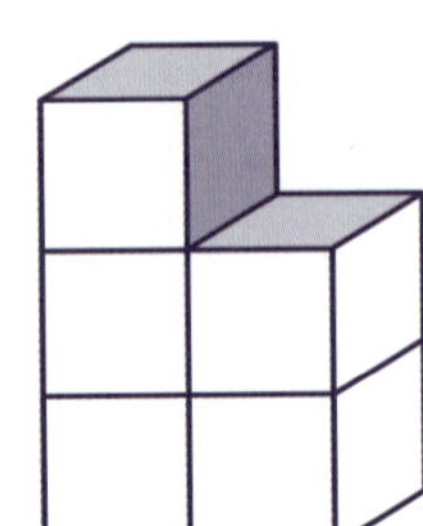

b ___ cm^3

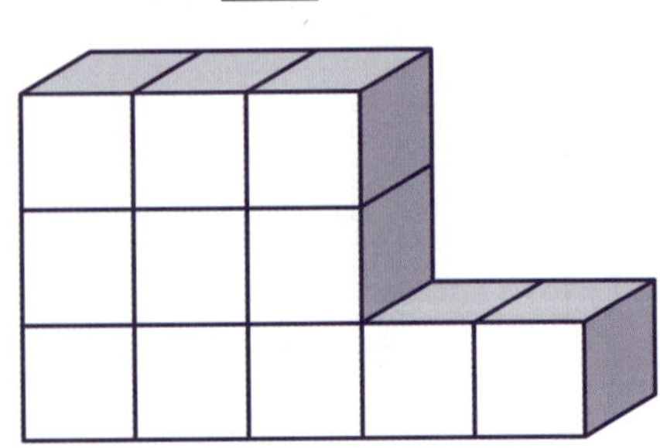

SELF CHECK Tick how you feel

Got it!	Need help...	I don't get it
☐	☐	☐

Check your answers
How many did you get correct? ☐

CATCH UP MATHS YEAR 5 BOOK B © PASCAL PRESS ISBN: 9781925726176

What is the volume? Circle the correct answer.

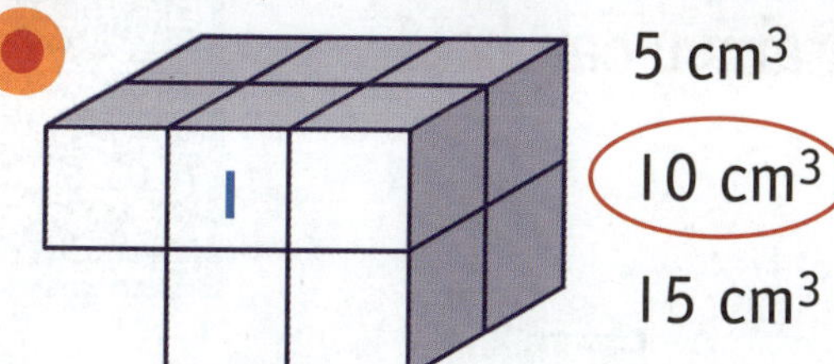

5 cm^3
10 cm^3 (circled)
15 cm^3

a
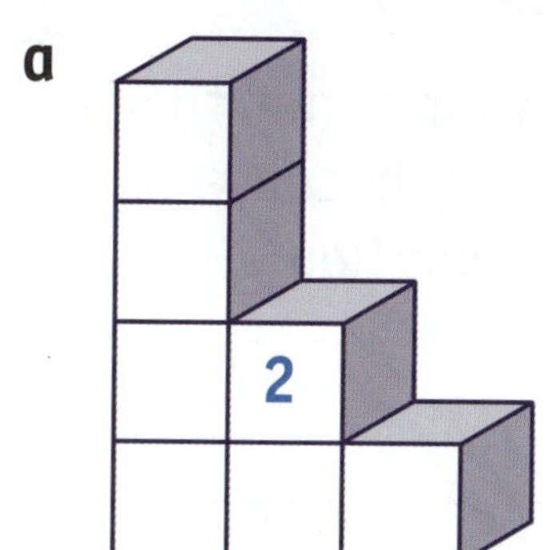

5 cm^3
6 cm^3
7 cm^3

b
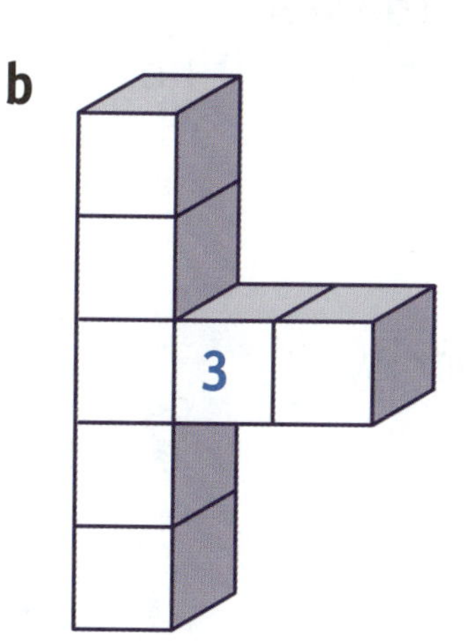

6 cm^3
7 cm^3
9 cm^3

c

16 cm^3 20 cm^3 32 cm^3

d
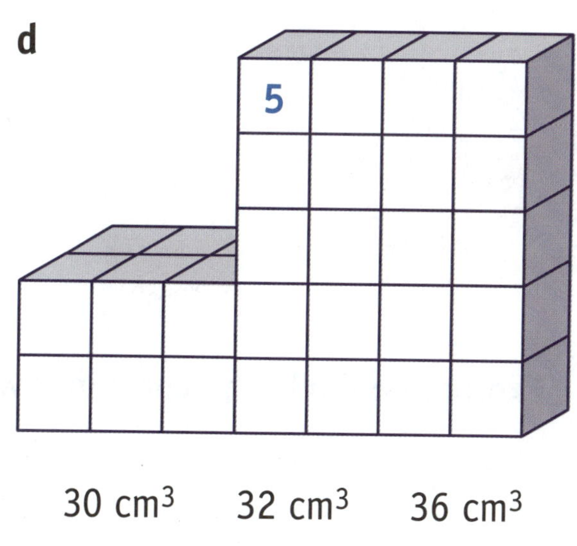

30 cm^3 32 cm^3 36 cm^3

Which objects in Question 1 have:

a the least volume? ______

b the greatest volume? ______

What is the difference in volume between:

a objects 2 and 3? ______

b objects 5 and 1? ______

How many cubic centimetres altogether were used to make the objects in Question 1?

____

CAPACITY

Capacity is the amount of liquid a container can hold.

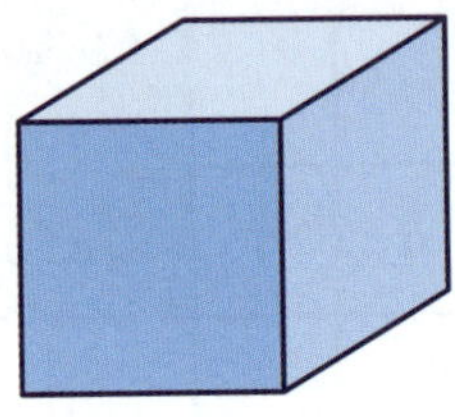
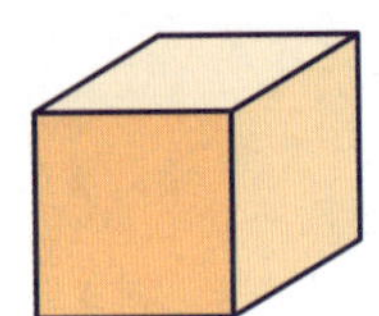
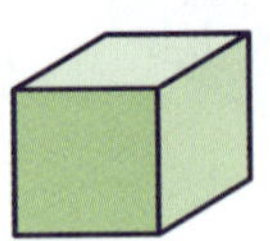

SCAN to watch video

The blue container has the largest capacity because it can hold the most liquid.

The green container has the smallest capacity because it can hold the least liquid.

Glass A

Glass B

Glass A has more water in it than Glass B. But Glass B can hold more water than Glass A, so it has a larger capacity.

Examples: Circle the container with the greater capacity.

a

b

c

Check the answers on the video!

Your turn

Draw a similar object with a smaller capacity.

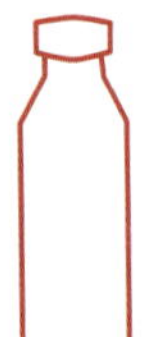

a

b

SELF CHECK Tick how you feel

Got it!	Need help...	I don't get it
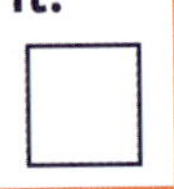 ☐	 ☐	 ☐

Check your answers

How many did you get correct? ☐

CATCH UP MATHS YEAR 5 BOOK B © PASCAL PRESS ISBN: 9781925726176

PRACTICE

Number the containers to order them from largest capacity (5) to smallest capacity (1).

● 5 3 2 1 4

a ☐ ☐ ☐ ☐ ☐

b ☐ ☐ ☐ ☐ ☐

c ☐ ☐ ☐ ☐ ☐

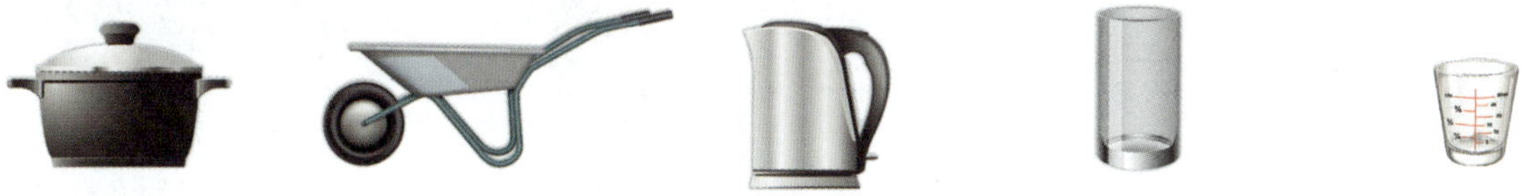

What is the total capacity of each set of containers?

● 10 L

b ________

d ________

a ________

c ________

e ________

LITRES

Litres (L) are used to measure liquids.
1 litre = 1000 millilitres
1 L = 1000 mL

This container can measure litres.

It can hold 3 litres (3 L).

Examples: How many litres can each container hold?

a 2 L **b** ___ L **c** ___ L **d** ___ L

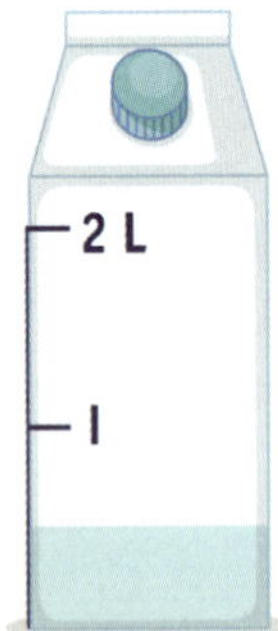

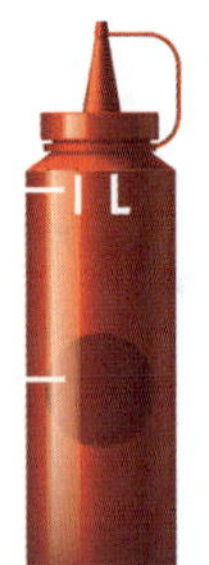

Check the answers on the video!

Your turn

Write the measurements using the symbol L for litres.

●	2 litres	2 L	**d**	85 litres	________
a	7 litres	________	**e**	105 litres	________
b	9 litres	________	**f**	1243 litres	________
c	22 litres	________	**g**	630 litres	________

SELF CHECK Tick how you feel

Got it!	Need help...	I don't get it
☐	☐	☐

Check your answers
How many did you get correct? ☐

CATCH UP MATHS YEAR 5 BOOK B © PASCAL PRESS ISBN: 9781925726176

PRACTICE

1 Match the equivalent measurements.

●	7 litres	98 L
a	15 litres	200 L
b	24 litres	7 L
c	98 litres	103 L
d	103 litres	15 L
e	200 litres	24 L

2 List five containers that would hold less than one litre.

juice box ______________________________

3 List five containers that would hold at least one litre.

bucket ______________________________

4 What is the total capacity of each group of containers?

● 6 L

2 L 1 L 3 L

a ___ L

b ___ L

5 Colour each container to show 2 litres.

●

a

b

c

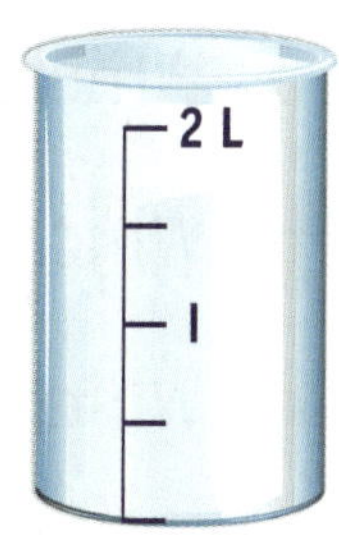

What amount of liquid is in each container?

 3 L

b ___ L

d ___ L

f ___ L

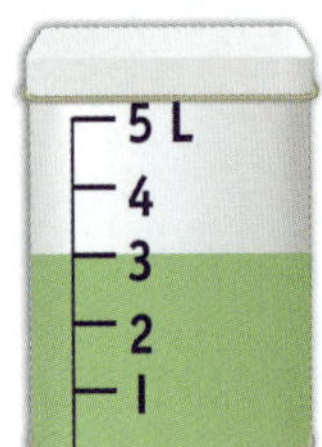

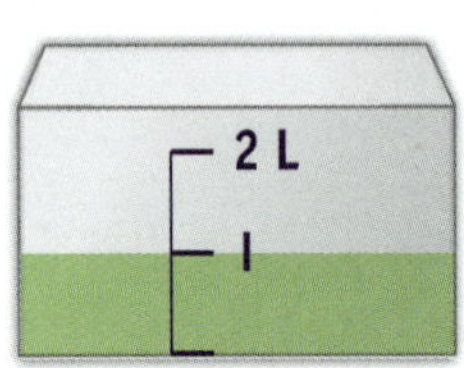

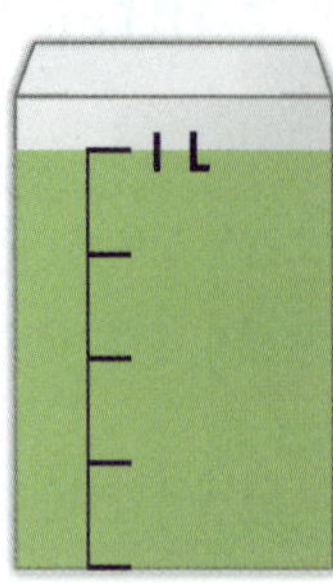

a ___ L

c ___ L

e ___ L

g ___ L

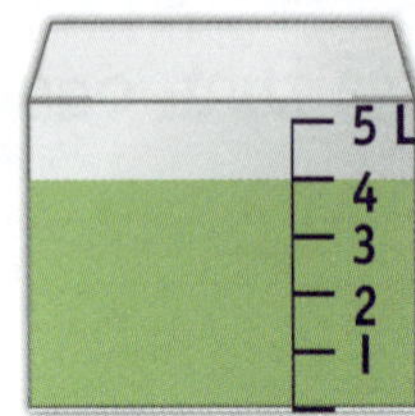

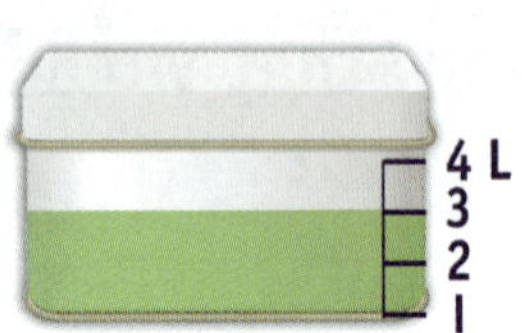

7

Colour the jugs to show the amounts.

 Jug 1: 3 L

a Jug 2: 1 L

b Jug 3: 4 L

c Jug 4: 5 L

8

Use the jugs in Question 7 to answer the following questions.

a Which jug has the most liquid in it? ___

b Which jug has the least liquid in it? ___

c Which jug has the greatest capacity? ___

d Which jug has the least capacity? ___

  ISBN: 9781925726176

9 Match the containers to the correct amounts.

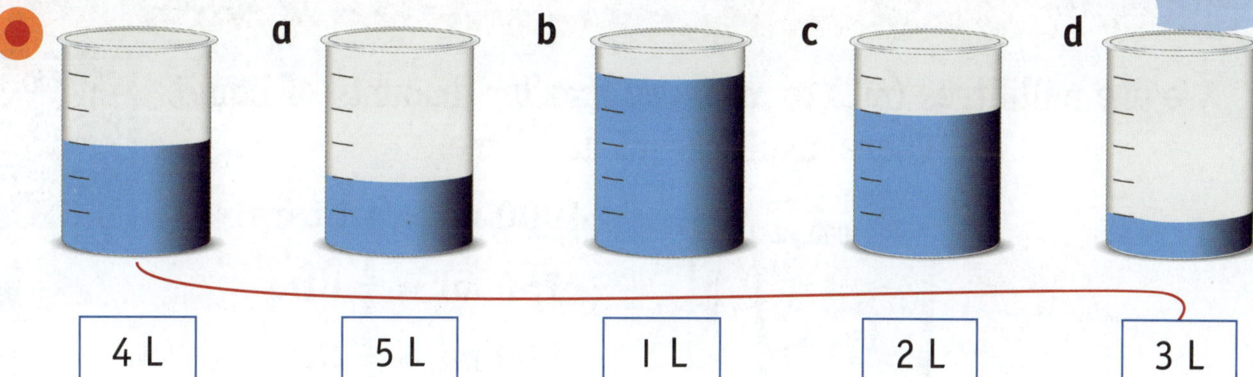

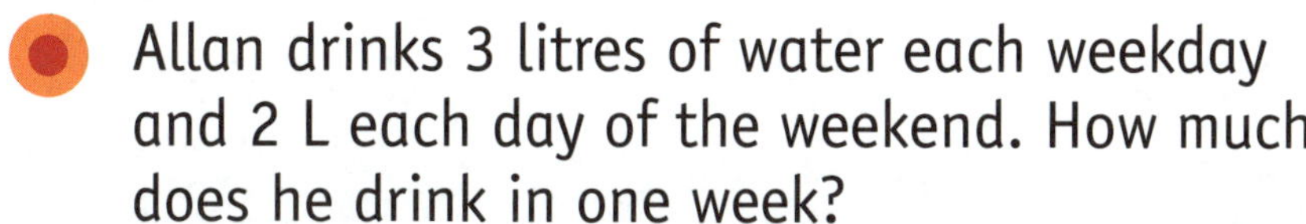

2 L

3 L

10 Solve.

		Working out
●	Allan drinks 3 litres of water each weekday and 2 L each day of the weekend. How much does he drink in one week?	3 L × 5 days = 15 L 2 L × 2 days = 4 L 15 L + 4 L = 19 L
a	Tony has a container that holds one litre. How many containers can Tony fill from a bucket that holds 20 litres?	
b	A bucket holds 5 litres. How many buckets of water will it take to fill a 150 L fish tank?	
c	Roslynn squeezes 8 oranges to make one litre of juice. How many oranges will she need to make 6 litres of juice?	
d	At Janice's juice stand she has these jugs with juice left over. If Janice had 16 L of juice to begin with, how much juice did she sell?	

  ISBN: 9781925726176

MILLILITRES

SCAN to watch video

We use millilitres (mL) to measure smaller amounts of liquid. There are 1000 mL in 1 litre.

1000 mL = 1 litre
750 mL = $\frac{3}{4}$ litre
500 mL = $\frac{1}{2}$ litre
250 mL = $\frac{1}{4}$ litre

Examples: Colour the containers to show the measurements.

a $\frac{1}{4}$ litre

b 1 litre

c $\frac{1}{2}$ litre

d $\frac{3}{4}$ litre

Check the answers on the video!

Your turn

Colour the containers to show the measurements. Hint: Each mark shows 100 mL.

300 mL

b 400 mL

d 750 mL

a 250 mL

c 700 mL

e 100 mL

SELF CHECK Tick how you feel

Got it!	Need help...	I don't get it
 ☐	 ☐	 ☐

Check your answers
How many did you get correct?

CATCH UP MATHS YEAR 5 BOOK B © PASCAL PRESS ISBN: 9781925726176

PRACTICE

1 Would you use millilitres (mL) or litres (L) to measure these?

- a glass of milk mL
- a a large carton of milk ___
- b a dose of medicine ___
- c drops in an eyedropper ___
- d water in a spa ___
- e petrol in a boat ___
- f a cup of coffee ___
- g juice in a juice box ___

2 Write as litres.

- 2500 mL $2\frac{1}{2}$ L
- a 1000 mL ______
- b 2000 mL ______
- c 9750 mL ______
- d 1500 mL ______
- e 8500 mL ______
- f 6000 mL ______
- g 10 000 mL ______
- h 1250 mL ______

3 Write as millilitres.

- 7 L 7000 mL
- a 4 L ______
- b $3\frac{1}{4}$ L ______
- c $3\frac{1}{2}$ L ______
- d $8\frac{1}{4}$ L ______
- e $1\frac{3}{4}$ L ______
- f $2\frac{1}{2}$ L ______
- g $6\frac{3}{4}$ L ______
- h 8 L ______

4 Write as litres and millilitres.

- 6200 mL 6 L 200 mL
- a 3800 mL ______
- b 1425 mL ______
- c 7500 mL ______
- d 9250 mL ______
- e 3800 mL ______
- f 1100 mL ______
- g 2750 mL ______
- h 4900 mL ______

5 Add these amounts.

- 3 L 650 mL + 3 L 210 mL = 6 L 860 mL

a 7 L 240 mL + 2 L 125 mL = ___ L ______ mL

b 4 L 250 mL + 3 L 740 mL = ___ L ______ mL

c 9 L 780 mL + 5 L 320 mL = ___ L ______ mL

6 Number the containers to order them from smallest capacity (1) to largest capacity (6).

☐ ☐ ☐ ☐ ☐ ☐

7 Colour the containers to show the measurements.

- 150 mL

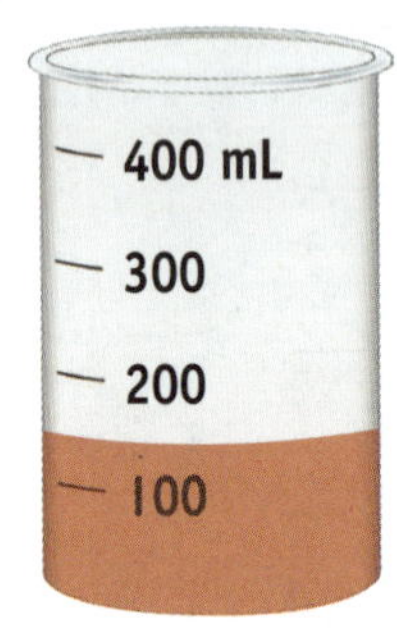

b 350 mL

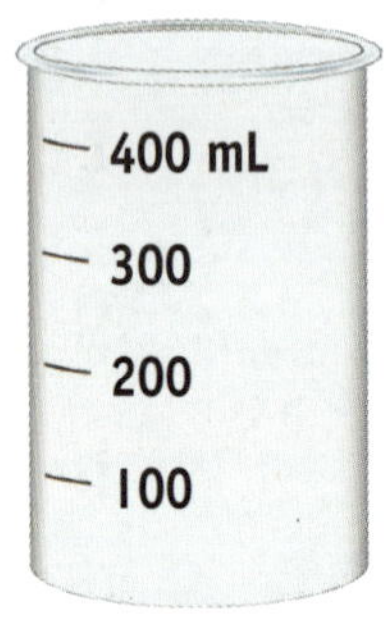

d 250 mL

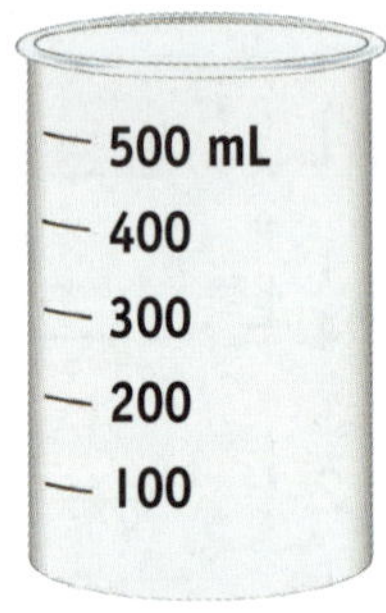

a 500 mL

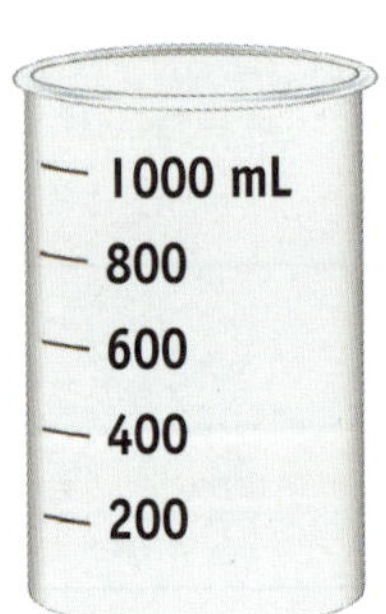

c 25 mL

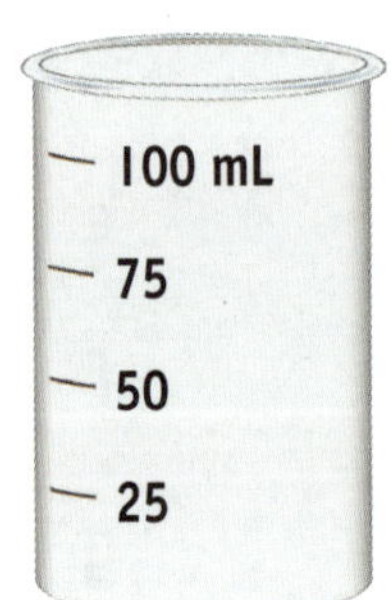

e 400 mL

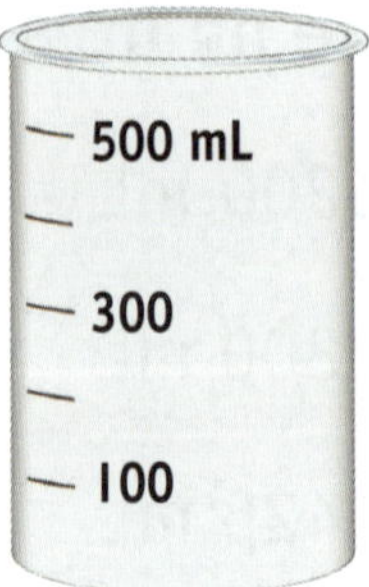

CATCH UP MATHS YEAR 5 BOOK B © PASCAL PRESS ISBN: 9781925726176

LITRES OR MILLILITRES

Litres (L) are used to measure large amounts of liquids.
Millilitres (mL) are used to measure small amounts of liquids.

This milk bottle holds 2 L.

15 mL has been measured in this medicine cup.

Examples: Rewrite these measurements using the short form of millilitres (mL) and litres (L).

a 220 millilitres 220 mL

b 6 litres ________

c 183 millilitres ________

d 20 millilitres ________

e 90 litres ________

f 1 litre ________

g 77 millilitres ________

h 587 litres ________

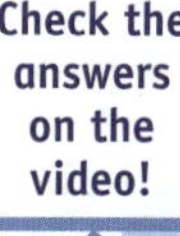

Your turn

Would the capacity be measured in mL or L?

 10 L

b 200 ___

d 180 ___

a 4 ___

c 20 ___

e 12 ___

SELF CHECK Tick how you feel

Got it!	Need help...	I don't get it
 ☐	 ☐	☐

Check your answers
How many did you get correct? ☐

  ISBN: 9781925726176

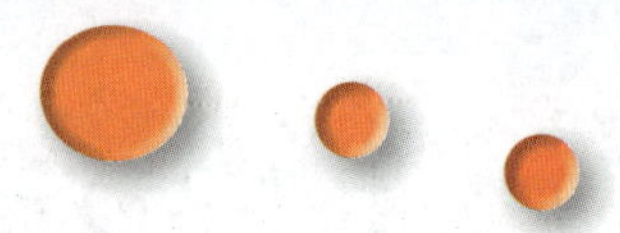

PRACTICE

1 Rewrite using the short form.

●	375 millilitres	375 mL	c	1000 millilitres	________
a	575 millilitres	________	d	900 millilitres	________
b	450 millilitres	________	e	150 millilitres	________

2 Write these amounts in millilitres.

●	6 L 235 mL	6235 mL	c	1 L 903 mL	________ mL
a	8 L 850 mL	________ mL	d	5 L 600 mL	________ mL
b	2 L 25 mL	________ mL	e	7 L 520 mL	________ mL

3 Convert into L.

●	8590 mL	8.590 L	c	9000 mL	________ L
a	7265 mL	________ L	d	7002 mL	________ L
b	3055 mL	________ L	e	10 420 mL	________ L

4 Convert into mL.

●	4.205 L	4205 mL	c	1.25 L	________
a	6.300 L	________	d	4.0 L	________
b	14.007 L	________	e	18.606 L	________

5 Number these amounts to order them from smallest (1) to largest (6).

●	3 L 530 mL	4 L 430 mL	1250 mL	600 mL	1000 mL
	4	5	3	1	2
a	7 L 522 mL	6290 mL	1560 mL	100 mL	4956 mL
	☐	☐	☐	☐	☐
b	9873 mL	8 L 430 mL	6580 mL	4 L 972 mL	3800 mL
	☐	☐	☐	☐	☐

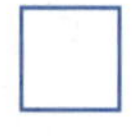

CATCH UP MATHS YEAR 5 BOOK B © PASCAL PRESS ISBN: 9781925726176

VOLUME AND CAPACITY REVIEW

Complete the sentence.

Volume is the amount of ___________ an object takes up.

What is the volume of each object?

a Volume = ___ cm^3

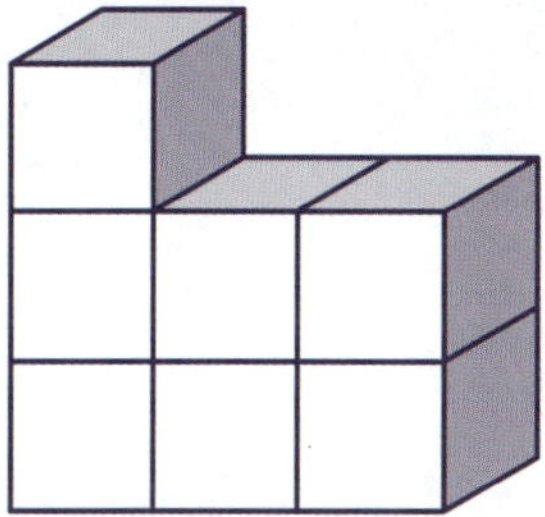

b Volume = ___ cm^3

c Volume = ___ cm^3

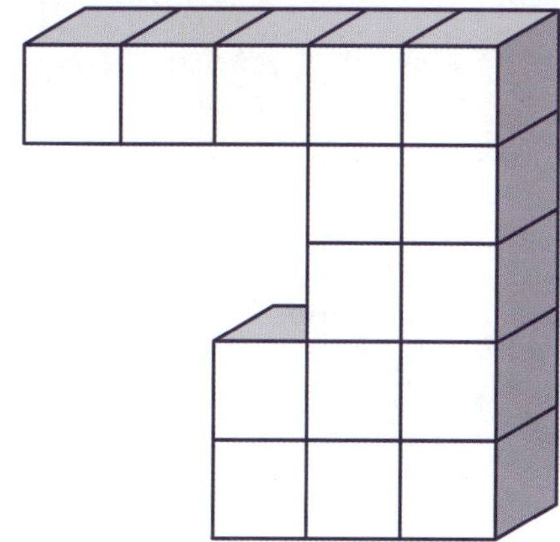

d Volume = ___ cm^3

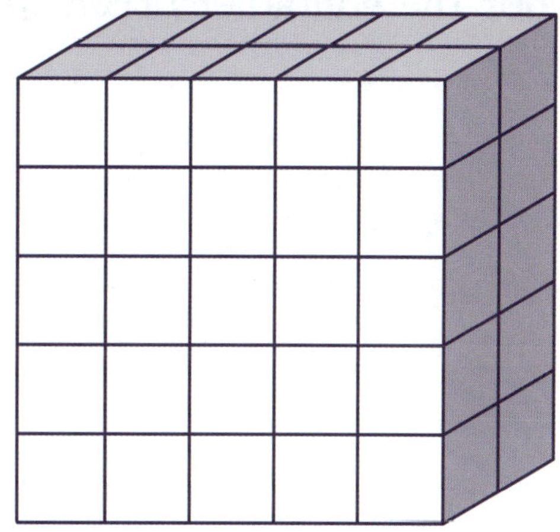

Which object in Question 2 has:

a the least volume? ______

b the greatest volume? ______

What is the difference in volume?

a object **b** and object **c** _______

b object **d** and object **a** _______

REVIEW

5 Complete the sentences.

Capacity is the amount of ____________ a container can _________.

1 L = _______ mL

6 Circle the container with the smaller capacity.

a

c

b

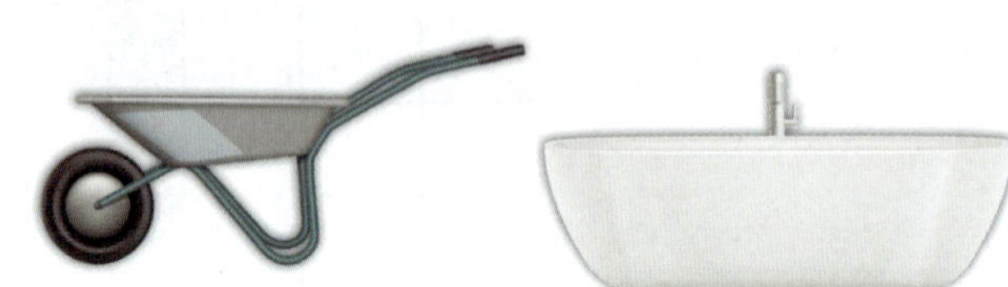

d

7 Order the containers from smallest capacity (1) to largest capacity (5).

a

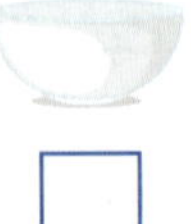 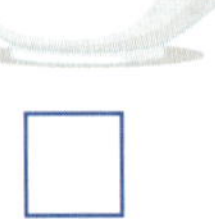  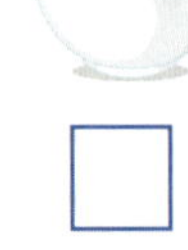

☐ ☐ ☐ ☐ ☐

b

☐ ☐ ☐ ☐ ☐

c

☐ ☐ ☐ ☐ ☐

  ISBN: 9781925726176

What is the total capacity of each set of containers?

a ___ L

b ___ L

c ___ L

Write the measurements using the symbol L.

a 8 litres ________

b 285 litres ________

c 20 litres ________

d 49 litres ________

e 1057 litres ________

f 10 litres ________

List three containers that would hold less than one litre.

____

List three containers that would hold one or more litres.

____

Colour each container to show one litre.

a

b

c

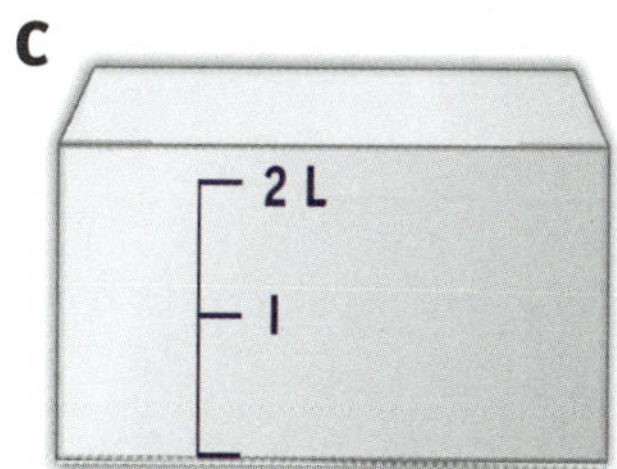

d

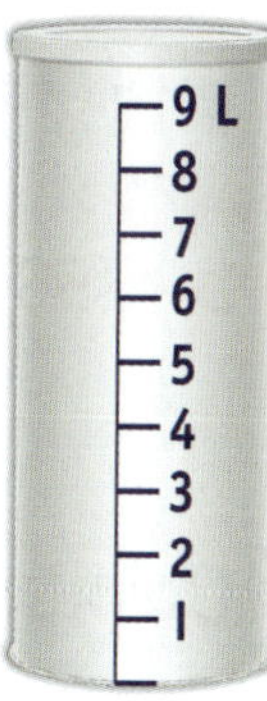

What amount of liquid is in each container?

a ________

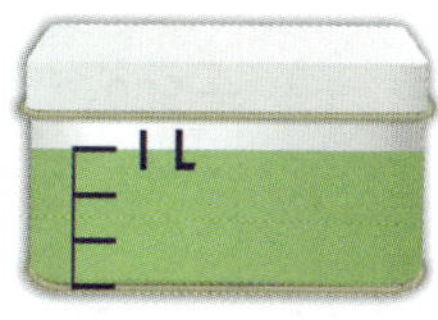

b ________

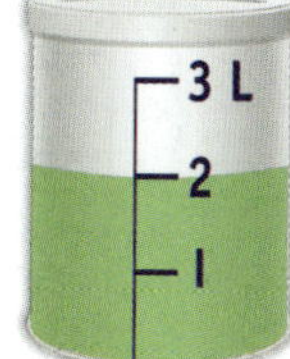

c ________

4 L
3
2
1

d ________

REVIEW

14 Solve. Show your working.

	Working out
a Danny drinks 1 litre of water on each day of the weekend and 2 litres of water every weekday, except for Wednesday when he drinks 3 litres of water. How much water does Danny drink in one week?	
b Li has a container that holds 3 litres. How many times does Li need to fill the container to fill a bucket that holds 33 litres?	
c A bucket holds 10 litres. How many buckets of water will it take to fill Antonia's fish tank, which holds 110 litres?	

15 Write the missing numbers.

a 1000 mL = ________ L

b 750 mL = ________ L

c ________ mL = $\frac{1}{2}$ L

d ________ mL = $\frac{1}{4}$ L

16 Colour the containers to show the amount.

a 500 mL

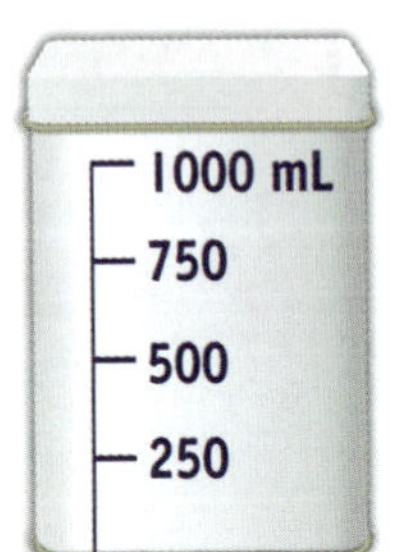

b 750 mL

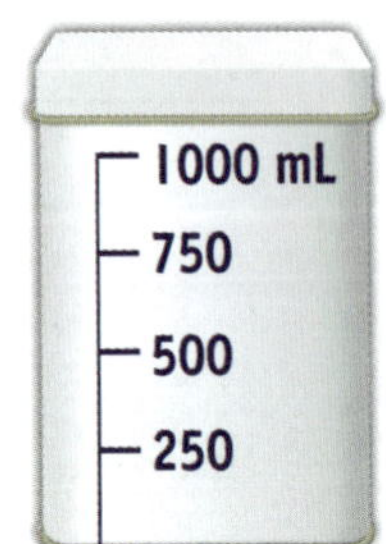

c 250 mL

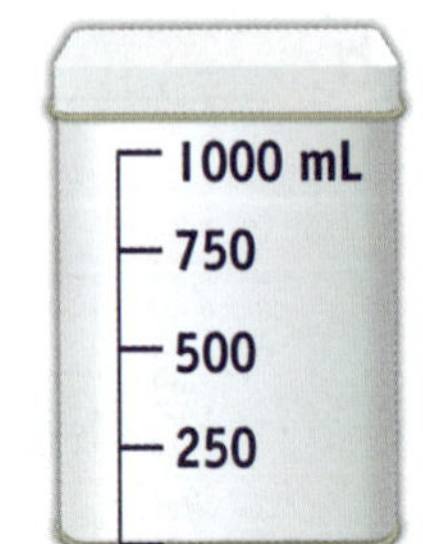

d 1000 mL

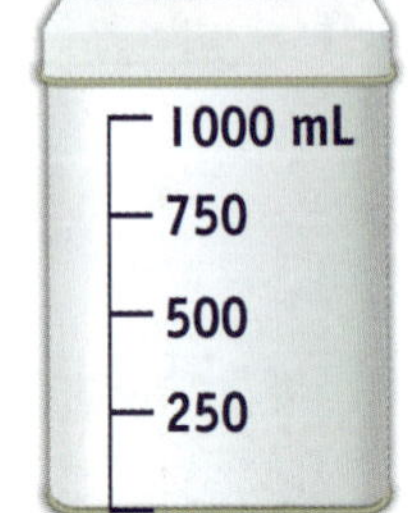

  ISBN: 9781925726176

17 Colour the containers to show the amount.

a 250 mL

b 300 mL

c 450 mL

d 650 mL

18 Would you use mL or L to measure these?

a milk in a baby's bottle ____

b petrol in a car ____

c a cup of tea ____

d a glass of milk ____

e drops of food colouring ____

f water in a swimming pool ____

19 Write as litres.

a 3500 mL __________

b 2000 mL __________

c 1500 mL __________

d 8750 mL __________

e 9250 mL __________

f 4750 mL __________

20 Write as millilitres.

a 8 L __________

b $3\frac{1}{4}$ L __________

c $6\frac{1}{2}$ L __________

d $2\frac{3}{4}$ L __________

e 9 L __________

f $10\frac{1}{4}$ L __________

21 Write as litres and millilitres.

a 3200 mL ______________

b 8400 mL ______________

c 9830 mL ______________

d 6840 mL ______________

e 6750 mL ______________

f 1875 mL ______________

REVIEW

Add these amounts.

a 4 L 250 mL + 2 L 530 mL = ______________

b 9 L 320 mL + 5 L 410 mL = ______________

c 6 L 140 mL + 8 L 560 mL = ______________

d 7 L 850 mL + 2 L 220 mL = ______________

e 9 L 33 mL + 4 L 125 mL = ______________

f 3 L 520 mL + 6 L 89 mL = ______________

Number the containers to order them from largest capacity (1) to smallest capacity (5).

a

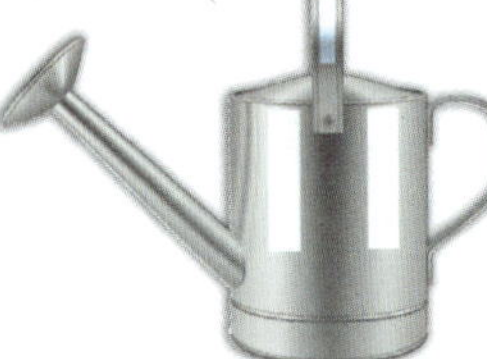

b

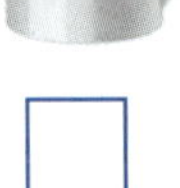

c

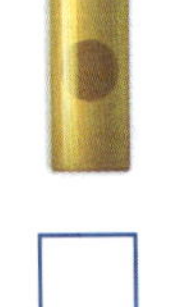

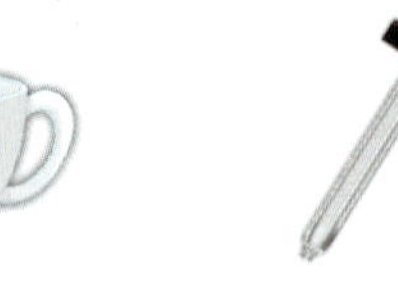

Write in short form.

a 250 millilitres ______________

b 580 millilitres ______________

c 120 millilitres ______________

d 50 millilitres ______________

e 2000 millilitres ______________

f 98 millilitres ______________

CATCH UP MATHS YEAR 5 BOOK B © PASCAL PRESS ISBN: 9781925726176

25 Write these amounts in millilitres.

a 8 L 535 mL __________ mL

b 6 L 530 mL __________ mL

c 1 L 680 mL __________ mL

d 5 L 55 mL __________ mL

e 22 L 100 mL __________ mL

f 11 L 25 mL __________ mL

26 Convert these mL into L. Use a decimal point.

a 7680 mL = __________ L

b 8003 mL = __________ L

c 13 303 mL = __________ L

d 16 001 mL = __________ L

e 3000 mL = __________ L

f 19 258 mL = __________ L

27 Convert these litres into millilitres.

a 5.204 L = __________ mL

b 9.237 L = __________ mL

c 3.75 L = __________ mL

d 12.25 L = __________ mL

e 6.0 L = __________ mL

f 8.52 L = __________ mL

28 Number these measurements from 1 (largest) to 5 (smallest) to order them in descending order.

a	5 L 220 mL ☐	3 L 650 mL ☐	1560 mL ☐	1 L 650 mL ☐	3400 mL ☐
b	2 L 755 mL ☐	4896 mL ☐	650 mL ☐	2570 mL ☐	1000 mL ☐
c	9550 mL ☐	6 L 530 mL ☐	4L 550 mL ☐	1382 mL ☐	2L 670 mL ☐
d	8750 mL ☐	8075 mL ☐	8000 mL ☐	8500 mL ☐	8975 mL ☐

KILOGRAMS

Heavy items are measured in kilograms (kg). 'Kilo' means 1000. Kilogram means 1000 grams.

A one-litre juice carton weighs about 1 kilogram (1 kg).

SCAN to watch video

Examples: Write the masses in short form.

a 4 kilograms 4 kg

b 12 kilograms ______

c 27 kilograms ______

d eleven kilograms ______

e one hundred kilograms ______

Check the answers on the video!

Circle the item that could be getting weighed.

 **1 kg**
- tennis ball
- bag of sugar

a **3 kg**
- an orange
- a watermelon

b **35 kg**
- a small child
- a bag of potatoes

c **1 kg**
- 1-litre carton of milk
- an apple

d **5 kg**
- a sandwich
- a full backpack

e **4 kg**
- a wallet
- a net of oranges

SELF CHECK Tick how you feel

Got it!	Need help...	I don't get it
☐	☐	☐

Check your answers

How many did you get correct? ☐

CATCH UP MATHS YEAR 5 BOOK B © PASCAL PRESS ISBN: 9781925726176

PRACTICE

1 Tick or cross the masses to show if they are written correctly.

● ☒ 7kg
a ☐ 42 Kg
b ☐ 107 KG
c ☐ 44k g
d ☐ 82 kg
e ☐ 303 kg
f ☐ 157 kG
g ☐ 4000kg

2 Use these items to answer the following questions.

Potatoes 10 kg
Oranges 3 kg
Cherries 1 kg
Watermelon 5 kg
Pineapple 2 kg

a Which item is heaviest? ____________

b Which item is lightest? ____________

c What is the mass of all the items together? ____________

3 How much do these quantities weigh?

● 3 nets of oranges $3 \times 3 = 9$ kg
a 5 bags of cherries ________
b 4 watermelons ________
c 2 pineapples ________
d 6 bags of potatoes ________
e 1 pineapple and 2 nets of oranges ________

4 Hany's cardboard box can hold 10 kg. How many of each item can he put into his box?

● potatoes 1
a oranges ___
b cherries ___
c watermelon ___
d pineapple ___

5 List three things that weigh each amount.

a less than a kilogram ____________________

b about a kilogram ____________________

c more than a kilogram ____________________

  ISBN: 9781925726176

GRAMS

Light objects are measured using grams (g). There are 1000 grams in 1 kilogram.

An egg weighs about 70 grams (70 g).

A paperclip weighs about 1 gram (1 g).

- In 5 kilograms, there are 5000 grams.
- In 2 kilograms, there are 2000 grams.
- In 4 kilograms 250 grams, there are 4250 grams.

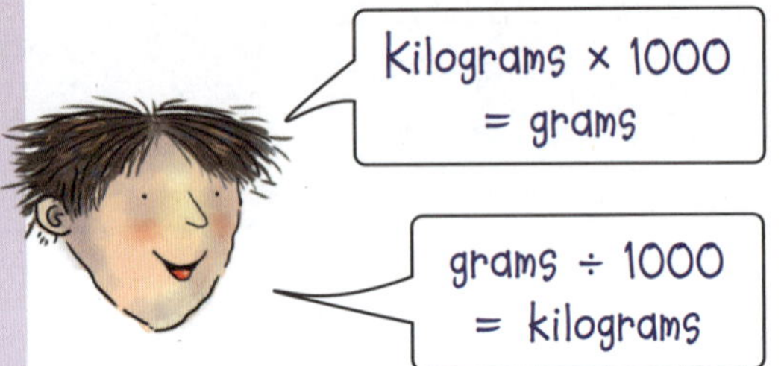

Examples: List six things that would be measured in grams.

a cherry ______ ______

______ ______

______ ______

Check the answers on the video!

Your turn

Circle the objects that would be measured in grams.

b 

d 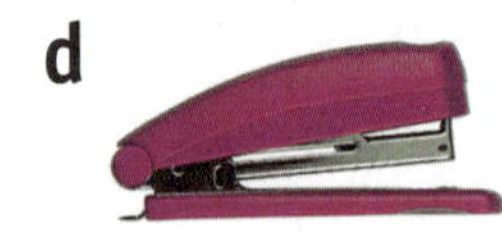

f 

a

c 

e 

g 

SELF CHECK Tick how you feel		
Got it! ☐	Need help... ☐	I don't get it ☐

Check your answers
How many did you get correct? ☐

CATCH UP MATHS YEAR 5 BOOK B © PASCAL PRESS ISBN: 9781925726176

PRACTICE

1 Use the short form to write the measurements.

- two hundred grams 200 g
- a four grams ______
- b sixty-four grams ______
- c one hundred and five grams ______
- d seven hundred and seventy grams ______

2 Tick or cross the labels to show if they are written correctly.

●	84 g ✔	b	203GR	d	320 g.	f	72g.
a	35G	c	45 g	e	235 G	g	700 g

3 Number the cans to order them from largest mass (1) to smallest (8).

● 8 

b ☐ 

d ☐ 

f ☐ 

a ☐ 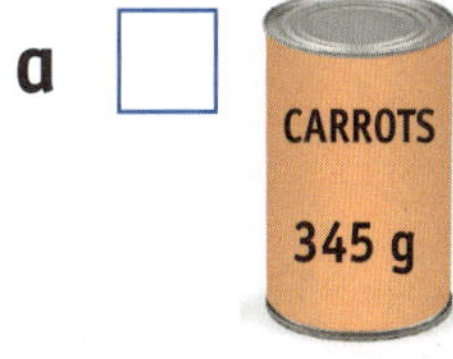

c ☐ 

e ☐ 

g ☐ 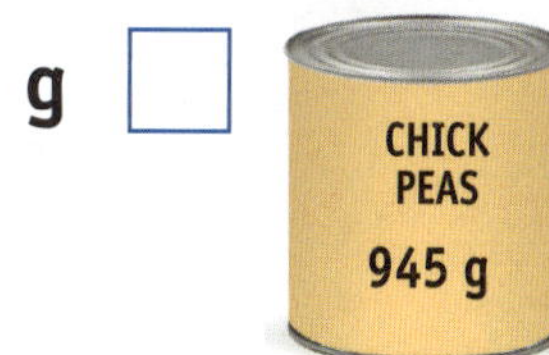

4 Use the cans in Question 3 to answer the following questions.

a Which cans have a mass more than 300 g? ______________________

b Which cans are 300 g or lighter? ______________________

5 How many more grams for these cans to make one kilogram?

- salmon 900 g
- a peas ______
- b jam ______
- c carrots ______
- d soup ______
- e corn ______
- f chickpeas ______
- g beans ______

6 How many of each can do you need to make 1 kg?

- salmon 10
- a peas ____
- b corn ____

PARTS OF A KILOGRAM

SCAN to watch video

One kilogram is 1000 g.
Parts of a kilogram can be measured in grams.
For example, $\frac{1}{2}$ kg = 500 g

FLOUR 1 kg = 1000 g

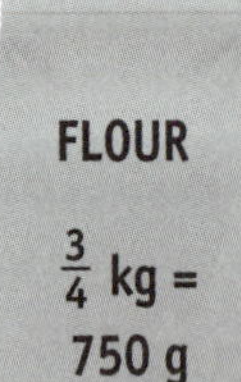

FLOUR $\frac{1}{2}$ kg = 500 g

FLOUR $\frac{1}{4}$ kg = 250 g

If I have $1\frac{1}{2}$ kg it equals 1500 grams.

If I have $3\frac{1}{4}$ kg it equals 3250 grams.

If I have $5\frac{3}{4}$ kg it equals 5750 grams.

Examples: Complete the equivalent measurements.

a	3250 g = $3\frac{1}{4}$ kg	**e**	750 g = ____ kg
b	5500 g = $5\frac{1}{2}$ kg	**f**	1250 g = ____ kg
c	3000 g = 3 kg	**g**	10 500 g = ____ kg
d	7750 g = $7\frac{3}{4}$ kg	**h**	500 g = ____ kg

Your turn

Convert the measurements in kilograms (kg) to grams (g).

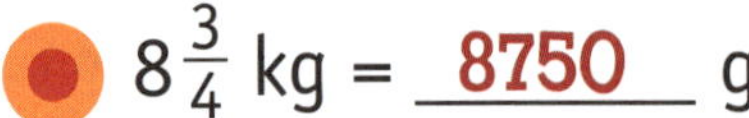

a $1\frac{1}{2}$ kg = ________ g

b 7 kg = ________ g

c $4\frac{1}{4}$ kg = ________ g

d $\frac{1}{4}$ kg = ________ g

e $5\frac{3}{4}$ kg = ________ g

f 2 kg = ________ g

g $3\frac{1}{2}$ kg = ________ g

h $9\frac{1}{4}$ kg = ________ g

SELF CHECK Tick how you feel

Got it! 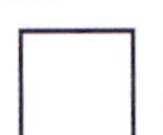 ☐

Need help...  ☐

I don't get it   ☐

Check your answers
How many did you get correct? ☐

CATCH UP MATHS YEAR 5 BOOK B © PASCAL PRESS ISBN: 9781925726176

PRACTICE

1 How many make up one kilogram?

- 50 g × 20 = 1 kilogram
- **a** 100 g × ______ = 1 kilogram
- **b** 200 g × ______ = 1 kilogram
- **c** 250 g × ______ = 1 kilogram
- **d** 500 g × ______ = 1 kilogram
- **e** 1000 g × ______ = 1 kilogram

2 Match the equivalent measurements.

a

kg	grams
$\frac{1}{4}$ kg	3000 g
$1\frac{1}{2}$ kg	250 g
3 kg	5000 g
5 kg	1500 g
$\frac{1}{2}$ kg	500 g

d

kg	grams
$2\frac{3}{4}$ kg	1000 g
10 kg	7500 g
1 kg	2750 g
$\frac{3}{4}$ kg	750 g
$7\frac{1}{2}$ kg	10 000 g

b

kg	grams
0.25 kg	2750 g
2.75 kg	500 g
17.31 kg	7500 g
7.5 kg	250 g
0.5 kg	17 310 g

e

kg	grams
0.75 kg	2280 g
8.5 kg	750 g
12.82 kg	12 820 g
2.28 kg	4710 g
4.71 kg	8500 g

c

kg	grams
2 kg 300 g	4300 g
4 kg 30 g	2003 g
4 kg 300 g	2300 g
2 kg 30 g	4030 g
2 kg 3 g	2030 g

f

kg	grams
8 kg 400 g	8040 g
8 kg 40 g	8004 g
8 kg 4 g	3008 g
3 kg 800 g	8400 g
3 kg 8 g	3800 g

COMPARING AND ORDERING MASS

Angelo	Adon	Conor
83 kg	79 kg	72 kg

When we compare masses, we work out which item is heaviest and which item is lightest. Then we can arrange the items in order of mass.

SCAN to watch video

Conor is the lightest person and Angelo is the heaviest.

Examples: Finish ordering these from lightest (1) to heaviest (4).

a

1 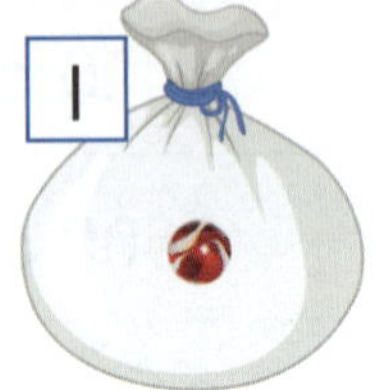
3 
2 
4 

b

☐ 
2 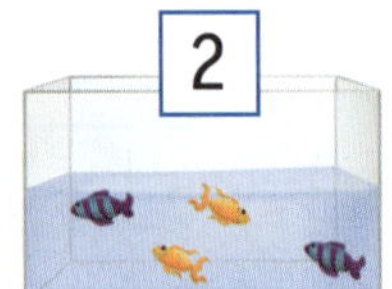
3 
☐ 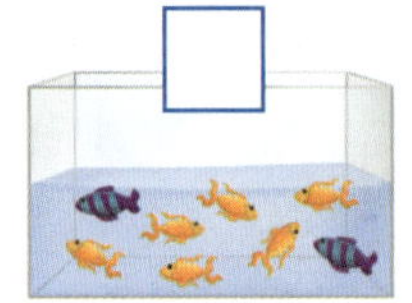

Check the answers on the video!

Your turn

Number the groups to order them from heaviest (1) to lightest (5).

1 
4 
3 
2 
5 

a

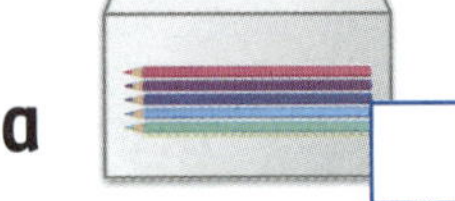 ☐
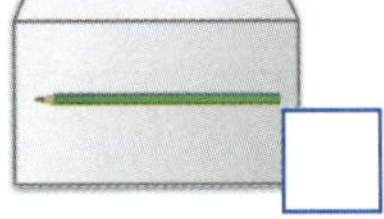 ☐
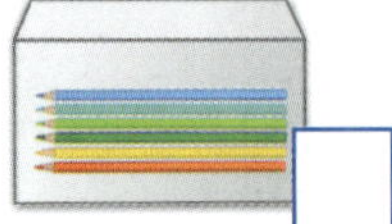 ☐
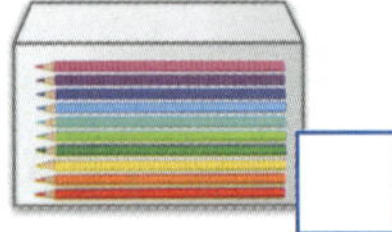 ☐
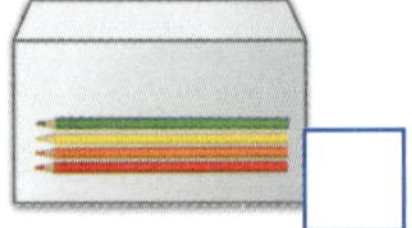 ☐

b

 ☐
 ☐
 ☐
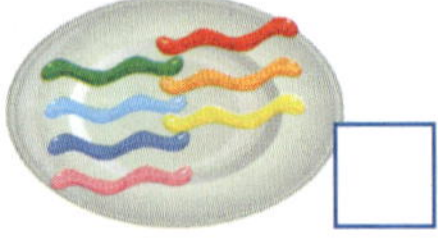 ☐
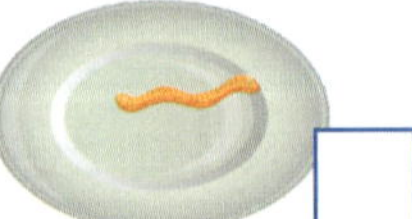 ☐

SELF CHECK Tick how you feel

Got it!	Need help...	I don't get it
 ☐	 ☐	 ☐

Check your answers

How many did you get correct? ☐

CATCH UP MATHS YEAR 5 BOOK B © PASCAL PRESS ISBN: 9781925726176

PRACTICE

1 Number the masses to order them from lightest (1) to heaviest (5).

●	1 kg 500 g	2536 g	2 kg 320 g	4589 g	5 kg 35 g
	1	3	2	4	5
a	5935 g	5 kg 355 g	5335 g	5 kg 593 g	9355 g
	☐	☐	☐	☐	☐
b	4 kg 205 g	5520 g	4 kg 520 g	5250 g	2550 g
	☐	☐	☐	☐	☐
c	3795 g	5975 g	7935 g	5 kg 759 g	3 kg 975 g
	☐	☐	☐	☐	☐

Jenai used balance scales to measure these items.

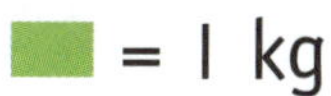
= 1 kg
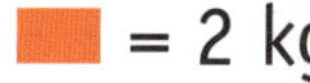
= 2 kg

2 How much do these items weigh?

● paint 5 kg   **a** bag ____   **b** ball ____   **c** gift ____

3 Write the missing numbers.

a You need ___ balls to balance the gift.

b You need ___ gifts to balance the bag.

4 What is the combined weight?

a paint and gift ___   **b** ball and bag ____   **c** all four items ____

5 Jenai put the ball on the same side of the scale as the bag.
How many paint cans does she need to balance the scale?

____ paint cans

READING SCALES

This is a balance scale.
We use it to measure small masses.
When it is level, the mass on one side is equal to the mass on the other side.

SCAN to watch video

The cake weighs 100 g.

This scale weighs up to 1 kg.

This scale can weigh up to 5 kg.

Examples: Read the balance scales and write the mass.

a 300 g

b ______ g

c ______ g

200 g
200 g

150 g
150 g

Check the answers on the video!

Your turn

Read the scale and write the mass.

● 200 g

a ______ kg

b ______ g

c ______ kg

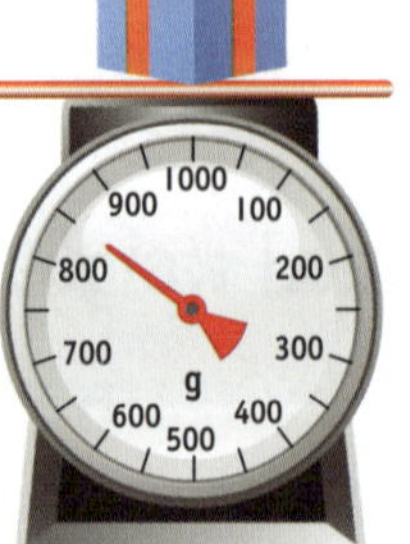

SELF CHECK Tick how you feel		
Got it!	Need help...	I don't get it

Check your answers

How many did you get correct?

CATCH UP MATHS YEAR 5 BOOK B © PASCAL PRESS ISBN: 9781925726176

PRACTICE

Read the scales and write the mass.

400 g b ______ d ______ f ______

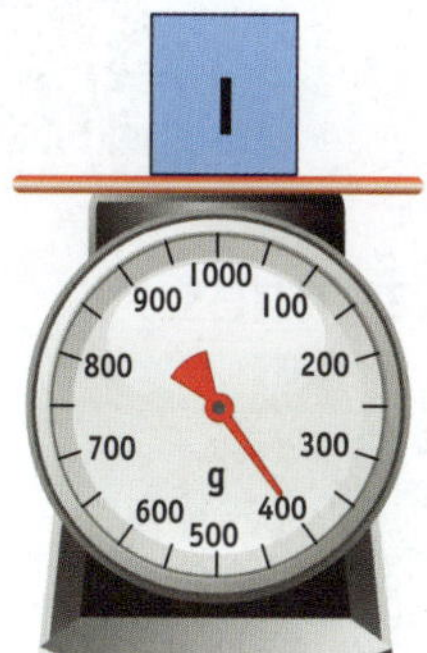

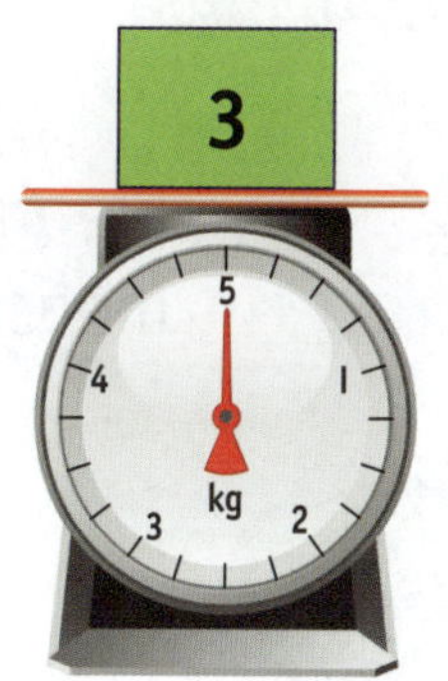

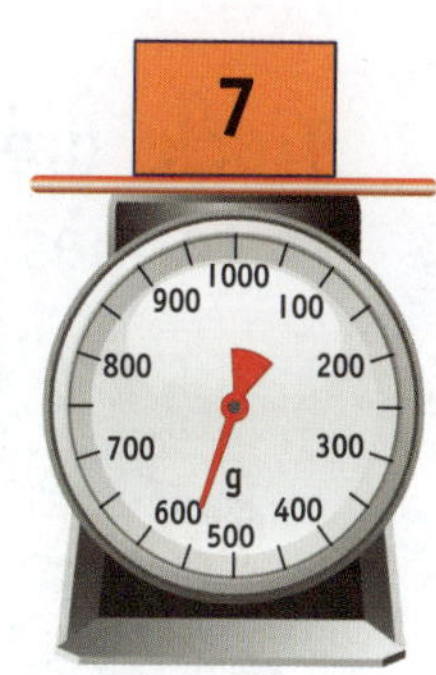

a ______ c ______ e ______ g ______

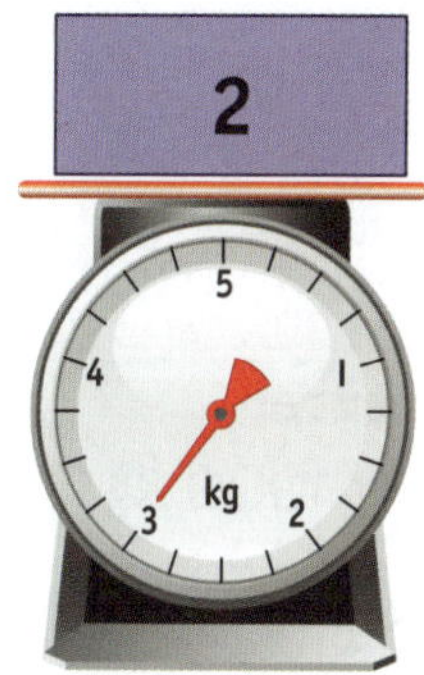

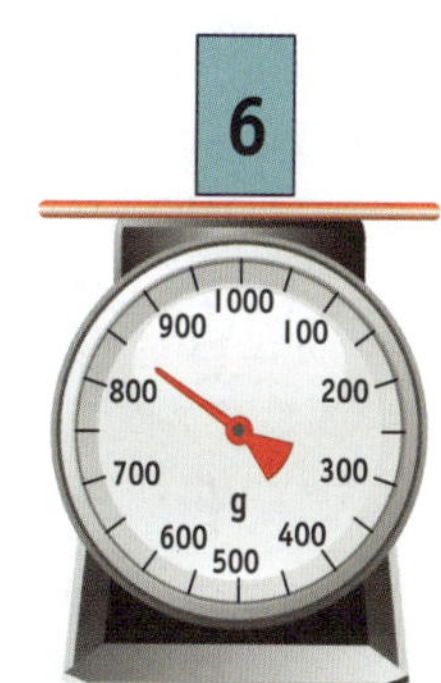

List the box numbers in Question 1 in order from lightest to heaviest.

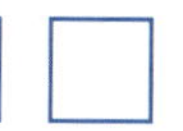

Which boxes weighed:

a more than 1 kg? ______ b less than 1 kg? ______

Draw a pointer on the scale to show the mass.

a
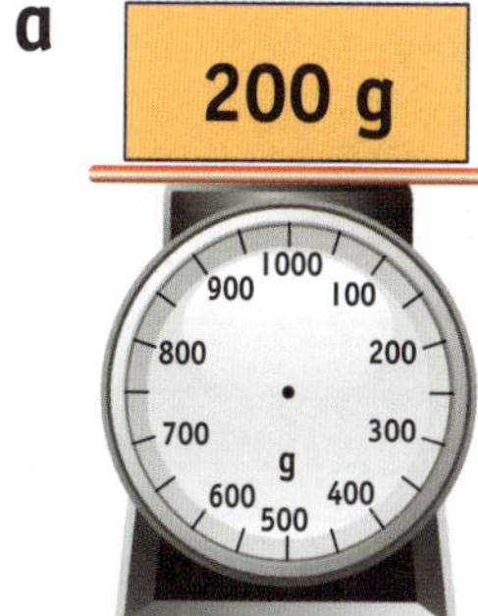

b
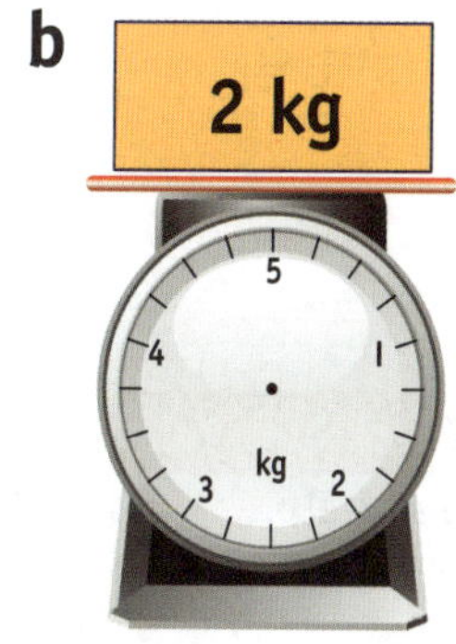

c
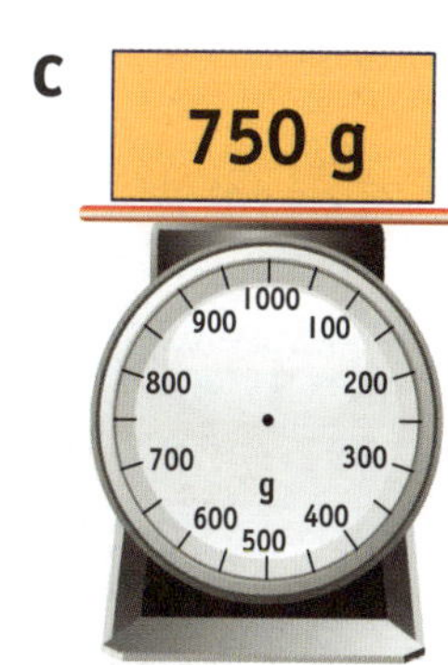

THE TONNE

Tonnes (t) are used for measuring very heavy items. One tonne is 1000 kilograms.

SCAN to watch video

Trucks are measured in tonnes.

Cars are measured in tonnes.

Elephants are measured in tonnes.

Examples:
Circle the things that would weigh one tonne or more.

a (bus) **e** whale
b person **f** taxi
c (hippo) **g** train
d lion **h** tuna fish

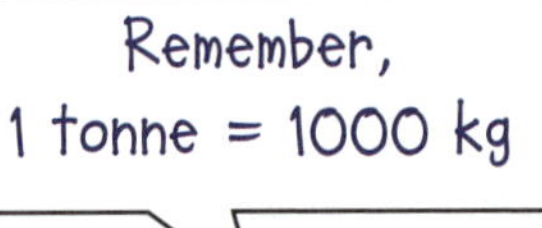

Your turn

Complete the equivalent measurements.

● $\frac{1}{2}$ tonne = 500 kilograms

a $\frac{1}{4}$ tonne = ________ kilograms

b $\frac{3}{4}$ t = ________ kg

c 3 t = ________ kg

Check your answers
How many did you get correct?

CATCH UP MATHS YEAR 5 BOOK B © PASCAL PRESS ISBN: 9781925726176

PRACTICE

1 How many make up one tonne?

- 250 kg × 4 = 1 tonne
- **a** 200 kg × ___ = 1 tonne
- **b** 50 kg × ___ = 1 tonne
- **c** 100 kg × ___ = 1 tonne
- **d** 500 kg × ___ = 1 tonne

2 Complete the table.

	Kilograms	Tonnes
	1200 kilograms	1.2 tonnes
a		9.73 tonnes
b	4500 kilograms	
c		4 tonnes
d		8.1 tonnes
e	3120 kilograms	
f	7000 kilograms	
g	6125 kilograms	
h		7.845 tonnes
i		12.9 tonnes
j	2400 kilograms	
k	16 040 kilograms	
l		16.25 tonnes
m		17.095 tonnes
n	21 090 kilograms	
o		1.003 tonnes
p		20.75 tonnes
q	14 001 kilograms	
r		33.589 tonnes
s		17.010 tonnes

GROSS MASS AND NET MASS

Gross mass is the mass of the container plus the mass of the contents. Net mass is the mass of the contents only.

The chocolates have a mass of 250 g.
The box has a mass of 50 g.

The net mass of the chocolates is 250 g.
The gross mass of the box of chocolates is 250 g + 50 g = 300 g.

Example: Write the missing numbers.

This tin has a mass of 20 g.

The peas inside have a mass of 90 g.

The gross mass = ____ g + 90 g

= ____ g

The net mass of the peas is ____ g.

Use the box of cereal to answer the questions.

● The box weighs 15 g.
What is the net weight of cereal?

515 – 15 = 500 g

a What is the gross weight of the box of cereal?

____

SELF CHECK Tick how you feel

Got it!	Need help...	I don't get it
☐	☐	☐

Check your answers
How many did you get correct? ☐

CATCH UP MATHS YEAR 5 BOOK B © PASCAL PRESS ISBN: 9781925726176

PRACTICE

1 Write the missing numbers.

The tin has a mass of 20 g.

The beans inside have a mass of 405 g.

The gross mass is 425 g.

The net mass of the beans is 405 g.

a 

The tin has a mass of 30 g.

The tomatoes inside have a mass of _____ g.

The gross mass is _____ g.

The net mass of the tomatoes is _____ g.

b TUNA 575 g

The tin has a mass of 15 g.

The tuna inside has a mass of _____ g.

The gross mass is _____ g.

The net mass of the tuna is _____ g.

2 Use the cereal box to answer the questions.

Box weight = 20 g

a What is the gross mass of the cereal? _______

b What is the net mass of the cereal? _______

3 Use the tin to answer the questions.

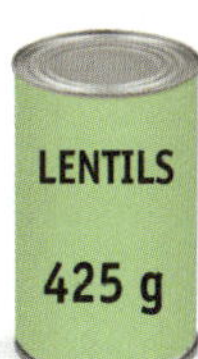

a The gross mass is _______.

b If the tin has a mass of 20 g, what is the net mass of the lentils? _______

4 The mass on the front of Allan's box of cookies has been rubbed off. If the box weighs 35 g and the net mass of the cookies is 840 g, what is the gross mass of the box of cookies? Show your working out.

____

MASS REVIEW

1 Complete the sentences.

Kilo means __________. Kilogram means __________ grams.

2 Write the masses in short form.

a 3 kilograms ________

b eighteen kilograms ________

c forty-six kilograms ________

d 89 kilograms ________

e two hundred and seven kilograms ________

f sixteen kilograms ________

g 17 kilograms ________

h 109 kilograms ________

i seventy-two kilograms ________

j eighty kilograms ________

3 Circle the item that could be getting weighed.

a **1 kg**
- bag of flour
- cricket ball

b **1 kg**
- ping pong ball
- net of apples

c **10 kg**
- a baby
- a book

d **10 kg**
- a net of oranges
- a bicycle

e **70 kg**
- a woman
- a giraffe

f **1 kg**
- a carton of juice
- a box of books

4 Tick or cross the masses to show if they are written correctly.

a ☐ 5kg

b ☐ 346 kg.

c ☐ 54 kg

d ☐ 183 KG

e ☐ 10kg

f ☐ 80 kG

g ☐ 8KG

h ☐ 27 Kg

i ☐ 49 kg

j ☐ 952 kg

k ☐ 1024 Kg

l ☐ 6397 kg.

CATCH UP MATHS YEAR 5 BOOK B © PASCAL PRESS ISBN: 9781925726176

5 Order these items from lightest (1) to heaviest (5).

a

2 kg

5 kg

4 kg

1 kg

☐ ☐ ☐ ☐ ☐

b

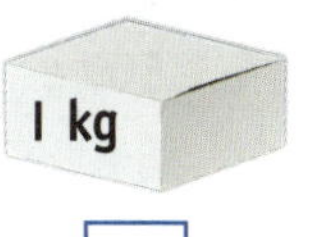

4 kg

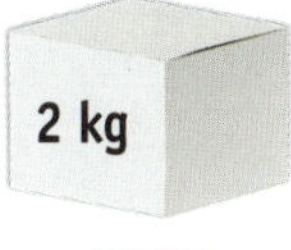

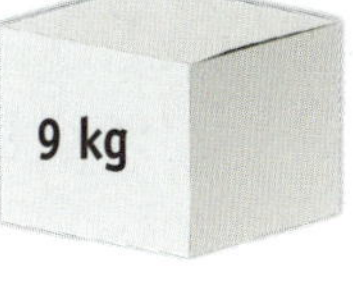

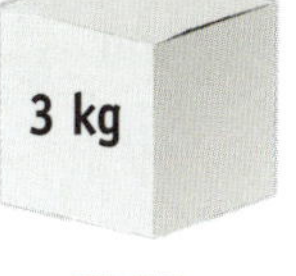

☐ ☐ ☐ ☐ ☐

c

250 g

750 g

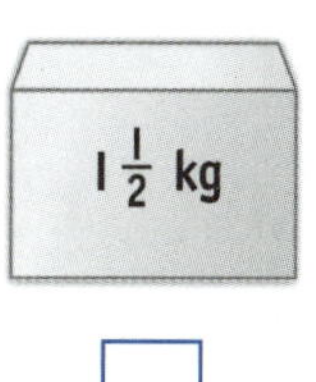

1 kg

☐ ☐ ☐ ☐ ☐

6 Complete the sentences.

Light masses are measured using ____________.

There are __________ grams in one kilogram.

7 Write the measurements in short form.

a forty-three grams ________

b two grams ________

c nine hundred and three grams ________

d fifteen grams ________

e seventy-one grams ________

f fifty-nine grams ________

8 Tick or cross the masses to show if they are written correctly.

a ☐ 42G

b ☐ 127 g

c ☐ 200 g

d ☐ 28 G

e ☐ 435 g.

f ☐ 73 g

g ☐ 340G

h ☐ 507g

i ☐ 20gr

j ☐ 62 g

k ☐ 17 gr

l ☐ 140 g

REVIEW

9 Order the boxes from smallest mass (1) to largest mass (8).

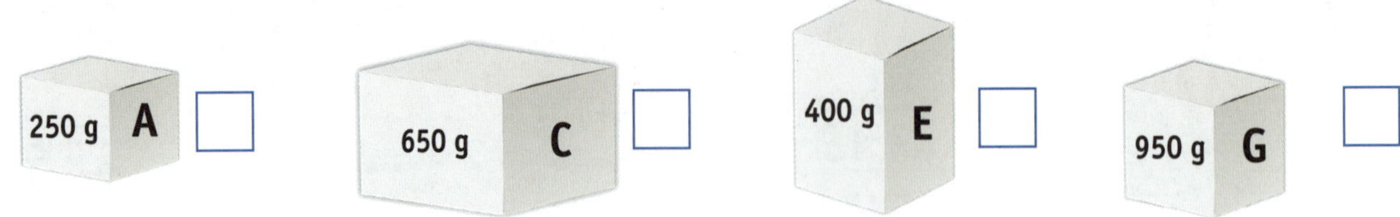

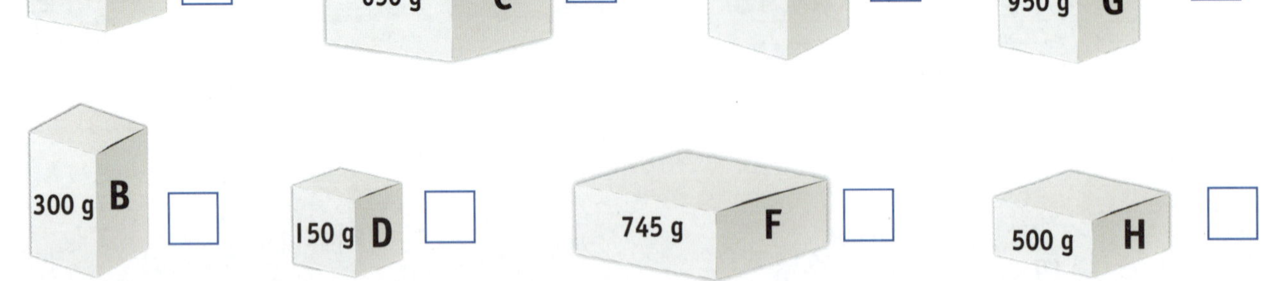

10 How many more grams for each box from Question 9 to make one kilogram?

a Box A ________ **c** Box C ________ **e** Box E ________ **g** Box G ________

b Box B ________ **d** Box D ________ **f** Box F ________ **h** Box H ________

11 Complete the equivalent measurements.

a 3500 g = ________ kg

b 1250 g = ________ kg

c 4000 g = ________ kg

d 7750 g = ________ kg

e 5500 g = ________ kg

f 6750 g = ________ kg

12 Convert these kilograms (kg) into grams (g).

a $1\frac{3}{4}$ kg = ________

b 6 kg = ________

c $2\frac{1}{4}$ kg = ________

d $10\frac{1}{2}$ kg = ________

e $3\frac{3}{4}$ kg = ________

f $7\frac{1}{2}$ kg = ________

g $3\frac{1}{2}$ kg = ________

h 8 kg = ________

i $6\frac{3}{4}$ kg = ________

j $1\frac{1}{4}$ kg = ________

  ISBN: 9781925726176

13 How many of each weight do you need to make up one kilogram?

a 200 g ____________

b 50 g ____________

c 500 g ____________

d 250 g ____________

e 100 g ____________

14 Complete the tables.

a

kg	grams
$\frac{3}{4}$ kg	
	250 g
$1\frac{1}{2}$ kg	
	2500 g
6 kg	
	3750 g
	10 000 g

b

kg	grams
0.5 kg	
	3750 g
15.42 kg	
	200 g
1.75 kg	
0.4 kg	
	14 125 g

c

kg	grams
3 kg 200 g	
	1008 g
	24 530 g
5 kg 755 g	
	8040 g
7 kg 10 g	
15 kg 5 g	

d

kg	grams
0.2 kg	
	17 250 g
6.3 kg	
10.89 kg	
	2672 g
2.4 kg	
3.07 kg	

REVIEW

15 Order these objects from heaviest (1) to lightest (5).

a

b

16 Draw a pointer on the scale to show the mass of the box.

a 450 g

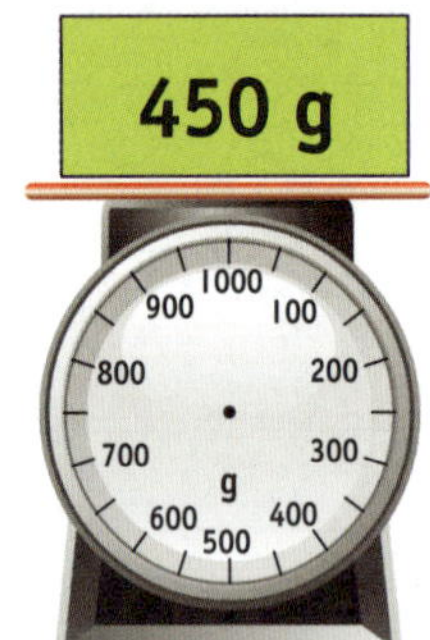

b 2 kg

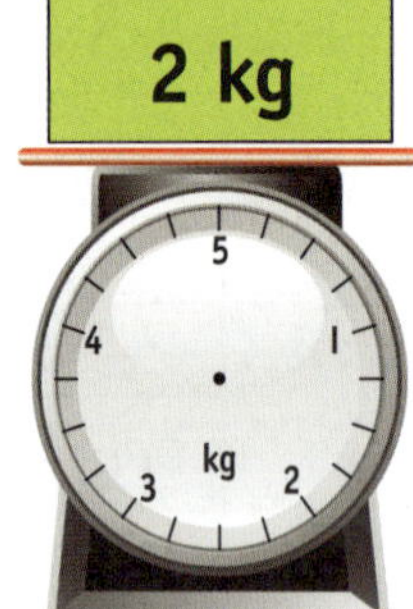

c 300 g

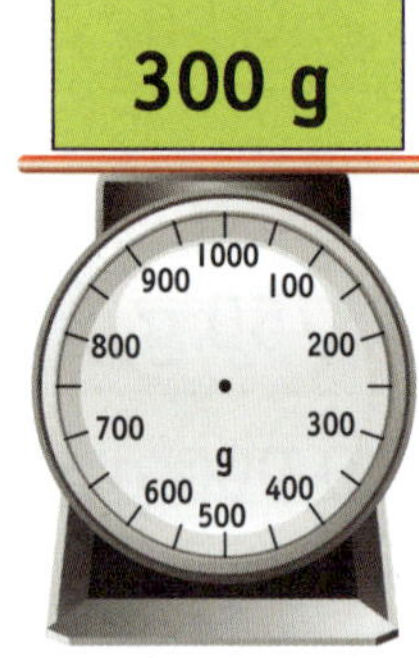

d 4 kg

17 Read the scale and write the mass.

a __________

b __________

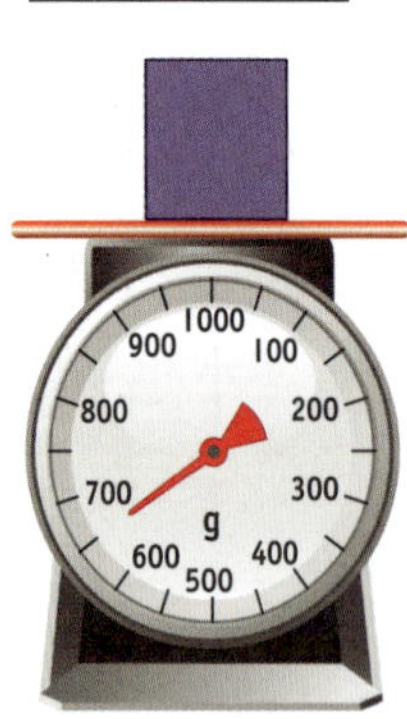

c __________

d __________

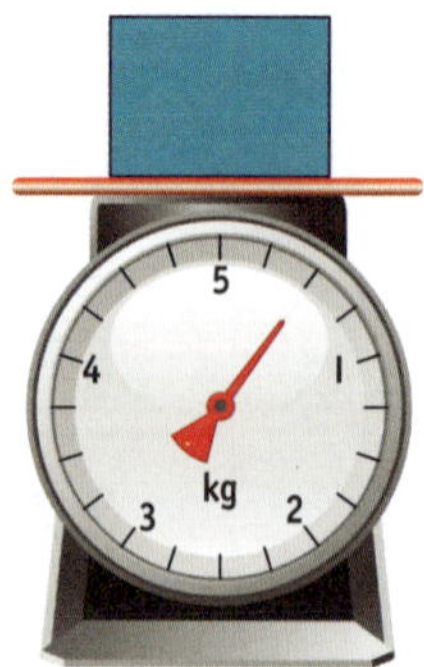

18 Complete the sentence.

One tonne is ________ kilograms.

  ISBN: 9781925726176

19 List five things that weigh more than one tonne.

____

____

20 Convert these measurements in tonnes to kilograms.

a $\frac{1}{4}$ tonne = ________ kg

c $\frac{3}{4}$ tonne = ________ kg

b $\frac{1}{2}$ tonne = ________ kg

d 1 tonne = ________ kg

21 How many of each weight would I need to make one tonne?

a 200 kg ______

c 100 kg ______

e 50 kg ______

b 500 kg ______

d 250 kg ______

f 10 kg ______

22 Complete the equivalent measurements.

	Kilograms	Tonnes
a	2400 kg	
b		8.462 t
c		10.5 t
d	6475 kg	

	Kilograms	Tonnes
e		9.603 t
f	14 020 kg	
g		5.004 t
h		17.55 t

23 Write the missing numbers.

The tin has a mass of 20 g.

The chickpeas inside have a mass of ______ g.

The gross mass is ______ g.

The net mass of the chickpeas is ______ g.

24 The mass on the front of George's box of cereal has rubbed off. The box weighs 15 g and the net mass of the cereal is 640 g.

What is the mass on the front of George's box of cereal? __________

O'CLOCK AND HALF PAST

Analogue clocks are divided into 12 parts. Each part is equal to one hour passing.

It is 4 o'clock.

Minute hand (big hand) points to the 12.

Hour hand (little hand) points to the 4.

It is half past 4.

Hour hand is halfway between the 4 and the 5 because it is half past 4.

Minute hand points to the 6.

Digital clocks use numbers to show the time. They have no hands.

The numbers on this side show the hours.

The numbers on this side show the minutes past the hour.

It is 4 o'clock.

4 hours

4:30

30 minutes past the hour

It is 4 thirty.

Examples: Match the times on the analogue and digital clocks.

a

b

c

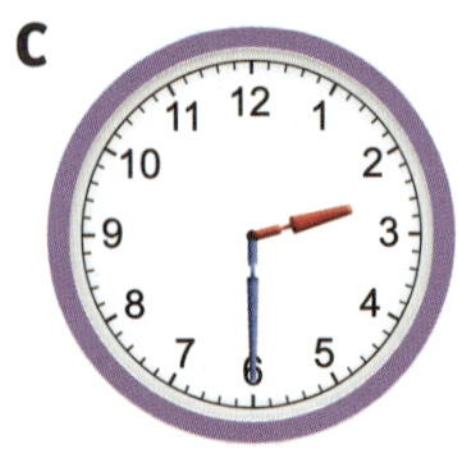

d

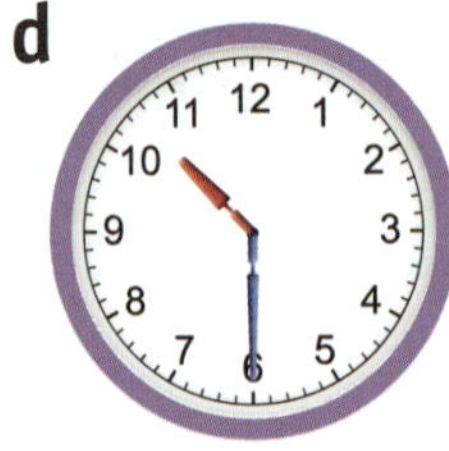

Check the answers on the video!

10:30

Your turn

Show the missing times.

a

b

c

4:30

SELF CHECK Tick how you feel

Got it! | Need help... | I don't get it

Check your answers

How many did you get correct?

 CATCH UP MATHS YEAR 5 BOOK B © PASCAL PRESS ISBN: 9781925726176

PRACTICE

1 Complete the times.

three o'clock

three o'clock

b

1 1:00

d

half past three

f

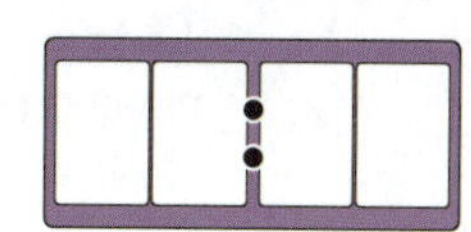

a

half past eight

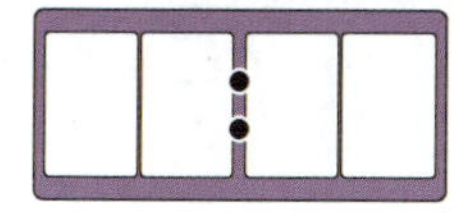

c

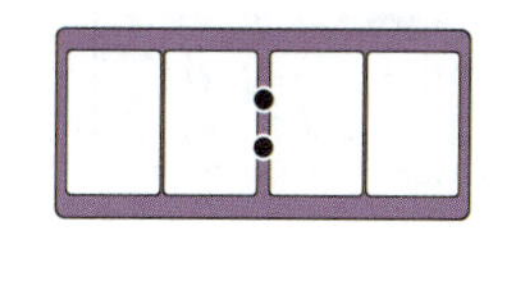

e

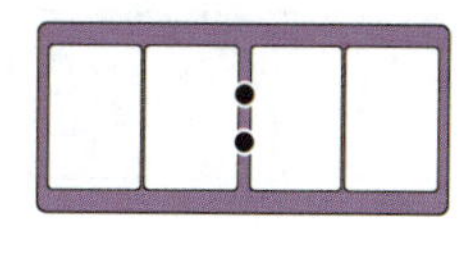

four thirty

g

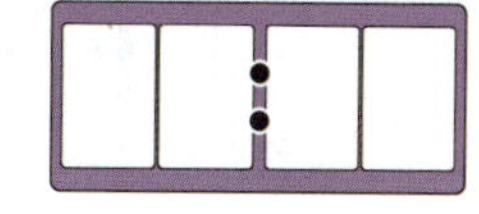

9 o'clock

2 Match the times.

a

b

c

half past 8 | half past 6 | two o'clock | three o'clock

Digital

3:00

8:30

6:30

2:00

two o'clock | three o'clock | six thirty | eight thirty

QUARTER PAST AND QUARTER TO

Analogue times

It is a quarter past 7.

The minute hand (big hand) is one-quarter of the way around the clock.

The hour hand is one-quarter of the way between the 7 and the 8.

It is a quarter to 8.

The minute hand is three-quarters of the way around the clock.

The hour hand is three-quarters of the way between the 7 and the 8.

Digital times

The numbers on this side show the hours. →

← The numbers on this side show the minutes past the hour.

It is 7 fifteen.

7 hours

45 minutes past the hour

It is 7 forty-five.

Examples: Match the times on the digital and analogue clocks.

a 

b 

c 

d 

Check the answers on the video!

Your turn

Show the missing times.

a

b

c

12:45

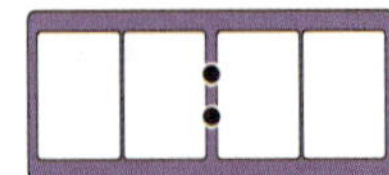

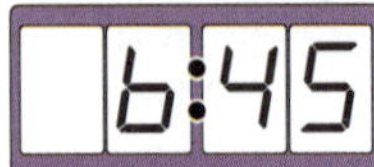

SELF CHECK Tick how you feel

Got it!	Need help...	I don't get it
 ☐	 ☐	☐

Check your answers

How many did you get correct?

CATCH UP MATHS YEAR 5 BOOK B © PASCAL PRESS ISBN: 9781925726176

PRACTICE

Complete the table.

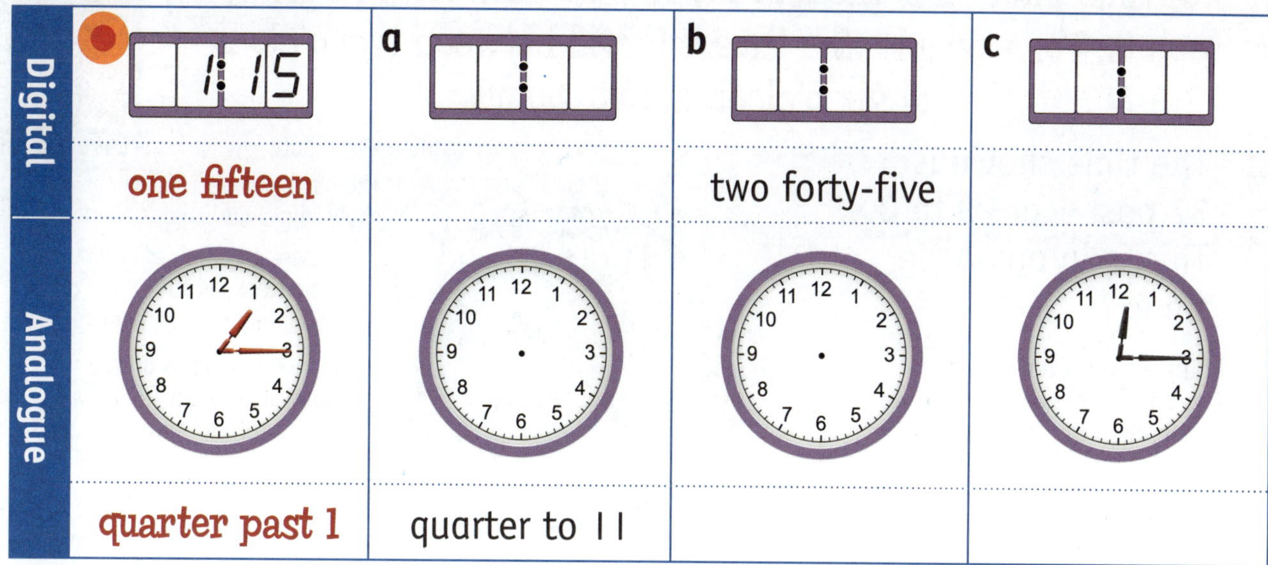

	Example	a	b	c
Digital	1:15	:	:	:
	one fifteen		two forty-five	
Analogue	(clock)	(clock)	(clock)	(clock)
	quarter past 1	quarter to 11		

Complete the times.

quarter to three

2:45

two forty-five

b

three fifteen

d

10:15

f

quarter to five

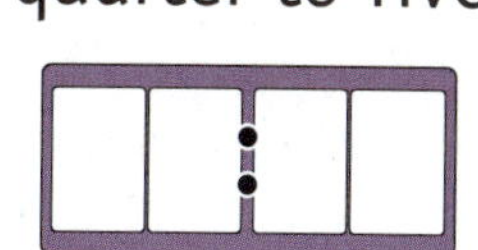

a

seven fifteen

c

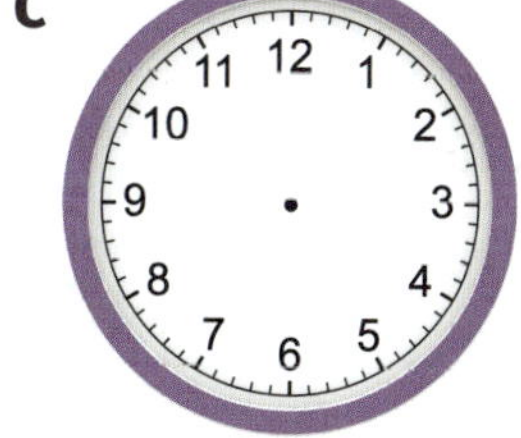

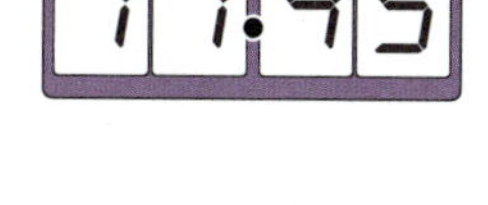

11:45

e

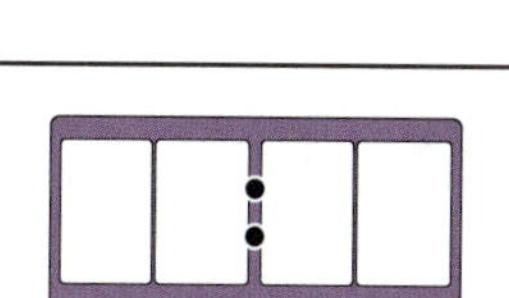

g

quarter to seven

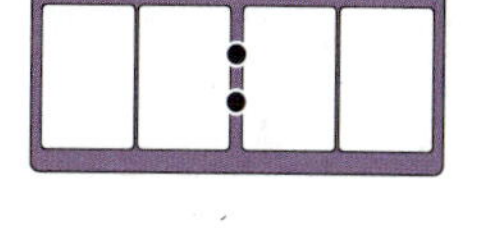

INTERVALS

A time interval is the amount of time between two given points. For example, the time interval between two o'clock and six o'clock is four hours.

SCAN to watch video

The time shown is 37 past 1 or 23 to 2. This is in analogue time.

If the big (minute) hand is on the 'to' side, count from the 12 to the **left**.

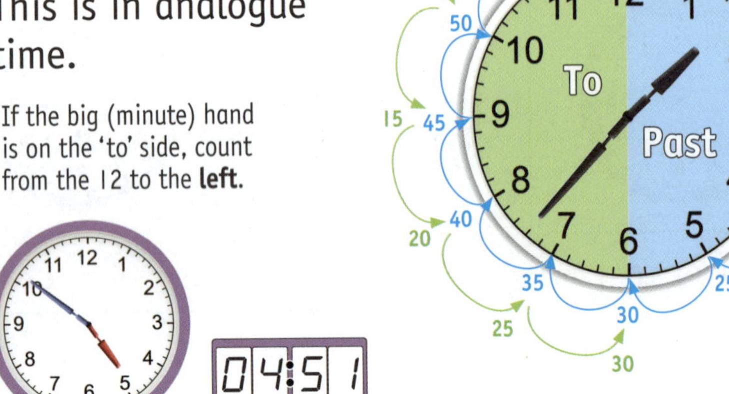

The black numbers show what hour it is.

There are 60 lines around the clock because there are 60 minutes in an hour.

If the big hand is on the 'past' side, count from the 12 to the **right**.

04:51

It is 9 to 5 or 51 past 4.

08:07

It is 7 past 8.

Important! When you convert to digital time, count from the 12 to the right, even if the big hand is on the 'to' side.

Examples:
Match the analogue clocks to the correct digital time.

Check the answers on the video!

a

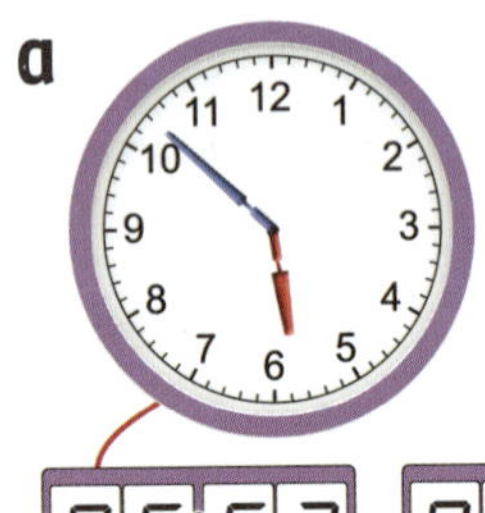

b

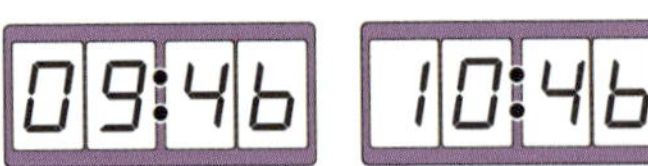

c

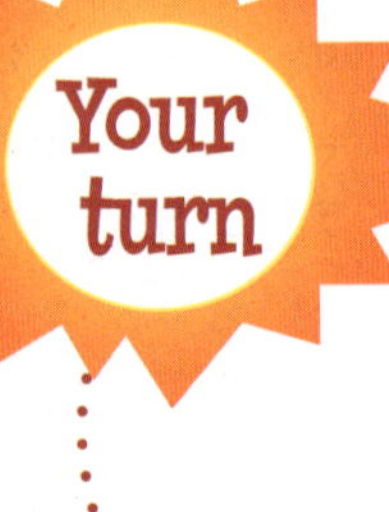

How would you read each clock in the Examples?

Clock	Analogue	Digital
A	8 to 6	five fifty-two
B		
C		

SELF CHECK Tick how you feel

Got it! ☐ Need help... ☐ I don't get it ☐

Check your answers

How many did you get correct? ☐

CATCH UP MATHS YEAR 5 BOOK B © PASCAL PRESS ISBN: 9781925726176

PRACTICE

1 Complete the table.

	Digital time	How we read it	What it means
●	10:16	ten sixteen	16 minutes past 10
a	8:23		
b		eleven forty-four	
c			22 minutes to 5
d		two eighteen	
e	4:56		

2 Draw the missing minute hand on each analogue clock to match the digital time.

● 
a 
b 
c 
d 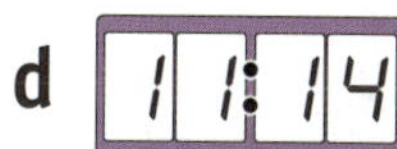

3 Show the time one minute later.

● 
a 
b 
c 
d 

4 Show the time five minutes later.

● 
a 
b 
c 
d 

5 Show the time two minutes earlier.

● 
a 
b 
c 
d 

CONVERTING TIMES

'Convert' means 'change into'. So when we convert times, we change them – for example, from minutes to hours or minutes to seconds.

60 seconds = 1 minute **60 minutes = 1 hour**

Convert 3 minutes to seconds.

3 minutes = 180 seconds

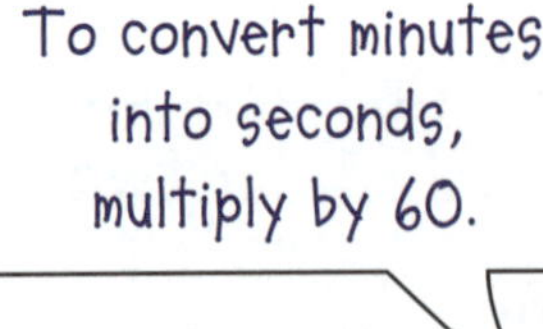

Convert 300 seconds to minutes.

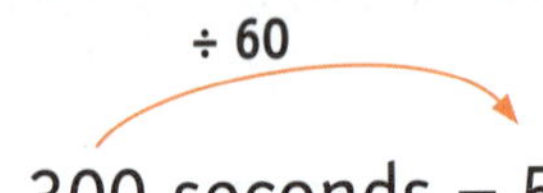

300 seconds = 5 minutes

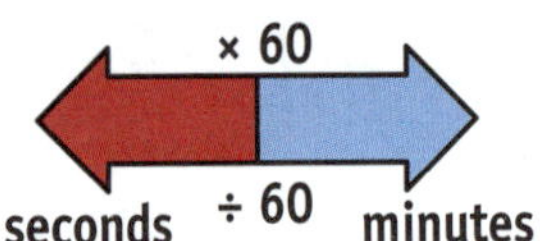

Examples: Write the missing numbers.

a 4 minutes = 240 seconds

c 600 seconds = ____ minutes

b 6 minutes = ____ seconds

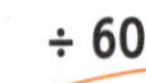

d 420 seconds = ____ minutes

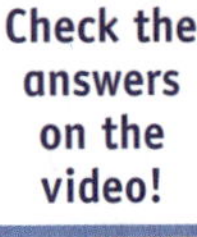

Convert the times.

- 2 minutes = 120 seconds

a 8 minutes = ____ seconds

b 9 minutes = ____ seconds

c 180 seconds = ____ minutes

Check your answers

How many did you get correct?

CATCH UP MATHS YEAR 5 BOOK B © PASCAL PRESS ISBN: 9781925726176

PRACTICE

1 Convert these seconds into minutes. Hint: Divide by 60.

- 60 seconds = 1 minute
- a 300 seconds = ___ minutes
- b 240 seconds = ___ minutes
- c 360 seconds = ___ minutes
- d 120 seconds = ___ minutes
- e 540 seconds = ___ minutes
- f 420 seconds = ___ minutes
- g 720 seconds = ___ minutes
- h 600 seconds = ___ minutes
- i 480 seconds = ___ minutes

2 Convert these minutes into seconds. Hint: Multiply by 60.

- 1 minute = 60 seconds
- a 3 minutes = _____ seconds
- b 6 minutes = _____ seconds
- c 2 minutes = _____ seconds
- d 8 minutes = _____ seconds
- e 12 minutes = _____ seconds
- f 16 minutes = _____ seconds
- g 18 minutes = _____ seconds
- h 14 minutes = _____ seconds
- i 10 minutes = _____ seconds

3 Convert these minutes into hours. Hint: Divide by 60.

- 60 minutes = 1 hour
- a 180 minutes = ___ hours
- b 300 minutes = ___ hours
- c 540 minutes = ___ hours
- d 660 minutes = ___ hours
- e 120 minutes = ___ hours
- f 4200 minutes = ___ hours
- g 1200 minutes = ___ hours

4 Convert these hours into minutes. Hint: Multiply by 60.

- 1 hour = 60 minutes
- a 5 hours = _____ minutes
- b 7 hours = _____ minutes
- c 4 hours = _____ minutes
- d 8 hours = _____ minutes
- e 11 hours = _____ minutes
- f 22 hours = _____ minutes
- g 30 hours = _____ minutes

24-HOUR TIME

There are 24 hours in a day. 12 hours are before midday and 12 hours are before midnight.

24-hour time has 4 digits.

The blue times are the 24-hour times you would read before midday (am).

The green times are the 24-hour times you would read after midday (pm).

Examples: Circle the 24-hour times before midday in blue and after midday in green.

a 08:00 **c** 01:00 **e** 0:00 **g** 07:00

b 19:00 **d** 12:00 **f** 23:00 **h** 04:00

Your turn

Is the time before midday or after midday?

● 21:00 after

b 05:00 ____________

d 17:00 ____________

a 02:00 ____________

c 20:00 ____________

e 03:00 ____________

SELF CHECK Tick how you feel

Got it!	Need help...	I don't get it
☐	☐	☐

Check your answers
How many did you get correct? ☐

CATCH UP MATHS YEAR 5 BOOK B © PASCAL PRESS ISBN: 9781925726176

PRACTICE

1 Convert these times to 24-hour time.

- ● 3:42 am 03:42
- a 7:30 am ______
- b 8:00 am ______
- c 6:14 am ______
- d 11:27 pm ______
- e 1:31 am ______
- f 2:22 am ______
- g 12:53 am ______
- h 7:43 pm ______
- i 7:12 am ______
- j 4:55 pm ______
- k 5:46 pm ______

2 Are these times am or pm?

- ● 02:36 am
- a 17:13 ___
- b 08:23 ___
- c 13:34 ___
- d 16:21 ___
- e 11:12 ___
- f 09:45 ___
- g 05:42 ___
- h 12:37 ___
- i 06:44 ___
- j 10:53 ___
- k 04:25 ___

3 Use 24-hour time to write the time one hour after.

- ● 7:01 am 08:01
- a 6:56 pm ______
- b 2:37 pm ______
- c 3:35 pm ______
- d 10:25 pm ______
- e 12:12 pm ______
- f 12:49 am ______
- g 1:14 am ______
- h 4:23 am ______
- i 4:05 am ______
- j 8:43 am ______
- k 10:28 am ______

4 Write T for True or F for False.

- ● [T] 7:25 am = 07:25
- a [ ] Midday is the same as 00:00.
- b [ ] 14:20 is the same as 20 past 4.
- c [ ] 15:30 is in the morning.
- d [ ] 17:46 is in the evening.
- e [ ] 13:02 would be a good time to eat lunch.
- f [ ] 07:11 is around dinner time.

5 List five places where 24-hour time is used.

airport
____

CALENDARS

A calendar shows the twelve months of the year in order. It tells us what year, month, date and day of the week it is.

Ashton's Calendar

July 2024						
S	**M**	**T**	**W**	**T**	**F**	**S**
	1	2 Haircut	3	4 Pete's B'day	5	6 Pete's Party
7	8	9	(10)	11	12	13
14	15 Deb's B'day	16	17	18 Aria's B'day	19	20
21	22	23	24	25	26 Excursion	27
28	29	30 Dentist	31			

The number 10 is circled on the calendar.
Read this date as Wednesday the 10th of July.

Days of the Week

S	Sunday
M	Monday
T	Tuesday
W	Wednesday
T	Thursday
F	Friday
S	Saturday

Months of the Year

Month	No. days
January	31
February	28 (29)
March	31
April	30
May	31
June	30
July	31
August	31
September	30
October	31
November	30
December	31

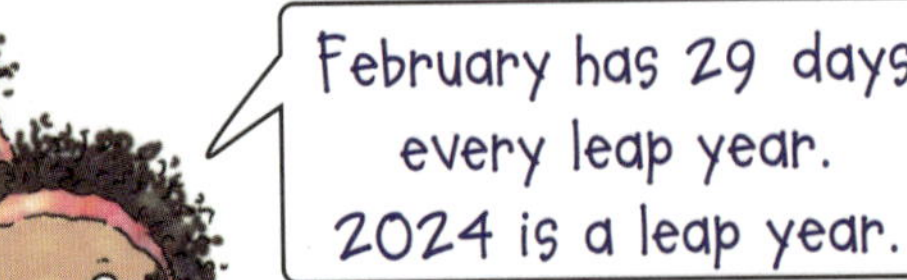

CATCH UP MATHS YEAR 5 BOOK B © PASCAL PRESS ISBN: 9781925726176

Example 1: Write the missing numbers.

a There are ___ months in one year.

b There are ___ days in one week.

Example 2: How many days in these months?

a June ___

b October ___

c January ___

d February ___

Your turn

Use Ashton's calendar to answer the questions.

● On Tuesday, 2nd July, Ashton will visit the hairdresser.

a On __________, _____ July, Ashton will visit the dentist.

b What day of the week is Deb's birthday?

____

c When are Pete's and Aria's birthdays?

__________, _____ July and __________, _____ July

d Ashton has an excursion on __________, _____ July.

e Put an X on Sunday, 14th July.

f Tick Tuesday, 23rd July.

g Draw a star on Saturday, 27th July.

SELF CHECK Tick how you feel

Got it!	Need help...	I don't get it
☐	☐	☐

Check your answers

How many did you get correct? ☐

PRACTICE

How many days are there in each season in the southern hemisphere?

- Spring 30 + 31 + 31 = 92
- b Autumn ____________
- a Summer ____________
- c Winter ____________

Which months have 31 days?

____

Which months have 30 days?

____

Use the calendars to complete the following questions.

August 2024

M	T	W	T	F	S	S
			1	2	3 ✗	4
5	6	7	8	9	10	11
12	13	14	15	16	17	18
19	20	21	22	23	24	25
26	27	28	29	30	31	

October 2024

M	T	W	T	F	S	S
	1	2	3	4	5	6
7	8	9	10	11	12	13
14	15	16	17	18	19	20
21	22	23	24	25	26	27
28	29	30	31			

Draw a ✗ on these dates on the calendars above.

- 3rd August
- b 16th August
- d 30th October
- a 2nd October
- c 17th October
- e 31st August

Write the day of each date in Question 4.

- Saturday
- b ____________
- d ____________
- a ____________
- c ____________
- e ____________

CATCH UP MATHS YEAR 5 BOOK B © PASCAL PRESS ISBN: 9781925726176

6 What day of the week are these dates on?

- 22nd August Thursday
- c 29th August ______
- a 11th October ______
- d 13th October ______
- b 26th October ______
- e 5th August ______

7 Complete the sentences.

- There are 4 Mondays in August and 4 Mondays in October.

a There are __ Saturdays in August and __ Saturdays in October.

b There are __ Wednesdays in August and __ Wednesdays in October.

c There are __ Thursdays in August and __ Thursdays in October.

d There are ___ weekdays in August and ___ weekdays in October.

8 Write the following events on the calendar.

a Haircut on 19th August

b Dentist on 4th October

c Doctor on 30 August

d Interview on 12th August

e Party on 23rd August

f Halloween on 31st October

g Football training on 1st August

h Max's birthday on 9th October

9 Use the calendar below to complete the activities.

a Write in the name and dates for the current month.

b Add in three important events.

c Write two pieces of information you can gather from the calendar.

- ______________________________
  ____
- ______________________________
  ____

Month and Year: ______________

M	T	W	T	F	S	S

TIMETABLES

A timetable is a chart that tells you when something is due to happen. Timetables are also called schedules.

SCAN to watch video

You can get timetables for buses, trains, cinemas, television channels, sports, classes ... Can you think of more?

Many timetables Use 24-hour time.

Craynar to Pilli Lilli Bus Timetable

Craynar	5:15	5:30	5:45	6:00	6:15
Banburg	5:27	5:42	5:57		6:27
Sans Ami	5:45	6:00			6:45
Pilli Lilli	6:01	6:16	6:39	6:34	7:01

Examples: Use the timetable above to answer the questions.

a The 5:30 bus has 3 stops.

b The 5:45 bus has __ stops.

c The 6:15 bus has __ stops.

d The fastest bus from Craynar to Pilli Lilli leaves Craynar at _______.

e What does [grey box] mean?

____

Your turn

Answer these questions using the timetable above.

● The 5:15 bus from Craynar arrives at Sans Ami at 5:45.

a The 5:45 bus from Craynar makes __ stops.

b If I get on the bus at Banburg at 6:27, what time will I get to Pilli Lilli? _______

c It takes ___ minutes to get from Craynar to Pilli Lilli on the 5:15 bus.

SELF CHECK Tick how you feel

Got it!	Need help...	I don't get it
☐	☐	☐

Check your answers

How many did you get correct? ☐

CATCH UP MATHS YEAR 5 BOOK B © PASCAL PRESS ISBN: 9781925726176

PRACTICE

TV Guide: Monday, 16 January

Channel	4:30 am	5:00 am	5:30 am	6:00 am	6:30 am	7:00 am
3	Take 3 4:30 am	Three Early News 5:00 am	This Day 5:30 am			
6	Jane Enjoying 4:30 am	Home Shop 5:00 am	Home Shop 5:30 am	Home Shop 6:00 am	Home Shop 6:30 am	Dollar Wars 7:00 am
9	Cartoon: BW 4:30 am; Toy Friends 4:50 am	Toy Jurassic 5:10 am	Monster Game! 5:30 am	Jungle Life 6:00 am	Spinner 6:30 am	Monster Game! 7:00 am
13	Holiday 4:30 am	Retro Reno 5:00 am		My Dream House 6:00 am	My Dream House 6:30 am	Homeward 7:00 am
15		Fish Tanks 5:00 am		Geri's Garage 6:00 am		Fix My Car 7:00 am
16		Cook with Max 5:00 am	Life Matters 5:30 am	Morning News 6:00 am		

1 Use the TV Guide to complete the following.

a The date of this TV Guide is ______________________.

b How many channels does this TV guide show? ___

c What time does the first *Monster Game!* start? _________

d What channel is the first *Monster Game!* on? ___

e How long does does the first *Monster Game!* go for? _______________

f The show *Fish Tanks* is on channel ___. It starts at ______ and finishes at _______.

g How long does *Home Shop* go for? _______________

2 Write the start times for these shows.

a *Fix My Car* ______ **b** *Life Matters* _____ **c** *Homeward* _____

3 How many times are these shows shown?

a *Monster Game!* ___

b *My Dream House* ___

c *Geri's Garage* ___

d *Cook with Max* ___

Check Point Ferry

Depart			Weekends and Public Holidays						Arrive
Check Point	Bo Drop Off	Canal	Eagle Point	Tennis Court	South Hall	Lovely Bay	Emma Bay	Bo Pick Up	Check Point
								8:20	8:25
8:30	8:35	8:36	8:38	8:40	8:45	8:50	8:55	9:05	9:10
9:30	9:35	9:36	9:38	9:40	9:45	9:50	9:55	10:05	10:10
10:30	10:35	10:36	10:38	10:40	10:45	10:50	10:55	11:05	11:10
11:30	11:35	11:36	11:38	11:40	11:45	11:50	11:55	12:05	12:10
12:30	12:35	12:36	12:38	12:40	12:45	12:50	12:55	1:05	1:10
1:30	1:35	1:36	1:38	1:40	1:45	1:50	1:55	2:05	2:10
2:30	2:35	2:36	2:38	2:40	2:45	2:50	2:55	3:05	3:10
3:30	3:35	3:36	3:38	3:40	3:45	3:50	3:55	4:05	4:10
4:30	4:35	4:36	4:38	4:40	4:45	4:50	4:55	5:05	5:10
5:30	5:35	5:36	5:38	5:40	5:45	5:50	5:55	6:05	6:10
6:30	6:35	6:36	6:38	6:40	6:45	6:50	6:55	7:00	7:05

4 Use the timetable above to answer the following questions.

a Would I use this timetable to catch a ferry to Check Point on Monday? Why or why not?

____

b How many ports does this ferry stop at? ___

c How long is a round trip on the Check Point ferry? ______________

d If I had a tennis game at Tennis Court at 1 pm, what time would I need to catch the ferry from Canal? _____

e What time does the earliest ferry leave Check Point? _____

f What time does the last ferry leave Check Point? _____

g What time does the last ferry arrive back at Check Point? _____

5 How long does it take to travel between these stops?

a Eagle Point, Emma Bay _____

b Canal, Lovely Bay _____

c South Hall, Check Point _____

d Check Point, Eagle Point _____

e Tennis Court, Bo Pick Up _____

CATCH UP MATHS YEAR 5 BOOK B © PASCAL PRESS ISBN: 9781925726176

Chris's High School Timetable

Term 1

Monday		Tuesday		Wednesday		Thursday		Friday	
Period 0	8:10–9:05	Period 0	8:10–9:05			Period 0	8:10–9:05	Period 0	8:10–9:05
Period 1	9:10–10:15	Period 1	9:10–10:10	Period 1	9:10–10:15	Period 1	9:10–10:00	Period 1	9:10–10:15
Period 2	10:15–11:15	Period 2	10:10–11:05	Recess	10:15–10:35	Period 2	10:00–10:45	Period 2	10:15–11:15
		Assembly	11:05–11:20						
Recess	11:15–11:35	Recess	11:20–11:40	Period 2	10:35–11:35	Recess	10:45–11:05	Recess	11:15–11:35
Period 3	11:35–12:35	Period 3	11:40–12:35	Period 3	11:35–12:35	Period 3	11:05–11:50	Period 3	11:35–12:35
Period 4	12:35–1:35	Period 4	12:35–1:35	Lunch 1	12:35–12:55	Period 4	11:50–12:35	Period 4	12:35–1:35
Lunch 1	1:35–1:55	Lunch 1	1:35–1:55	Lunch 2	12:55–1:15	Lunch 1	12:35–12:55	Lunch 1	1:35–1:55
Lunch 2	1:55–2:15	Lunch 2	1:55–2:15	Period 4	1:15–2:15	Lunch 2	12:55–1:15	Lunch 2	1:55–2:15
Period 5	2:15–3:15	Period 5	2:15–3:15	Teacher planning session		Sport		Period 5	2:15–3:15

6 Use the timetable above to answer the following questions.

a Which days does Chris finish school at 3:15?

____

b What day and time is the teacher planning session?

____

c How long does lunch go for on Thursdays? _____

d When does Chris have Sport? _________________________

7 What time does recess start?

a Monday _____ **c** Wednesday _____ **e** Friday _____

b Tuesday _____ **d** Thursday _____

8 How many periods does Chris have on these days?

a Monday __ **c** Wednesday __ **e** Friday __

b Tuesday __ **d** Thursday __

TIMELINES

A timeline is a number line that shows special dates in order, starting from the earliest date.

Anthony's Timeline

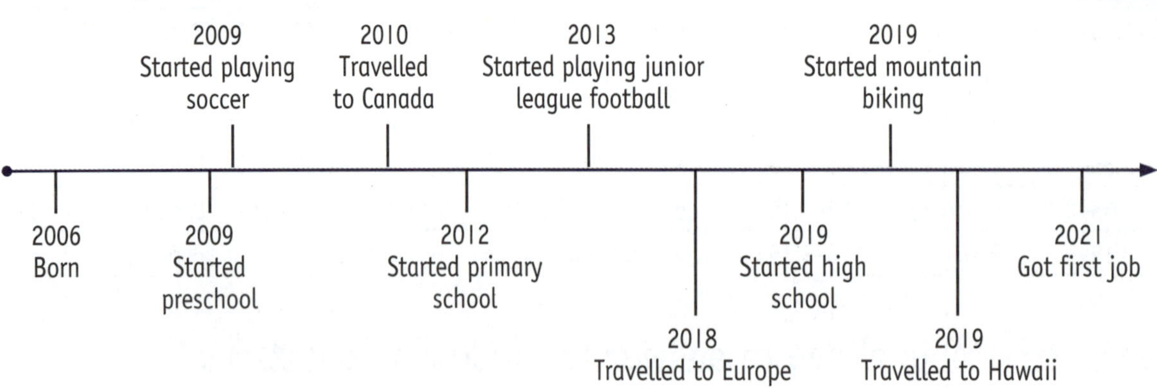

Examples: Use Anthony's timeline to answer the questions.

a Anthony was born in 2006.

b He started preschool when he was ___ years old.

c He started primary school in the year ______.

d In what year did Anthony start these sports?

- soccer ______
- mountain biking ______
- junior league football ______

e How old was Anthony when he competed in these sports?

- soccer ______
- mountain biking ______
- junior league football ______

f How many holidays has Anthony been on? ___

g What were Anthony's holiday destinations?

____

CATCH UP MATHS YEAR 5 BOOK B © PASCAL PRESS ISBN: 9781925726176

On the timeline below, record the year you were born, the year you started primary school, and the year you started Year 5.

Then add other special events in your life.

Write three questions you could ask about the information on your timeline.

- ______________________________
  ____
- ______________________________
  ____
- ______________________________
  ____

SELF CHECK Tick how you feel

Got it!	Need help...	I don't get it
☐	☐	☐

Check your answers

How many did you get correct? ☐

PRACTICE

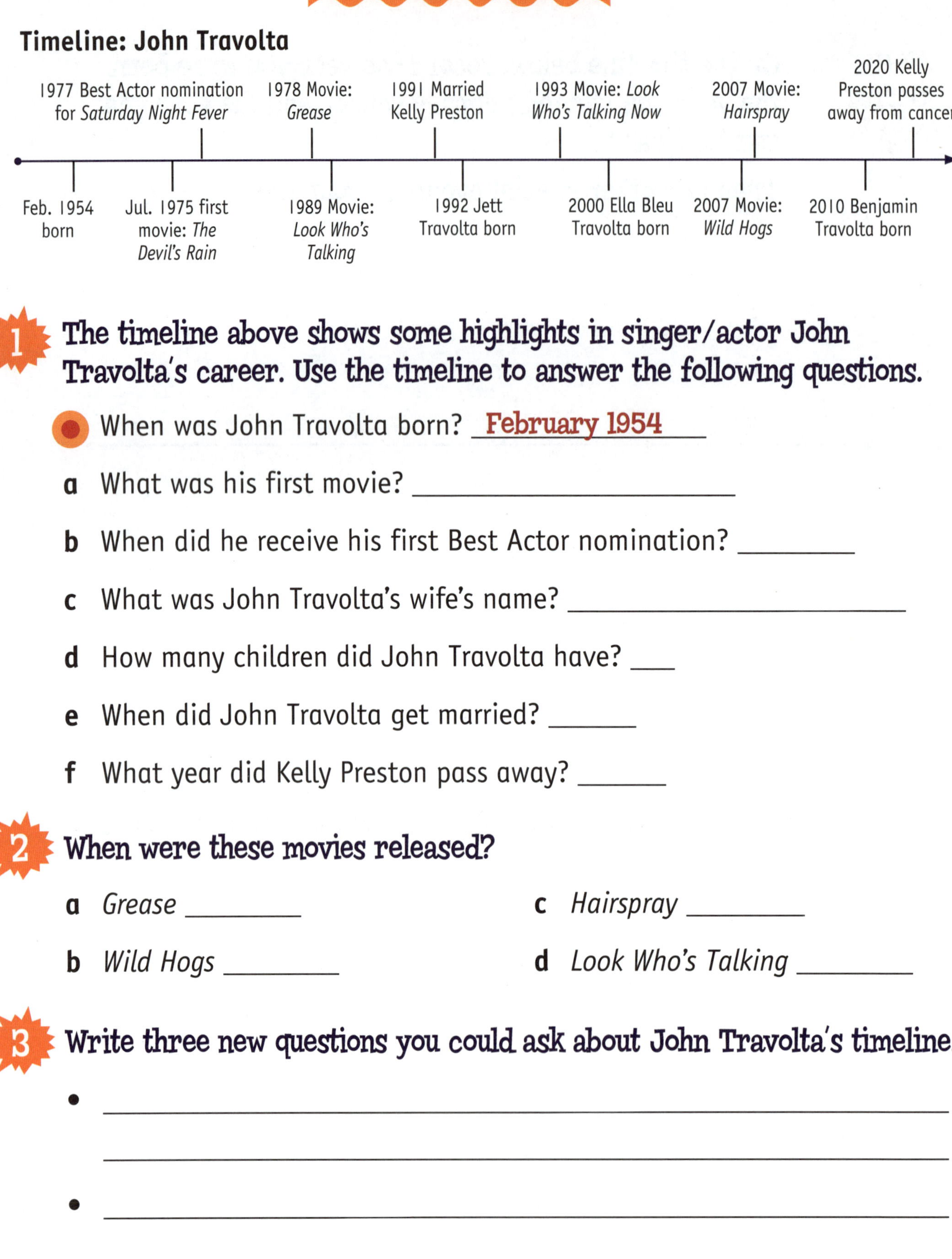

1 The timeline above shows some highlights in singer/actor John Travolta's career. Use the timeline to answer the following questions.

- When was John Travolta born? **February 1954**

a What was his first movie? ______________________

b When did he receive his first Best Actor nomination? ________

c What was John Travolta's wife's name? ______________________

d How many children did John Travolta have? ___

e When did John Travolta get married? ______

f What year did Kelly Preston pass away? ______

2 When were these movies released?

a *Grease* ________

b *Wild Hogs* ________

c *Hairspray* ________

d *Look Who's Talking* ________

3 Write three new questions you could ask about John Travolta's timeline.

- ____
- ____
- ____

CATCH UP MATHS YEAR 5 BOOK B © PASCAL PRESS ISBN: 9781925726176

Timeline: Cruise Ships

1900:
The first luxury cruise ship, called *Prinzessin Victoria Luise*, was built in Germany.

1912:
Titanic disaster

1939–45:
Cruise ships were repurposed to transport soldiers and supplies during World War 2.

1950s:
Interest in cruise ships dropped as many people chose commercial flights instead.

1966:
Norwegian Capricorn Lines launched their first ship, *Norwegian Sunward*.

1970s:
Cruise ships were redesigned so people enjoyed the experience onboard, rather than using them just for transport.

1970:
Royal Caribbean's first ship, *Song of Norway*, is launched.

1972:
Carnival Cruise Lines's maiden voyage of first ship, TSS *Mardi Gras*.

2018:
Largest-ever cruise ship built: *Symphony of the Seas*.

2022:
Royal Caribbean's first journey of the world's largest cruise ship, *Wonder of the Seas*.

4 **Use the timeline above to answer the following questions.**

a When was the *Prinzessin Victoria Luise* built? ________

b When was the *Titanic* disaster? ________

c The largest cruise ship in the world is ____________________________.

d When were cruise ships redesigned? ________

e When was Carnival Cruise Lines's first voyage? _______

f *Norwegian Sunward* was ____________________________ first cruise ship.

g What was the name of Carnival Cruise Line's first ship?

____

5 **Write three new questions you could ask about the cruise timeline.**

- ____
____
- ____
____
- ____
____

TIME REVIEW

1 Cross out the incorrect words.

a The minute hand is the **big** / **small** hand on an analogue clock.

b The hour hand is the **big** / **small** hand on an analogue clock.

2 Complete the times.

a ______________

b half past four

c 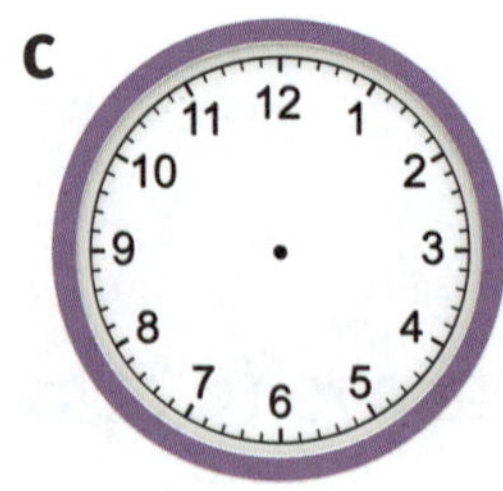 ______________

d ______________

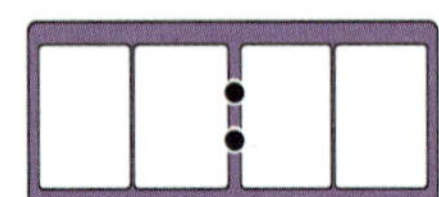

a ______________

b ______________

c ______________

d twelve thirty

3 Join the matching times.

Analogue

a 

b 

c 

d 

half past 6	five o'clock	seven o'clock	half past 11

Digital

6:30

7:00

seven o'clock	six thirty	five o'clock	eleven thirty

CATCH UP MATHS YEAR 5 BOOK B © PASCAL PRESS ISBN: 9781925726176

4 Write the time you would say.

a ____________________

c ____________________

e ____________________

b ____________________

01:15

d ____________________

03:45

f ____________________

5 Complete the table.

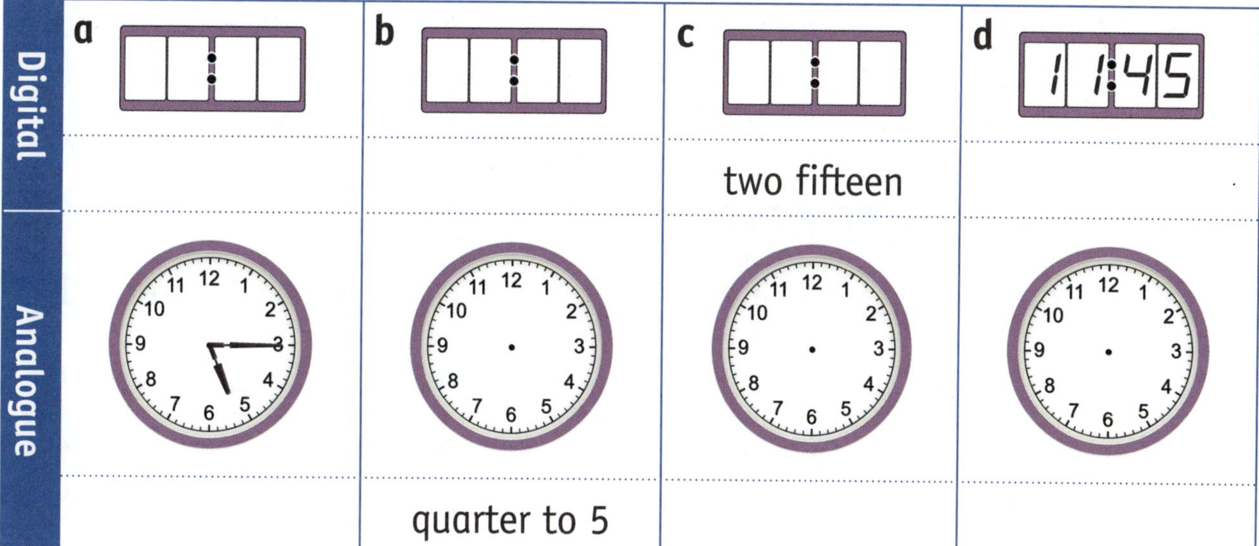

	a	b	c	d
Digital	[:]	[:]	[:]	11:45
			two fifteen	
Analogue				
		quarter to 5		

6 Draw the missing minute hand on each analogue clock to match the digital time.

a 

b 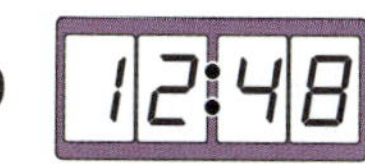

c 

d 

e 

REVIEW

7 Complete the table.

	Digital time	How we read it	What it means
a	11:29		
b		nine forty-nine	
c			26 minutes to 11
d			17 minutes past 1
e	4:11		

8 Convert these times in seconds into minutes.

a 240 seconds = ___ minutes
b 720 seconds = ___ minutes
c 480 seconds = ___ minutes
d 540 seconds = ___ minutes
e 6600 seconds = ___ minutes
f 1200 seconds = ___ minutes

9 Convert these times in minutes into seconds.

a 2 minutes = ____ seconds
b 10 minutes = ____ seconds
c 5 minutes = ____ seconds
d 3 minutes = ____ seconds
e 30 minutes = ____ seconds
f 40 minutes = ____ seconds

10 Convert these minutes to hours.

a 480 minutes = ___ hours
b 360 minutes = ___ hours
c 60 minutes = ___ hour
d 180 minutes = ___ hours
e 7200 minutes = ___ hours
f 6000 minutes = ___ hours

11 Convert these hours into minutes.

a 2 hours = _____ minutes
b 6 hours = _____ minutes
c 12 hours = _____ minutes
d 4 hours = _____ minutes
e 32 hours = _____ minutes
f 15 hours = _____ minutes

  ISBN: 9781925726176

12 Convert these times to 24-hour time.

a 5:09 am ______ f 8:35 am ______ k 2:19 am ______

b 11:53 pm ______ g 1:52 am ______ l 7:58 am ______

c 10:44 am ______ h 6:23 pm ______ m 8:05 am ______

d 12:18 pm ______ i 6:46 pm ______ n 9:21 pm ______

e 4:37 pm ______ j 3:46 pm ______ o 12:17 am ______

Use the calendars to complete the following questions.

June 2025

S	M	T	W	T	F	S
30						1
2	3	4	5	6	7	8
9	10	11	12	13	14	15
16	17	18	19	20	21	22
23	24	25	26	27	28	29

July 2025

S	M	T	W	T	F	S
	1	2	3	4	5	6
7	8	9	10	11	12	13
14	15	16	17	18	19	20
21	22	23	24	25	26	27
28	29	30	31			

13 Draw a ✗ on these dates on the calendars above.

a 9th June c 6th July e 29th July

b 24th July d 20th June f 3rd June

14 Circle the first day of each month in green and the last day of each month in blue.

15 What day of the week are these dates on?

a 16th July ______________ c 27th July ______________

b 13th June ______________ d 2nd June ______________

REVIEW

16 Complete the sentences.

- **a** There are ___ Mondays in June and ___ Mondays in July.
- **b** There are ___ Thursdays in June and ___ Thursdays in July.
- **c** There are ___ Sundays in June and ___ Sundays in July.
- **d** There are ___ weekdays in June and ___ weekdays in July.
- **e** There are ___ weekend days in June and ___ weekend days in July.

17 Write the following on the calendar.

- **a** Dentist on 5th June
- **b** Doctor on 30th July
- **c** Excursion 12th July
- **d** Party on 21st June
- **e** Haircut on 4th July
- **f** Movies on 10th July

18 How many days in each of these?

- **a** June ___
- **b** September ___
- **c** December ___
- **d** February 2024 ___
- **e** winter ___
- **f** autumn ___
- **g** one fortnight ___
- **h** one leap year ___
- **i** one week ___

Use the gym timetable to answer the following questions.

Weekday	Mon	Tue	Wed	Thu	Fri	Weekend	Sat	Sun
6:00 am	**Box** Brigid	**Circuit** Ari	**Box** Mitch	**Circuit** Karen	**Circuit** Emma	7:00 am	**Circuit** Geri	
8:15 am	**Mid Pace** Jen	**Tai Chi** Anya	**Light** Ari	**Mid Pace** Nia	**Balance** Lachlan	7:30 am		**Circuit** Jo
9:15 am	**Circuit** Brigid	**Circuit** Nia	**Circuit** Ari	**Circuit** Emma	**Circuit** Matt	8:15 am	**Light** Geri	
10:30 am	**Gentle** Emma	**Light** Nia	**Mid Pace** Ari	**Light** Shen	**Mid Pace** Shen	9:00 am		
11:30 am					**Gentle** Kim	9:15 am	**Box** Geri	
2:30 pm				**Burn** Matt		10:00 am		
4:30 pm	**Circuit** Ari	**Box** Mel	**Circuit** Shen	**Circuit** Leo	**Circuit** Jo	4:00 pm		**Pilates** Shen
5:45 pm	**Circuit** Ari	**Circuit** Matt	**Circuit** Karen	**Box** Matt				**Circuit** Ari
7:00 pm	**Box** Mitch	**Metafit** Mitch	**Box** Mat					

  ISBN: 9781925726176

19 How many classes are on each of these?

a Monday ___ c Wednesday ___ e Friday ___

b Tuesday ___ d Thursday ___ f the weekend ___

20 How many circuit classes are held on Monday? ___

21 When is the metafit class? ______________

22 Who leads these classes?

a Circuit, Monday at 9:15 am ______________

b Mid pace, Wednesday at 10:30 am ______________

c Burn, Thursday at 2:30 pm ______________

23 Fill in the gaps.

a The Balance class starts at ___________.

b The Balance class goes for ______________.

c The last class on the weekend is ____________ on ____day and it is led by _________.

24 How many classes does each person lead?

a Shen ___ c Anya ___ e Lachlan ___

b Nia ___ d Emma ___ f Geri ___

25 Use the soccer timeline to answer the questions.

a When and where was the first Football Association founded?

____

b What happened in 1872?

____

____

c In what year and country was FIFA formed?

____

d What important event occurred in 1908?

____

____

e Where and when was the first World Cup held?

____

f What year were substitutions first allowed in soccer?

____

Timeline: History of Soccer

- 2010 First World Cup tournament played in Africa, in Johannesburg, South Africa.
- 2009 Cristiano Ronaldo became the most expensive player in the world, joining Real Madrid for £85m.
- 2007 FIFA has 208 members.
- 1998 World Cup expanded to 32 teams.
- 1992 First FIFA rankings published. Germany is #1.
- 1991 First women's World Cup in China.
- 1982 World Cup has 24 teams for the first time.
- 1958 Rules changed to allow substitution for the first time.
- 1930 First World Cup in Montevideo, Uruguay.
- 1908 First successful international competition in Belgium at the Olympics.
- 1904 FIFA formed in Paris, France.
- 1888 Football League established in England.
- 1872 First international game held in Glasgow, Scotland.
- 1863 Football Association founded in London, England.

g How many teams were in the World Cup in 1982? ____

h How many teams were in the World Cup in 1998? ____

i In what year was Germany number 1? ________

j How many members did FIFA have in 2007? _____

k Write a question you could ask about the History of Soccer timeline.

____

____

CATCH UP MATHS YEAR 5 BOOK B © PASCAL PRESS ISBN: 9781925726176

DIRECTIONS

Directions use position words to tell us the location of something.

Here are some words to help you describe location.

above	down	left	south
below	east	north	up
behind	in front of	right	west

Check the answers on the video!

Example: Write the words **up**, **down**, **right** and **left** to complete the directions along the red path.

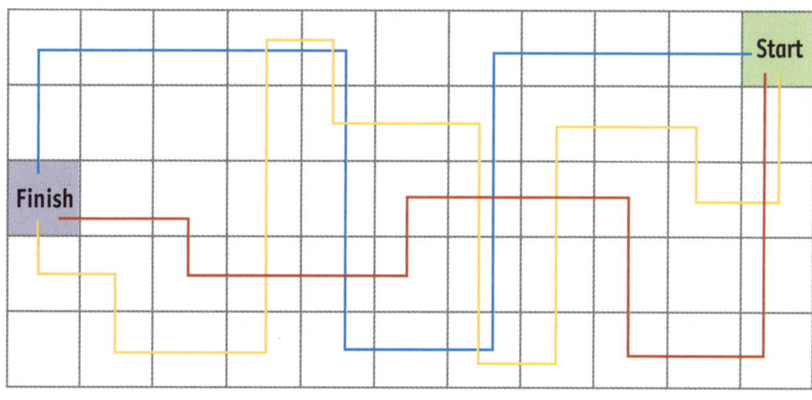

4 down ________,
2 left ________,
2 ________,
3 ________,
1 ________,
3 ________,
1 ________,
2 ________

Your turn

Using numbers and words, write directions to follow each path.

a Blue path ____

____

b Yellow path ____

____

____

Check your answers
How many did you get correct?

1 **Here is Roisin's home.**

- Write directions for her friend Ann to find Roisin in Allan's bedroom when she arrives.

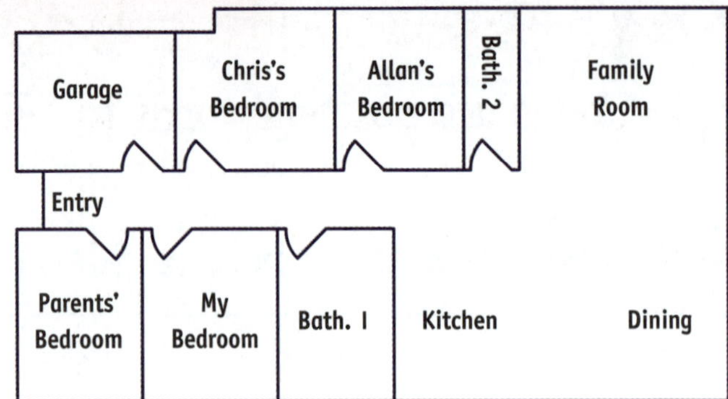

Walk down the hallway past my parents' room on the right and the garage on the left. Continue walking past my bedroom on the right and Chris's bedroom on the left. Walk past the bathroom on the right and Allan's bedroom will be on the left.

a Write directions from the entry to the dining area.

____

____

____

____

b Write directions from bathroom 2 to the entry.

____

____

____

____

2 **Describe the position of each place.**

- Chris's bedroom Between the garage and Allan's bedroom

a Roisin's bedroom ______________________________

b Bathroom I ______________________________

c Dining area ______________________________

d Family room ______________________________

CATCH UP MATHS YEAR 5 BOOK B © PASCAL PRESS ISBN: 9781925726176

3 Use the map to write directions for the following.

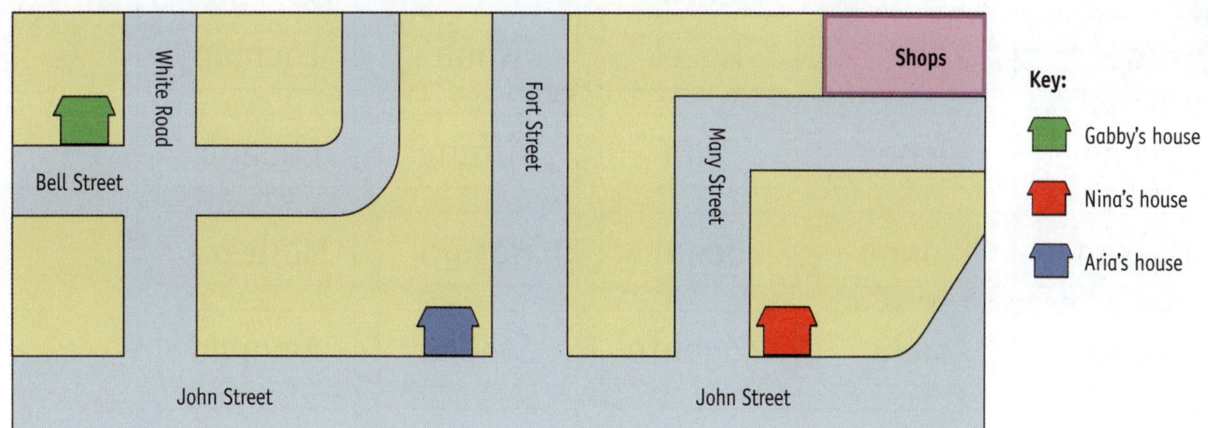

a Gabby's house to Nina's house ______________________

____

____

b Aria's house to Gabby's house ______________________

____

____

c Aria's house to Nina's house ______________________

____

____

4 Write directions to get to the shops from each of the houses.

a Gabby's house to the shops ______________________

____

____

b Nina's house to the shops ______________________

____

____

c Aria's house to the shops ______________________

____

____

  ISBN: 9781925726176

5T's Tote Tray Stand

Zac	Will	Hazel	Noah	Lachlan
Michael	Jade	Tina	Tiffany	Phoenix
Hassan	Rianne	Jenai	Hosain	Marleece
Nevaeh	Amelia	Magenta	Stelios	Amanda
Rose	Velia	Marianthi	Leo	Gabby
Mia	Arty	Felicity	Xavier	Mrs White

5 Write the name on the tote tray that matches the location.

- Below Rianne's Amelia
- **a** Above Stelios's ____________
- **b** To the left of Jade's ____________
- **c** To the right of Tina's ____________
- **d** In the bottom right corner ____________
- **e** In the top left corner ____________
- **f** In the top right corner ____________
- **g** In the bottom left corner ____________

6 Describe the position of each person's tray.

- Velia's second row from the bottom, second from the left
- **a** Hassan's ____________
- **b** Rose's ____________
- **c** Xavier's ____________
- **d** Michael's ____________
- **e** Noah's ____________

CATCH UP MATHS YEAR 5 BOOK B © PASCAL PRESS ISBN: 9781925726176

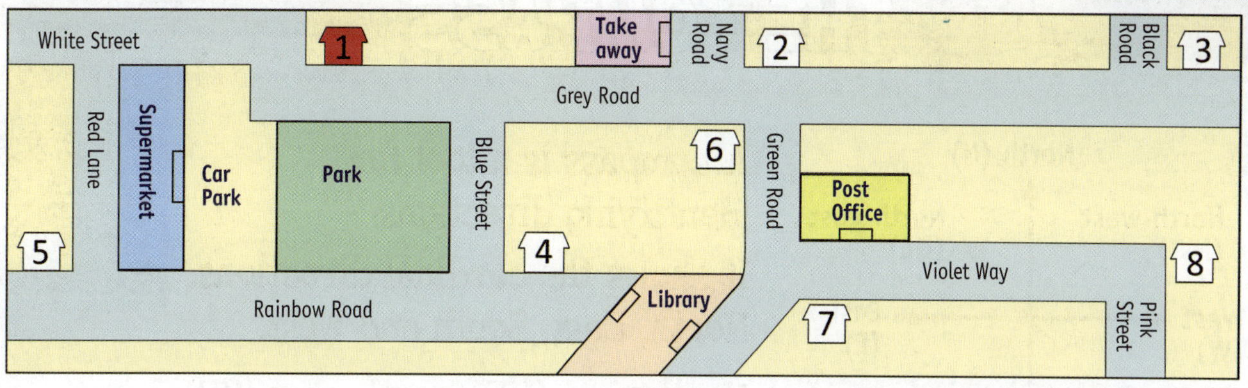

7 **Locate the houses on the map. Colour the houses to match.**

1 Fabio	3 David	5 Anthony	7 Alex
2 Mikayla	4 Rani	6 Ashton	8 Paula

8 **Write directions for:**

Rani to get to the supermarket

Turn right onto Rainbow Road, continue past Blue Street and the park, turn right into the car park and then left into the supermarket.

a David to get to the park

b Fabio to get to the library

c Alex to get to the take away

THE COMPASS

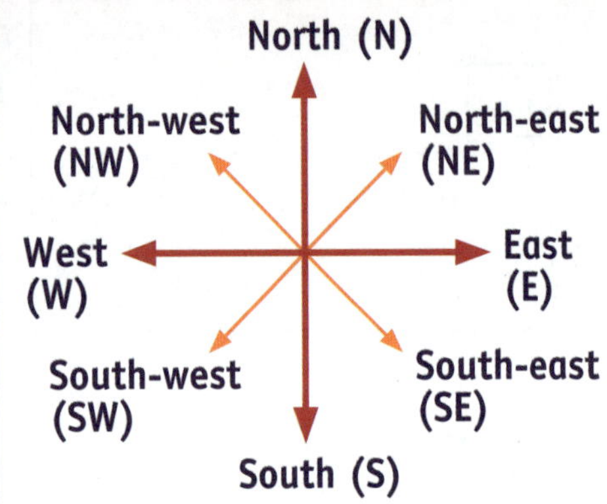

A compass is a tool for identifying directions.

It shows the cardinal directions: North, East, South and West.

North-east (NE), south-east (SE), south-west (SW) and north-west (NW) are intercardinal directions.

SCAN to watch video

Examples:
Write the missing information on the compass.

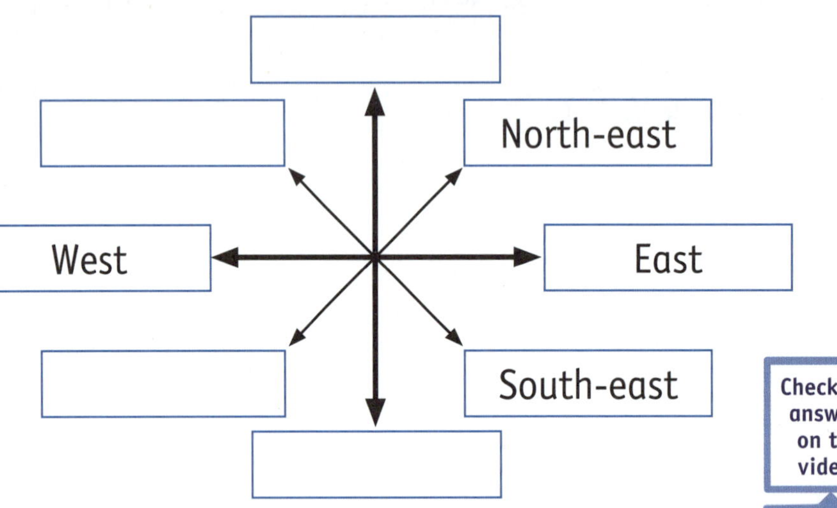

Check the answers on the video!

1 Start from the ✗.
Draw the shape that is:

● North ●

a West

b South-east

c South-west

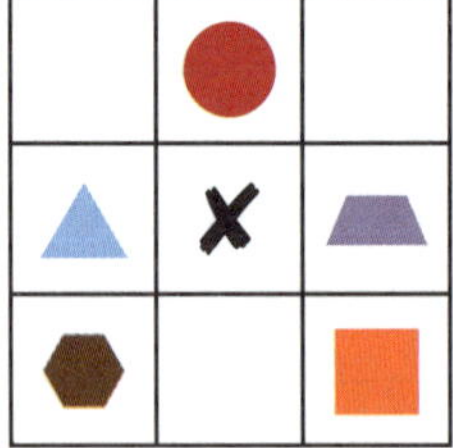

2 Draw these things on the grid.

● ▬ East of X

a ★ North-west of X

b ☾ North-east of X

c ✔ South of X

Check your answers

How many did you get correct?

CATCH UP MATHS YEAR 5 BOOK B © PASCAL PRESS ISBN: 9781925726176

PRACTICE

1 Use the directions to label the places on the map.

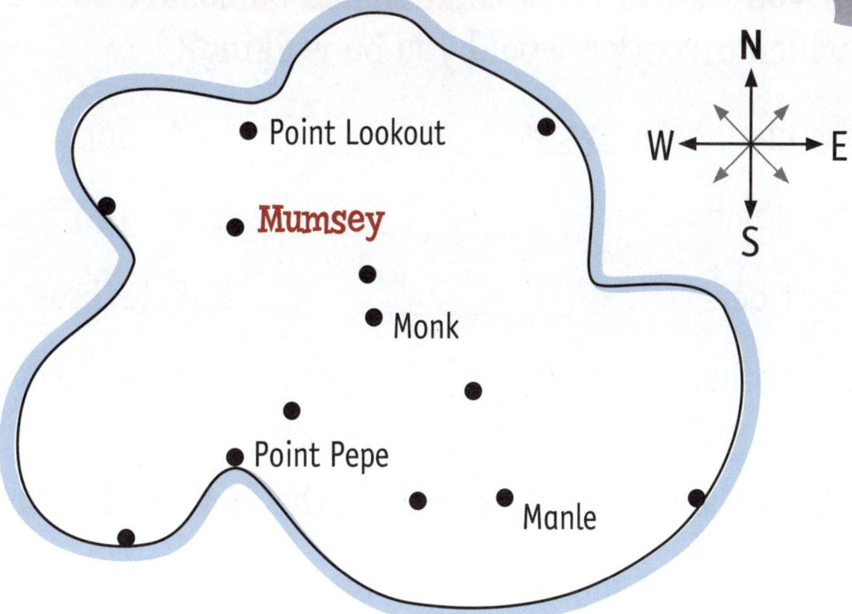

- Mumsey: S of Point Lookout
- **a** Barton: N of Monk
- **b** Shoutout: W of Manle
- **c** Cat Point: E of Manle
- **d** Mount Pete: SE of Monk
- **e** Martha's Point: NE of Monk
- **f** Dale: SW of Monk
- **g** Tom's Point: SW of Point Lookout
- **h** Al's Point: SW of Point Pepe

2 Look at the zoo map.

If you are standing at the ★, what animals will you find when you walk in these directions?

- SE monkeys
- **a** W ____________
- **b** NW ____________
- **c** S ____________

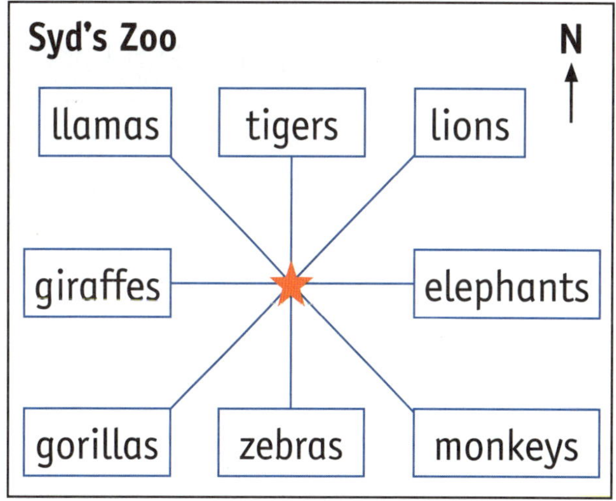

3 From the ★, what direction should you walk to find these animals?

- tigers north
- **a** gorillas ____________
- **b** lions ____________
- **c** elephants ____________

If you walked from each animal enclosure to the ★, which direction would you be walking?

- elephants west
- a llamas ______
- b tigers ______
- c gorillas ______
- d lions ______
- e giraffes ______
- f zebras ______
- g monkeys ______

5 Draw these rides to complete the map of the Funaround Fun Park.

- The ferris wheel is SE of the food hall.
- a The dodgem cars are NW of the food hall.
- b The ice-cream stand is north of the food hall.
- c The roller-coaster is east of the food hall.
- d The silly slides are NE of the food hall.
- e The crazy clowns are south of the food hall.
- f The pirate ship is west of the food hall.

Funaround Fun Park

N

Food Hall

CATCH UP MATHS YEAR 5 BOOK B © PASCAL PRESS ISBN: 9781925726176

GRID REFERENCES

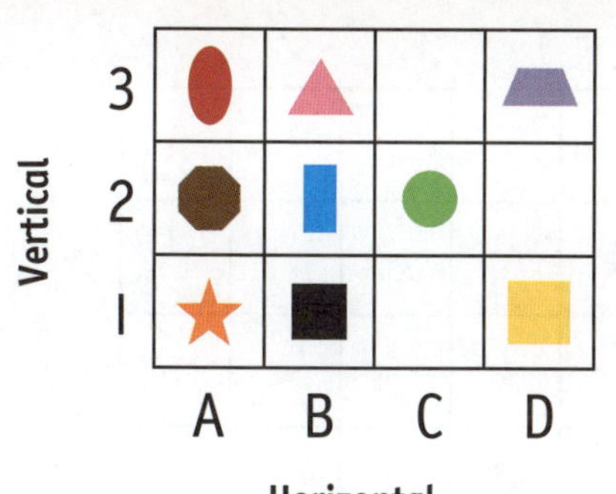

This is a grid.
Each space on this grid is named by a letter and a number.

SCAN to watch video

When you give grid references, say the horizontal grid label first and then the vertical grid label.

This grid has 4 columns: A, B, C, D.
It has three rows: 1, 2 and 3

The ● has the grid reference (C,2).

The ■ has the grid reference (D,1).

Put grid references in brackets with a comma between them.

Examples:

Complete the grid references using the grid above.

a ▮ has the grid reference (B,<u>2</u>).

b ▲ has the grid reference (B,__).

c ★ has the grid reference (A,__).

Your turn

1 Using the grid above, write the grid references for these shapes.

■ <u>(B,1)</u> **a** ▰ _____ **b** ⬮ _____

2 Draw the shapes at these grid references in the grid above.

⬣ (A,2) **a** ☾ (C,3) **b** ☺ (C,1) **c** ♥ (D,2)

SELF CHECK Tick how you feel

Got it!	Need help...	I don't get it

Check your answers
How many did you get correct?

PRACTICE

1 Write the grid references of the coloured stars.

● ★ (1,G)

a ★ ______

b ★ ______

c ★ ______

d ★ ______

e ★ ______

f ★ ______

g ★ ______

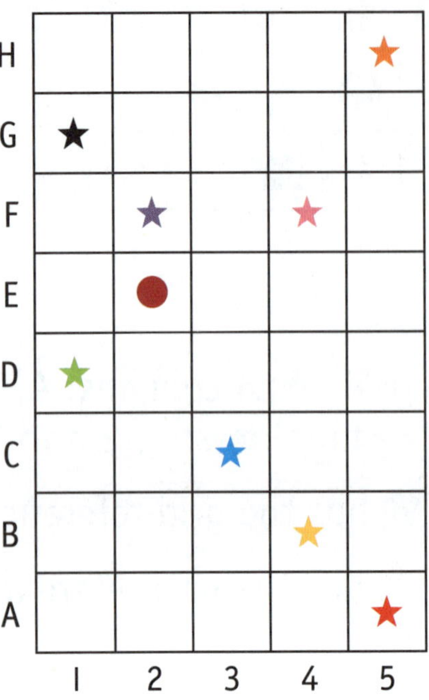

2 Draw the coloured circles at the given grid references.

● ● (2,E)

a ● (1,A)

b ● (5,E)

c ● (2,H)

d ● (3,G)

e ● (1,E)

f ● (2,C)

g ● (4,A)

3 This grid has been named using two sets of numbers. Write the grid reference for each smiley.

Remember, the horizontal grid label comes first and then the vertical grid label.

● ☺ (3,6)

a ☺ ______

b ☺ ______

c ☺ ______

d ☺ ______

e ☺ ______

f ☺ ______

g ☺ ______

h ☺ ______

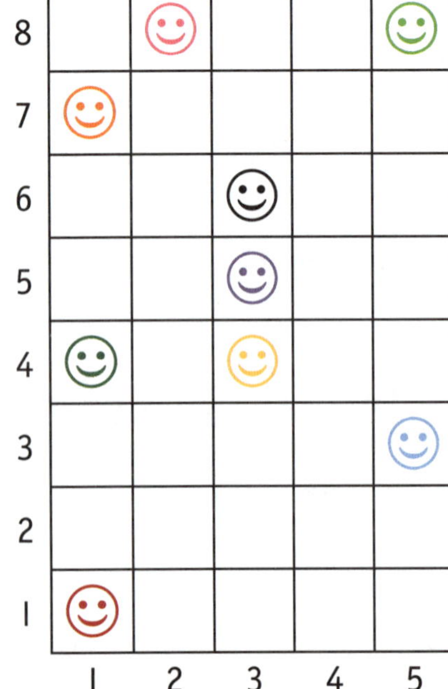

4 Draw ▲ in (4,2) and ▲ in (2,4).

CATCH UP MATHS YEAR 5 BOOK B © PASCAL PRESS ISBN: 9781925726176

COORDINATES

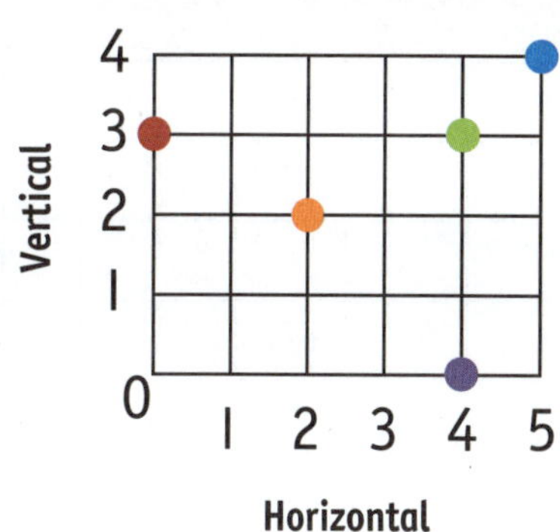

This is a grid where each line is named by a number.
A pair of coordinates names a point where lines meet.

SCAN to watch video

This grid has 6 vertical lines: 0, 1, 2, 3, 4, 5.
It has 5 horizontal lines: 0, 1, 2, 3, 4.

The ● has the coordinates (0,3).

The ● has the coordinates (5,4).

When you give coordinates, say the horizontal grid label first and then the vertical grid label.

Examples:
Complete the coordinates using the grid above.

a ● has the coordinates (4,3).

b ● has the coordinates (__,__).

c ● has the coordinates (__,__).

Put coordinates in brackets with a comma between them.

1 The grid at the right is named with letters and numbers.
Write the coordinates for the stars.

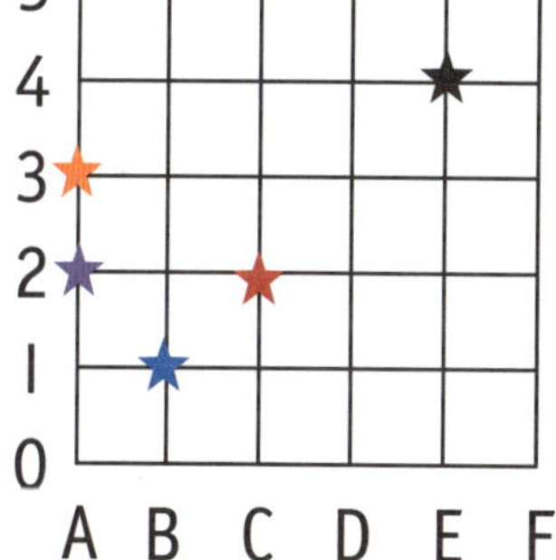

● ★ (A,3)

a ★ ______

b ★ ______

c ★ ______

2 Draw the stars at the coordinates.

● ★ (A,2) **a** ★ (C,3) **b** ★ (C,1) **c** ★ (D,2)

SELF CHECK Tick how you feel

Got it!	Need help...	I don't get it
☐	 ☐	 ☐

Check your answers
How many did you get correct?

PRACTICE

1 Using the grid, write the coordinates of these circles.

- ● (0,3)
- a ● ______
- b ● ______
- c ● ______
- d ● ______
- e ● ______
- f ● ______
- g ● ______

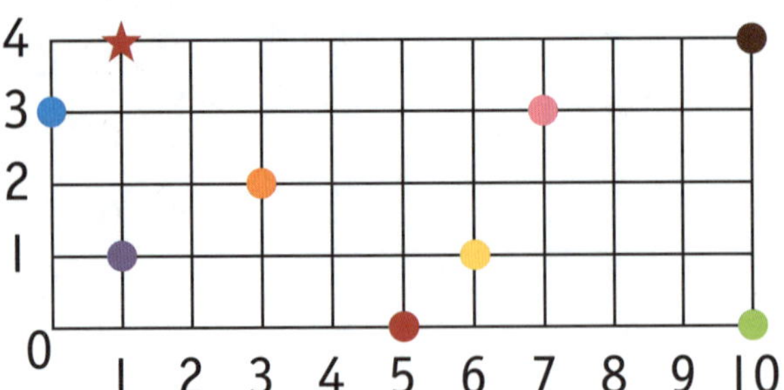

2 Draw the stars at the coordinates on the grid above.

- ★ (1,4)
- a ★ (4,0)
- b ★ (6,3)
- c ★ (8,1)
- d ★ (9,4)
- e ★ (0,2)
- f ★ (2,2)
- g ★ (7,0)

3 Write the coordinates of the vertices of each shape.

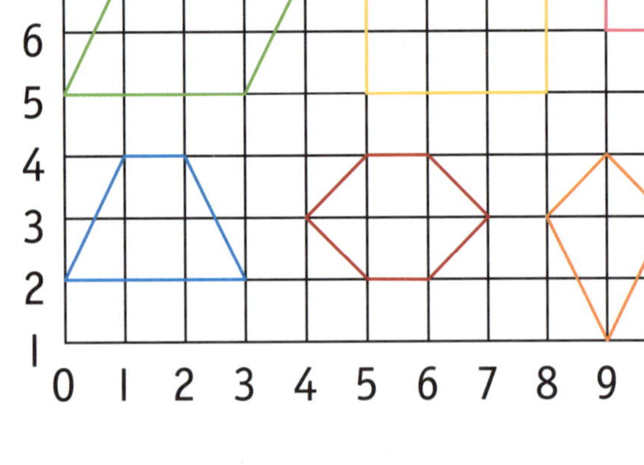

- kite (9,1) (10,3) (9,4) (8,3)
- a rectangle ______________________
- b trapezium ______________________
- e square ______________________
- c parallelogram ______________________
- d hexagon ______________________

4 Plot the set of points. Join them, then name the shape you made.

- (B,3) (B,5) (D,3) (D,5) square
- a (E,5),(E,7) (H,7) (I,5) ____________
- b (A,1) (A,2) (D,1) (D,2) ____________
- c (I,2) (I,3) (J,4) (K,3) (K,2)

  ____
- d (G,1) (G,3) (E,4) (E,2) ____________

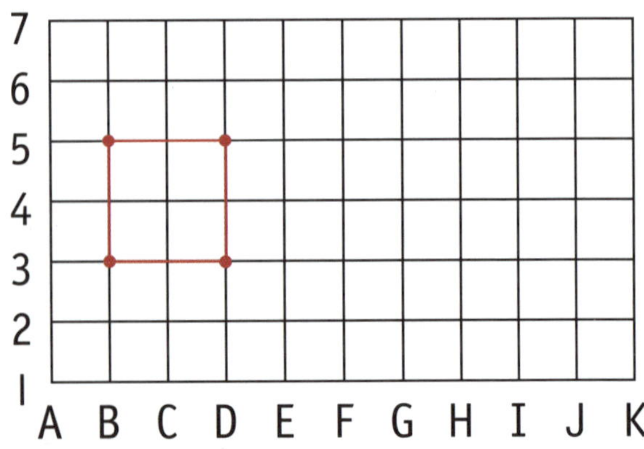

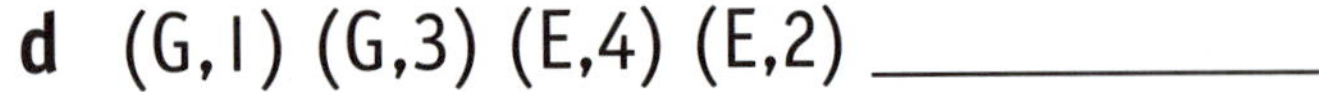

CATCH UP MATHS YEAR 5 BOOK B © PASCAL PRESS ISBN: 9781925726176

LEGENDS

Legends show where important places or landmarks are on a map. Legends are also called keys.

SCAN to watch video

This map shows the town of Vacay.

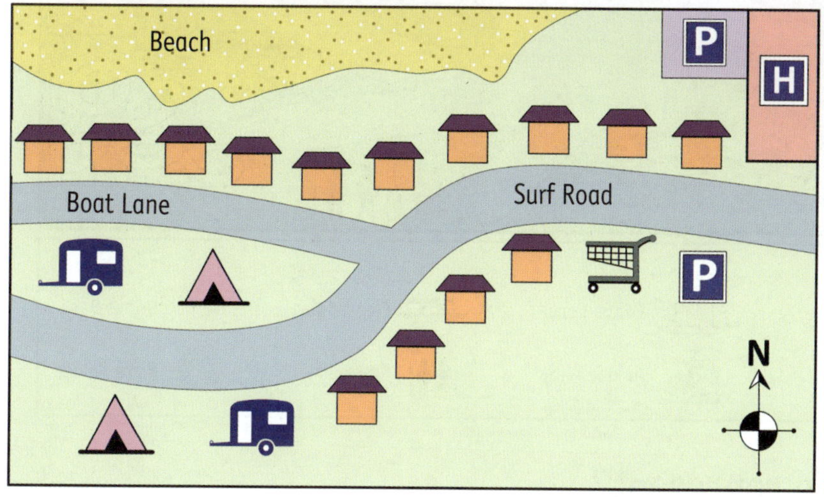

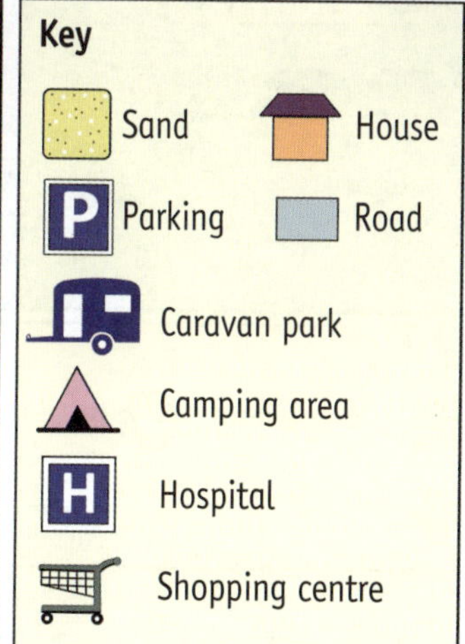

Examples: Look at the key in the map of Vacay. What do the pictures represent?

Keys or legends can have pictures, colours and lines.

a camping area

b ______

c ______

d ______

e  ______

f 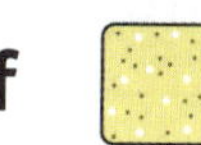 ______

g  ______

h  ______

Your turn

1 How many places in Vacay can you:

- camp? 2

a shop? ___

b stay in a caravan park? ___

2 How many houses are there in Vacay? ___

SELF CHECK Tick how you feel		
Got it!	Need help...	I don't get it
☐	☐	☐

Check your answers

How many did you get correct? ☐

PRACTICE

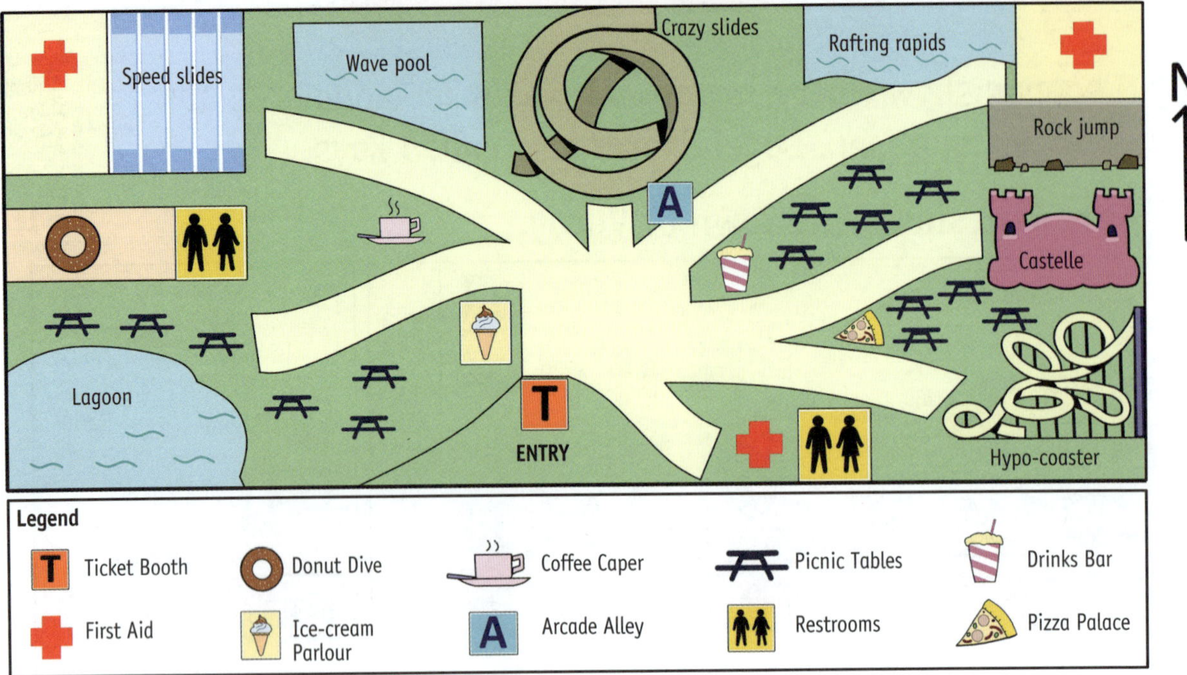

1 What do these symbols represent?

b T ____________

a ____________

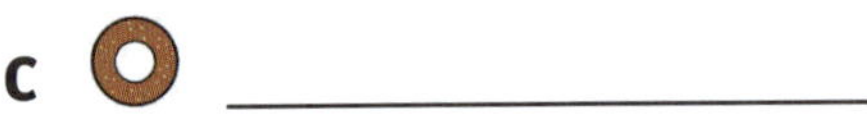

c ____________

2 Draw the symbol for each place.

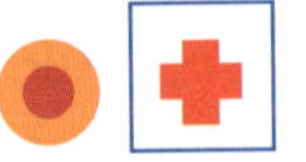

First Aid

a Pizza Palace

b Restrooms

c Drinks Bar

d Arcade Alley

e Ice-cream Parlour

3 Draw these symbols on the map.

Coffee in 1 place

a Restrooms in 3 places

b F Food in 4 places

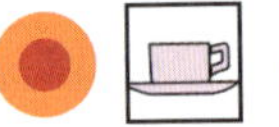

c First Aid in 2 places

d T Ticket booth near the Entry

e Ice cream in one place

f Drinks bar in 3 places

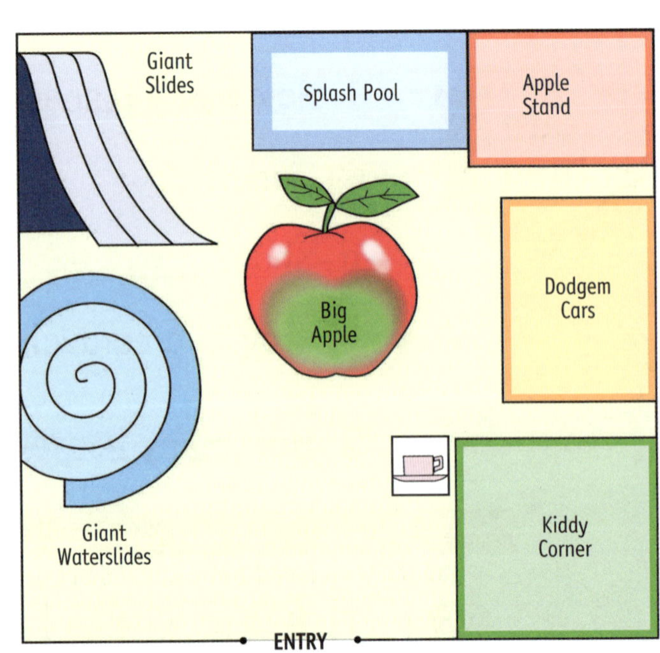

  ISBN: 9781925726176

POSITION REVIEW

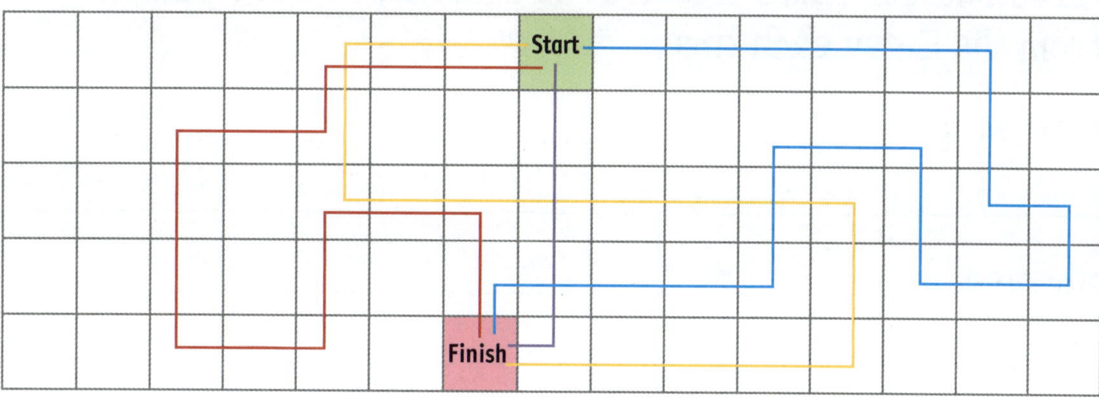

1 Write directions to follow each path from Start to Finish.

a red path ______________________________

____

____

b blue path ______________________________

____

____

c purple path ______________________________

____

____

d yellow path ______________________________

____

____

Use this diagram to answer the following questions.

First Floor of Anh's home

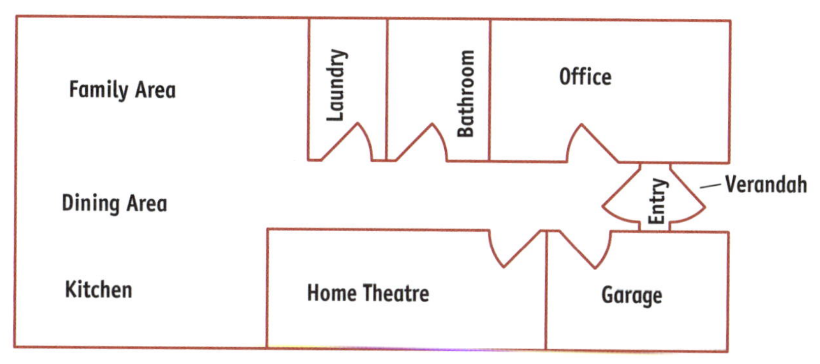

2 **Write directions for Anh's friend Jo to find him in these places. Start from the Entry each time.**

a Home theatre ______________________________

____

b Family area ______________________________

____

c Laundry ______________________________

____

3 **Describe the position of each place in Anh's house.**

a Dining area ______________________________

b Bathroom ______________________________

c Garage ______________________________

d Kitchen ______________________________

4 **Who sits at the desk that matches the description?**

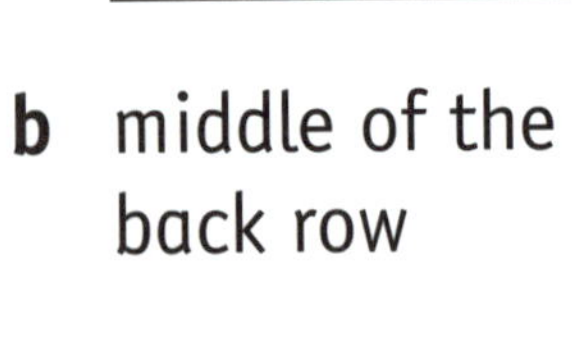

a front row, second from the left

____

b middle of the back row

____

Kindergarten Red Classroom

		Back		Teacher's desk
Lev	Lisa	Kira	Joe	Hani
Annie	Ravi	Peter	Pasha	Adam
Kylie	Sam	Jake	Elise	Zola
Whiteboard		Front		

c middle row, second from the left ______________

d back row, second from the right ______________

e back left corner ______________

f front right corner ______________

CATCH UP MATHS YEAR 5 BOOK B © PASCAL PRESS ISBN: 9781925726176

5 Describe the position of these students' desks.

a Annie ______

b Lev ______

c Jake ______

d Lisa ______

e Pasha ______

f Hani ______

6 Label the compass.

a ______

b ______

c ______

d ______

e ______

f ______

g ______

h ______

7 Use the grid at the right to answer the following questions.

✗ (orange)		
✗ (red)	★	✗ (blue)
	✗ (purple)	

8 Start from the star. What colour cross is:

a west? ______

b east? ______

c south? ______

d NW? ______

9 Draw these things on the grid above.

a ✗ SW of ★

b ✗ N of ★

c ✗ NE of ★

d ✗ SE of ★

REVIEW

Use the map of the fun park to answer the following questions.

10 If you are standing at the information stand (Info), which attraction is in each of these directions?

a N ____________________

b E ____________________

c SW ____________________

d NE ____________________

e S ____________________

f W ____________________

g SE ____________________

h NW ____________________

Map of Fun Park

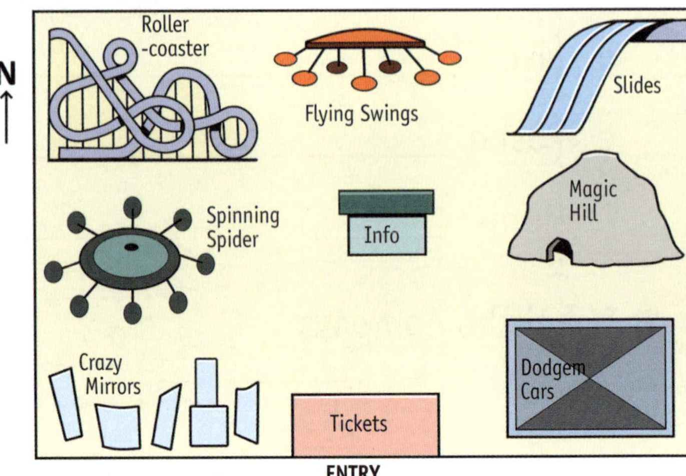

11 If you walked from each attraction to the information stand, which direction would you be walking?

a Spinning spiders to info stand: ____________________

b Slides to info stand: ____________________

c Entry to info stand: ____________________

d Dodgem cars to info stand: ____________________

e Roller-coaster to info stand: ____________________

f Flying swings to info stand: ____________________

g Crazy mirrors to info stand: ____________________

h Magic hill to info stand: ____________________

CATCH UP MATHS YEAR 5 BOOK B © PASCAL PRESS ISBN: 9781925726176

Use the grid to answer the following questions.

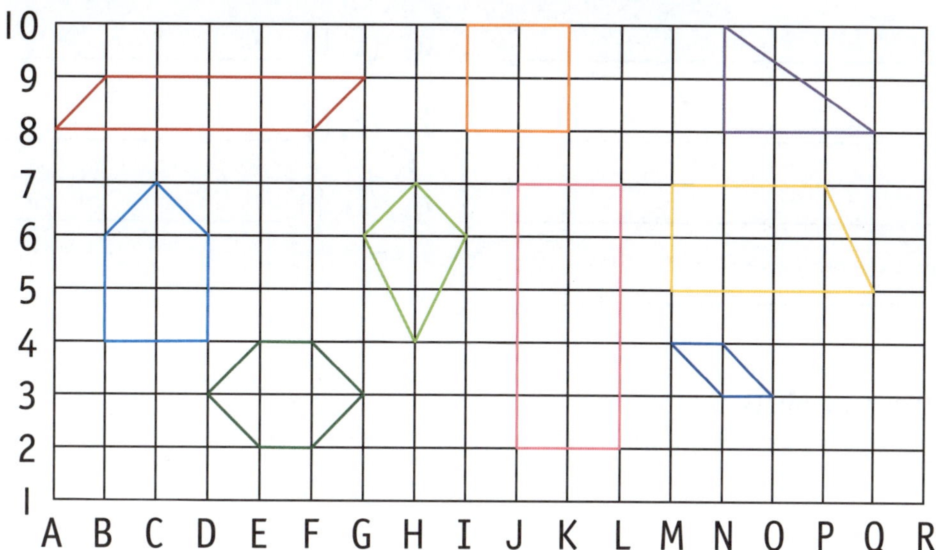

12 Write the coordinates of the vertices for each shape.

a pentagon ________________

b kite ________________

c hexagon ________________

d trapezium ________________

13 Write the name of the shape formed by these coordinates.

a (M,4) (N,3) (N,4) (O,3) ________________

b (A,8) (B,9) (G,9) (F,8) ________________

c (I,8) (I,10) (K,10) (K,8) ________________

d (N,8) (N,10) (Q,8) ________________

e (J,2) (J,7) (L,2) (L,7) ________________

Use the map at the right to answer the following questions.

14 Write the grid reference for these cities.

a Alice Springs ______

b Port Hedland ______

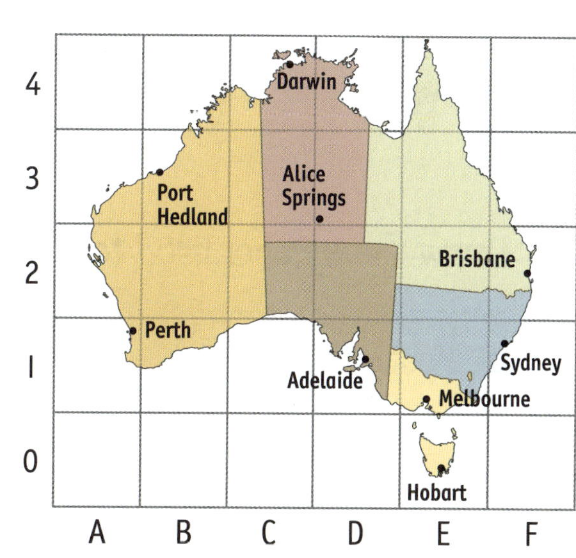

Write the cities with these grid references.

a (F,2) ______________________

b (E,0) ______________________

c (A,1) ______________________

d (C,4) ______________________

e (E,1) ______________________

f (F,1) ______________________

g (D,1) ______________________

Harty Community Map

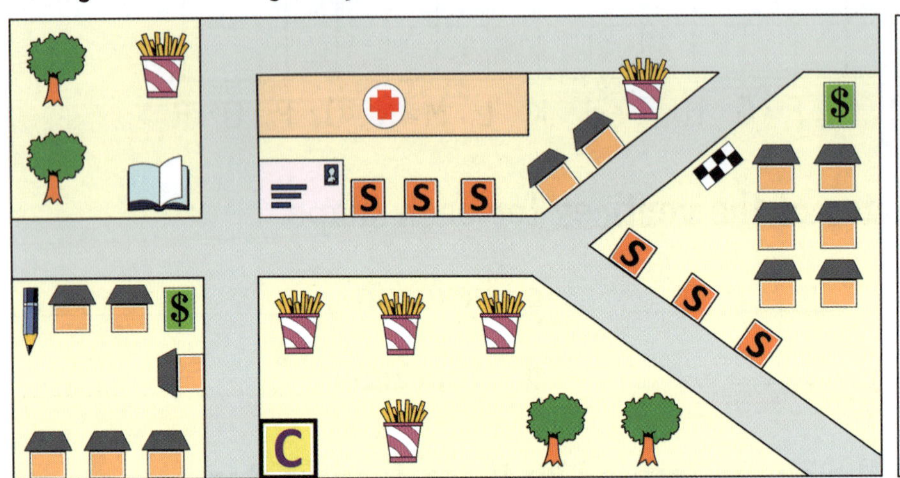

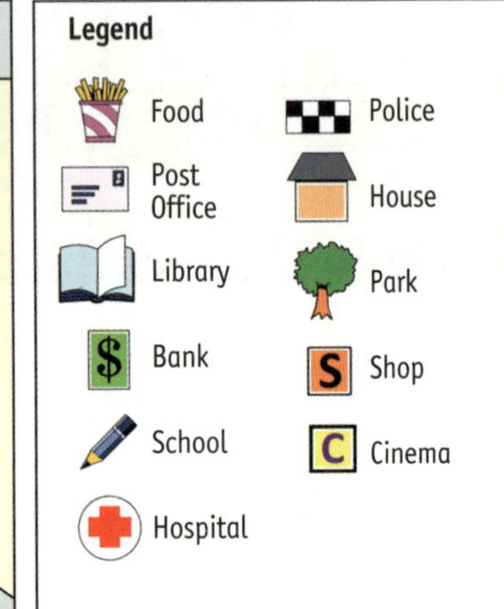

Look at the legend on the Harty Community Map.
What do these pictures represent?

a ______________________

b ______________________

c ______________________

d ______________________

e ______________________

f ______________________

g ______________________

h ______________________

Draw the symbol used to represent these things.

a Post office ☐

b Police ☐

c House ☐

CATCH UP MATHS YEAR 5 BOOK B © PASCAL PRESS ISBN: 9781925726176

ANSWERS

1. LENGTH

Metres

Page 1 – Your Turn

Adult to check

Page 2 – Practice

1 a 528 cm b 317 cm c 231 cm d 859 cm e 644 cm f 309 cm g 710 cm h 920 cm i 699 cm j 404 cm k 808 cm

2 a 6 m 37 cm b 6 m 1 cm c 9 m 52 cm d 4 m 81 cm e 5 m 45 cm f 8 m 90 cm g 3 m 30 cm h 2 m 6 cm i 6 m 60 cm j 1 m 7 cm k 7 m 15 cm

3 a 210 cm b 8 m 65 cm c 989 cm d 4 m e 637 cm f 3 m 2 cm g 520 cm h 4 m 29 cm i 603 cm j 6 m 46 cm k 711 cm l 8 m 1 cm m 990 cm

4 1: 1 m 69 cm 2: 2 m 10 cm 3: 3 m 2 cm 4: 4 m 5: 4m 29 cm 6: 5 m 20 cm 7: 6 m 3 cm 8: 6 m 37 cm 9: 6 m 46 cm 10: 7 m 11 cm 11: 8 m 1 cm 12: 8 m 65 cm 13: 9 m 89 cm 14: 9 m 90 cm

5 a $2\frac{1}{2}$ m b 45 m c 300 m d 2 m e $\frac{1}{4}$ m f 1 m g 3 m

6 a 625 cm b 275 cm c 125 cm d 550 cm e 725 cm f 975 cm g 1025 cm h 150 cm i 1225 cm j 425 cm k 859 cm

Centimetres

Page 4 – Your Turn

a 6
b 9

Page 5 – Practice

1 Adult to check

2 a 11 cm b 9 cm c 1 cm d 7 cm

3 a 9 cm b 12 cm c 8 cm

4 a green b orange c orange, blue, pink, green

Millimetres

Page 6 – Your Turn

1 a 45 mm b 83 mm c 126 mm

2

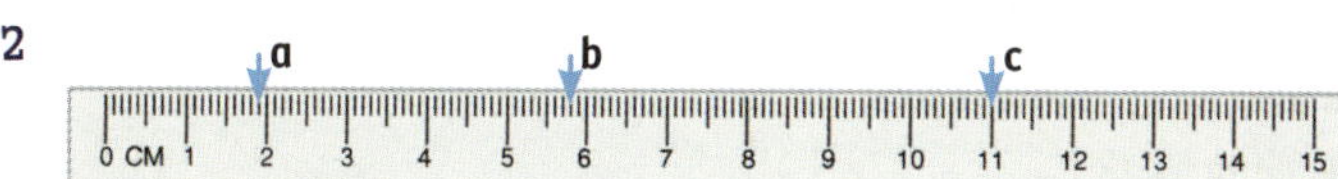

Page 7 – Practice

1 Adult to check

2 a mm b cm c mm d mm e cm

3 a 44 mm b 126 mm c 13 mm

4 13 mm, 44 mm, 72 mm, 126 mm

5 a 7 cm 1 mm b 5 cm c 69 mm d 8 cm 3 mm e 91 mm f 144 mm g 2 cm 8 mm

6 a 80 mm b 110 mm c 20 mm d 200 mm e 800 mm

Millimetres, Centimetres and Metres

Page 8 – Your Turn

a 20 cm b 30 m c 13 mm d 10 cm

Page 9 – Practice

1 c

2 a 300 cm b 50 mm c 12 cm d 4 cm e 10 cm f 8 m g 17 cm h 6 cm i 1300 cm

3 a No b Yes c Yes d Yes e No f Yes g No h Yes i Yes

4

a mm to cm	
60	6
130	13
280	28
1490	149

b cm to mm	
14	140
67	670
133	1330
552	5520

c cm to m	
200	2
550	5.5
600	6
1400	14

d m to cm	
24	2400
56	5600
740	74 000
153	15 300

Kilometres

Page 10 – Your Turn

1 a 7000 m b 12 000 m c 14 000 m d 26 000 m e 315 000 m

2 a 9 km b 16 km c 34 km d 25 km e 537 km

Page 11 – Practice

1 Adult to check

2 a 5025 b 7047 c 1009 d 4653 e 6814 f 24 102 g 96 012 h 10 146 i 38 031

3 a 2 km 829 m b 5 km 945 m c 6 km 3 m d 9 km 12 m e 14 km 388 m f 21 km 510 m g 35 km 7 m h 46 km 30 m i 71 km 10 m j 74 km 136 m k 847 km 528 m

4 a 625 cm, 0.006 25 km b 200 000 cm, 2000 m c 300 cm, 0.003 km d 9.14 m, 0.009 14 km e 5 374 000 cm, 53.74 km

Perimeter

Page 12 – Your Turn

a P = 5 cm + 1 cm + 5 cm + 1 cm = 12 cm
b P = 1 m + 2 m + 3 m + 4 m = 18 m

Page 13 – Practice

1 a P = 10 cm b P = 20 cm c P = 8 m d P = 15 m e P = 12 cm f P = 32 cm

2 Adult to check

3 a 26 m b 20 m c 18 m

4 a Yes b No c Yes d Yes e No

5 1 Trees & Herbs 2 Palms 3 Succulents 4 Citrus 5 Roses 6 Pots and Planters 7 Outdoor Furniture 8 Indoor Plants

6 a 4 m b 30 mm c 8 m d 18 cm e 8 m f 18 m g 17 mm

7 a P = 25 m b P = 28 cm c P = 20 cm d P = 40 cm

ANSWERS

1. LENGTH CONTINUED

Decimal Notation

Page 16 – Your Turn

a 6.34 b 1.02 c 4.00 d 5.20 e 1.35 f 1.18 g 4.85 h 0.94 i 14.253 j 27.595 k 43.682

Page 17 – Practice

1 a 1.58 m b 6.92 m c 1.09 m d 9.20 m e 3.01 m f 30.30 m g 16.41 m h 39.87 m i 62.03 m j 0.43 m k 0.64 m

2 a 0 m 16 cm b 3 m 79 cm c 5 m 20 cm d 4 m 5 cm e 8 m 17 cm f 13 m 6 cm g 52 m 80 cm h 0 m 8 cm i 0 m 2 cm

3 a 2 m 16 cm, 2.16 m b 810 cm, 8.1 m c 3683 cm, 36 m 83 cm d 154 m 53 cm, 154.53 cm e 24 738 cm, 247.38 cm

Temperature

Page 18 – Your Turn

a 13 °C, 21 °C, 23 °C, 53 °C, 82 °C
b 300 °C, 330 °C, 401 °C, 650 °C, 910 °C
c 0 °C, 9 °C, 37 °C, 90 °C, 102 °C

Page 19 – Practice

1 a 42 °C b 99 °C c 8 °C d 16 °C e 173 °C

2 a fifteen degrees Celsius
b ninety-eight degrees Celsius
c sixty-one degrees Celsius
d four hundred and four degrees Celsius

3 a 7 °C b 100 °C c 10 °C

4 Adult to check

Length Review Page 20

1 a 100 cm b 50 cm c 25 cm

2 a 253 cm b 807 cm c 616 cm d 808 cm e 1516 cm f 2485 cm

3 a 3 m 57 cm b 8 m 5 cm c 6 m 70 cm d 4 m 79 cm e 0 m 38 cm f 1 m 15 cm

4 a 10 cm b 14 cm c 2 cm d 8 cm

5 2 cm, 8 cm, 10 cm, 14 cm

6 Adult to check

7 a cm b cm c mm d cm e mm f mm

8 a 14 b 42 c 95 d 133

9 Adult to check

10 a 11 cm b 6 m c 100 m d 1 mm e 25 cm

11 a 7 b 40 c 2000 d 1 e 600 f 13.6 g 25 h 15 i 560 j 29 k 14 l 17 200 m 8 n 157 o 593.5 p 12.5 q 6.7 r 69 000

12 a 6000 b 8000 c 17 000 d 24 000 e 37 000 f 243 000 g 146 000 h 549 000 i 6 847 000 j 8 625 000

13 a 7 b 11 c 2 d 13 e 43 f 612 g 33 h 81 i 549 j 642

14 a 5 m, 0.005 km b 225 cm, 0.002 25 km c 500 000 cm, 5 km d 6270 cm, 62.7 m e 3 352 000 cm, 33.52 km

15 a P = 12 m b P = 16 cm c P = 15.5 cm d P = 17 cm e P = 16 m f P = 22 cm

16 Adult to check

17 a 11 b 28 c 16 d 10 e 13 f 12 g 70 h 17 i 10 j 102

18 a P = 28 m b P = 18 cm

19 a 7.49 b 3.07 c 2.00 d 9.95 e 3.82 f 0.58 g 0.11 h 0.06 i 4.70 j 415.32 k 555.95 l 249.30

20 a 7.09 b 4.93 c 2.07 d 8.93 e 5 f 6.60 g 0.83 h 0.72 i 0.03 j 0.08 k 232.65 l 834.51

21 a 7 m 3 cm b 8 m 30 cm c 2 m 19 cm d 1 m 0 cm e 0 m 8 cm f 60 m 36 cm g 0 m 53 cm h 153 m 45 cm i 1 m 4 cm j 245 m 63 cm k 346 m 20 cm l 657 m 6 cm m 908 m 42 cm n 421 m 87 cm

22 a 8.33 m b 4.56 m c 7.57 m d 0.20 m e 0.59 m f 0.06 m

23 a 1.85 m b 17.62 m c 50.15 m

24 a 34 °C, 21 °C, 10 °C, 8 °C, 3 °C
b 43 °C, 32 °C, 24 °C, 13 °C, 5 °C
c 19 °C, 14 °C, 7 °C, 2 °C, 0 °C
d 22 °C, 18 °C, 17 °C, 11 °C, 4 °C
e 63 °C, 45 °C, 41 °C, 15 °C, 10 °C

25 a 19 °C b 26 °C c 748 °C

26 a six degrees Celsius
b eighty-three degrees Celsius
c four hundred and ten degrees Celsius

27 a 25 °C b 20 °C c 70 °C d 100 °C

28 Adult to check

2. ANGLES

Page 28 – Your Turn

Adult to check

Page 29 – Practice

1 Adult to check

2 a 4, 3, 5, 2, 1 b 2, 1, 4, 5, 3 c 1, 5, 2, 4, 3

3 Adult to check

  ISBN: 9781925726176

2. ANGLES CONTINUED

Right Angles

Page 30 – Your Turn

Adult to check

Page 31 – Practice

1 Adult to check
2 Adult to check
3 a = c > e = g <
b < d = f =
4 a

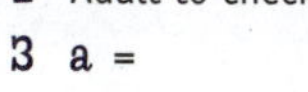
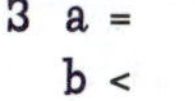

c e b d

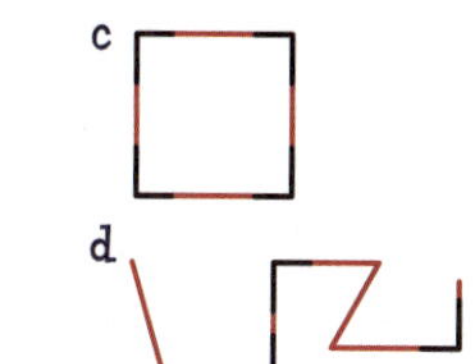
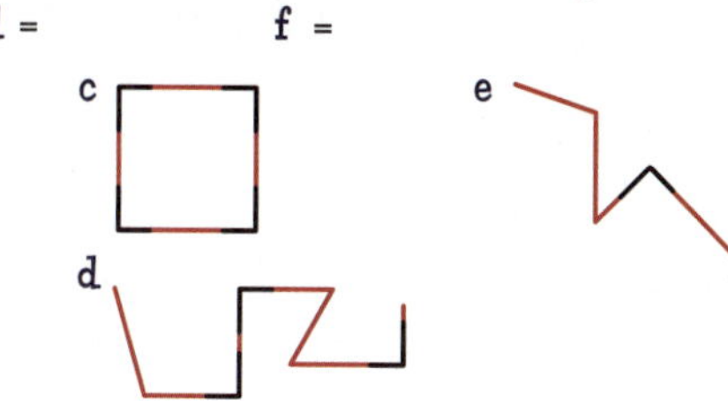

Acute and Obtuse Angles

Page 32 – Your Turn

Adult to check

Page 33 – Practice

1 a ∠DEF or ∠FED, Acute
b ∠PQR or ∠RQP, Obtuse
c ∠CAT or ∠TAC, Obtuse
d ∠PMN or ∠NMP, Acute
e ∠IHG or ∠GHI, Obtuse

2 a 4, 3, 2, 1, 5 b 5, 1, 2, 3, 4 c 4, 3, 5, 1, 2

3 a

b

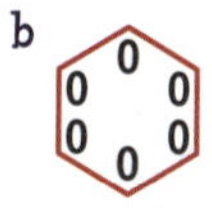

c

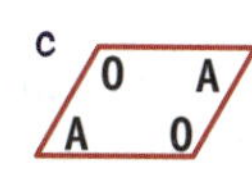

d

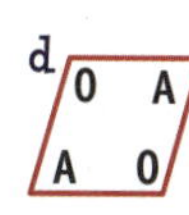

Straight, Reflex and Revolution Angles

Page 34 – Your Turn

Adult to check

Page 35 – Practice

1 Reflex angles: c, d, g
Straight angles: a, f
Revolutions: b, e
2 Adult to check
3 a A b E c B d D e C
4 Adult to check
5 Adult to check

Using a Protractor

Page 36 – Your Turn

a Inside
b Inside
c Outside

Page 37 – Practice

1 a 75° b 170° c 15° d 115° e 35°
2 a 25° c 160° e 30° g 155°
b 45° d 117° f 142°
3 Adult to check

Angle Sum of a Triangle

Page 39 – Your Turn

a 70°
b 50°
c 50°

Page 40 – Practice

1 a 40°, isosceles c 60°, equilateral e 60°, right angled
b 50°, right angled d 115°, scalene
2 a 200 ✗ b 190 ✗ c 180 d 180 e 180

Angle Sum of Quadrilaterals

Page 41 – Your Turn

1 a 135° b 85° c 90° d 80° e 75°

Page 42 – Practice

1 a 55° d 90° g 100° j 30°
b 61° e 242° h 120° k 112°
c 34° f 42° i 78°

Angles Review Page 43

1 a turn, rays, arms
b right, ninety
2 Adult to check
3 Adult to check
4 a c e

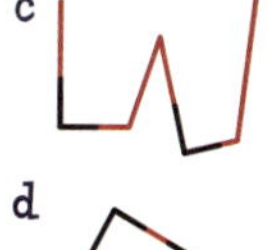

b d f

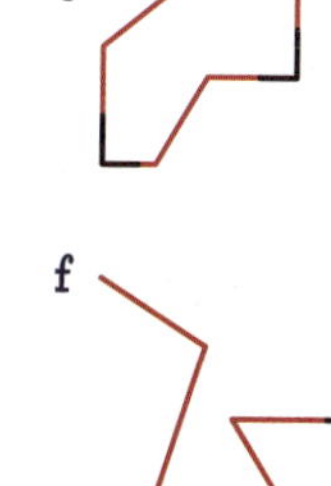

5 Adult to check
6 a ∠ABC or ∠CBA, Acute
b ∠CDE or ∠EDC, Obtuse
c ∠PQR or ∠RQP, Obtuse
d ∠FGH or ∠HGF, Reflex
e ∠STU or ∠UTS, Obtuse
f ∠VWX or ∠XWV, Straight
7 Acute: a, f. Obtuse: c, h. Reflex: e. Straight: d. Revolution: b, g
8 a A b C c D d B e E
9 Adult to check
10 Adult to check
11 Adult to check
12 a 50° b 165° c 140° d 90°
13 a 135° b 170° c 90° d 115° e 45°
15 Adult to check
16 a 60°, equilateral c 90°, right angle e 30°, isosceles
b 50°, scalene d 60°, scalene f 20°, scalene
17 a 131° c 90° e 105°
b 141° d 67° f 233°

3. 2D SHAPES

2D Shapes

Page 48 – Your Turn

a square (regular quadrilateral)
b irregular pentagon
c irregular triangle
d rectangle (irregular quadrilateral)
e regular triangle
f irregular hexagon

Page 49 – Practice

1 a regular pentagon
b rectangle (irregular quadrilateral)
c irregular hexagon
d irregular octagon
e regular hexagon

2

	Name	Letters that help identify shape	No. angles	No. sides	No. corners
a	triangle	tri	3	3	3
b	pentagon	pent	5	5	5
c	hexagon	hex	6	6	6
d	heptagon	sept	7	7	7
e	octagon	oct	8	8	8
f	nonagon	non	9	9	9
g	decagon	deca	10	10	10

Types of Lines

Page 50 – Your Turn

1 Vertical lines: b, e. Horizontal lines: f, i. Parallel lines: a, c, h. Perpendicular lines: d, g.

Page 51 – Practice

1 Adult to check
2 Adult to check
3 Adult to check

Circles

Page 52 – Your Turn

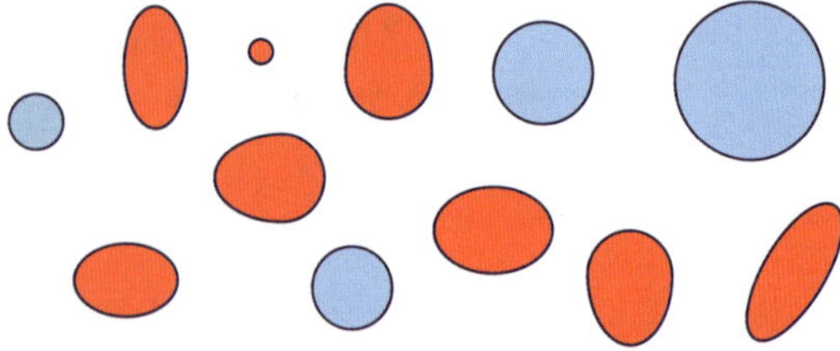

Page 53 – Practice

1 Adult to check
2 a Diameter
b Radius
c Chord
d Circumference
e Sector
f Arc
3 Sample answers:
a G
b CF
c E

Triangles

Page 54 – Your Turn

Equilateral triangles: c, d, g.
Isosceles triangles: f.
Scalene triangles: a.
Right-angled triangles: b, e.

Page 55 – Practice

1 a equilateral
b scalene
c right angled
d scalene
e right angled
f equilateral
g isosceles
2 Adult to check
3 Adult to check

Quadrilaterals

Page 56 – Your Turn

a trapezium
b kite
c square
d parallelogram

Page 57 – Practice

1

2 Adult to check
3 Adult to check

Polygons

Page 58 – Your Turn

a 7 (irregular heptagon)
b 10 (irregular decagon)
c 9 (regular nonagon)
d 12 (regular dodecagon)

Page 59 – Practice

1 Adult to check
2 a regular pentagon
b irregular hexagon
c regular hexagon
d regular triangle
e irregular hexagon
f regular pentagon
g irregular pentagon
3 a regular heptagon
b irregular octagon
c irregular heptagon
d regular octagon
e regular heptagon
f irregular octagon
4 a regular decagon
b regular dodecagon
c irregular decagon
d irregular nonagon
e regular nonagon
f regular decagon
g irregular dodecagon

Transformations

Page 60 – Your Turn

a reflection
b rotation

Page 61 – Practice

1 a rotation
b reflection
c translation

2 a

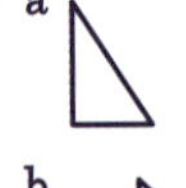

b

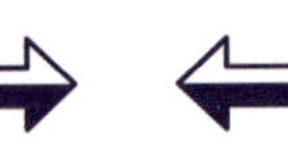

c

3 a

b

c

4 a b c

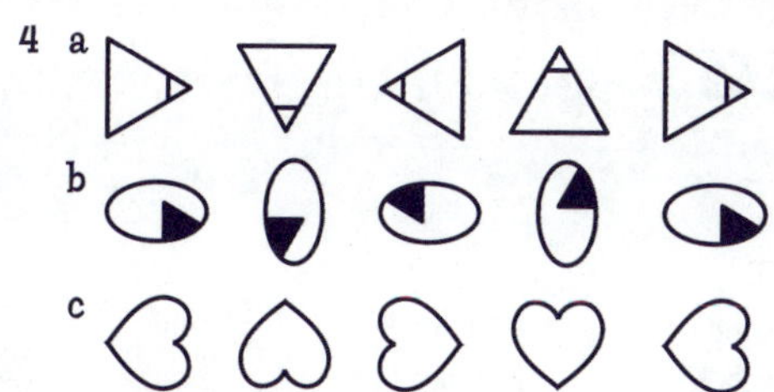

Tessellations

Page 62 – Your Turn

a ✔ b ✔ c ✘ d ✘

Page 63 – Practice

1 a 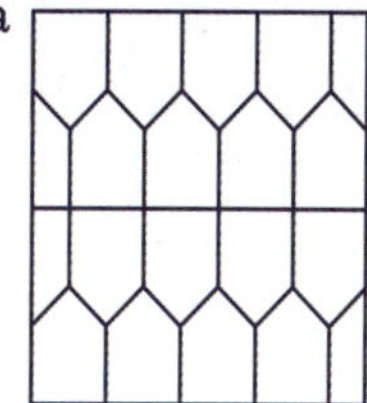 b 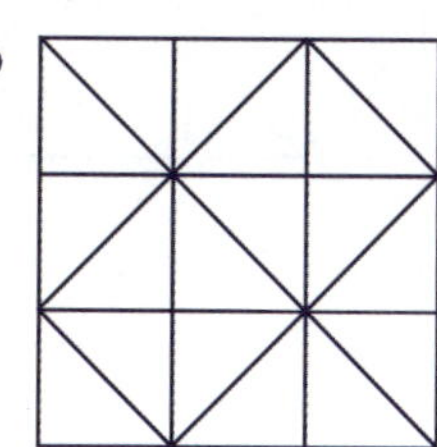

2 No because the shapes are overlapping.

3 a 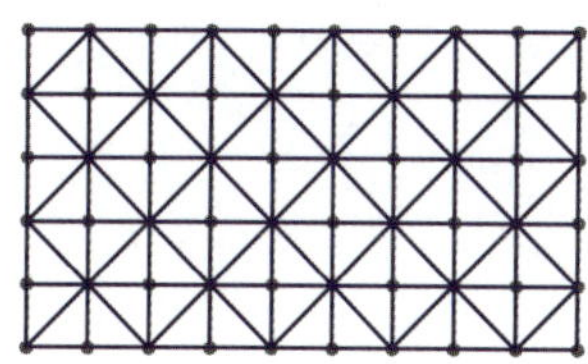 b 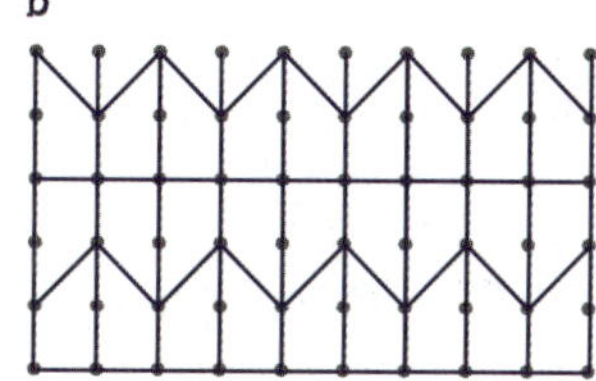

4 Adult to check

Symmetry

Page 64 – Your Turn

a 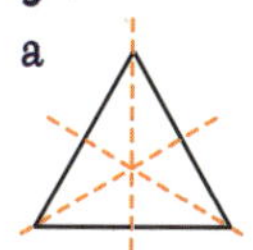 b 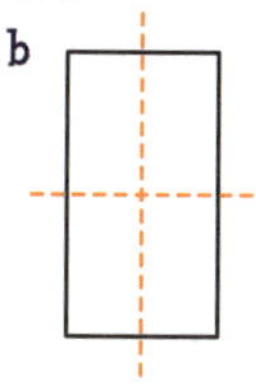 c 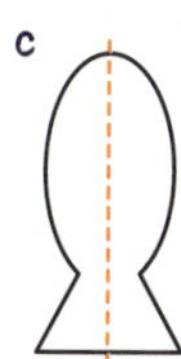

Page 65 – Practice

1 Adult to check

2 Adult to check

3

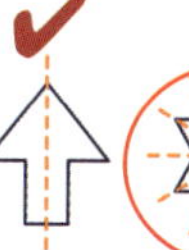

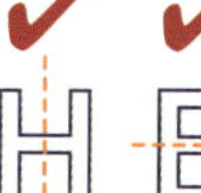

4 a b

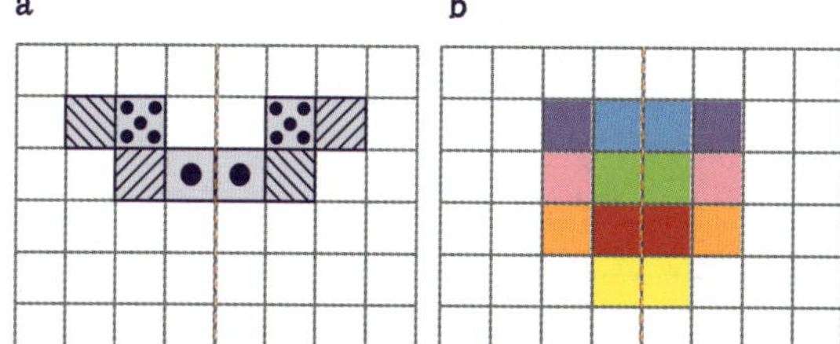

2D Shapes Review Page 66

1 a irregular octagon
b regular hexagon
c regular dodecagon
d irregular dodecagon
e regular decagon
f regular triangle
g rectangle (irregular quadrilateral)
h regular heptagon
i regular nonagon
j irregular octagon
k regular pentagon
l square (regular quadrilateral)
m irregular pentagon
n irregular heptagon
o regular octagon

2 a horizontal
b parallel
c horizontal
d perpendicular
e vertical
f parallel
g vertical
h perpendicular
i parallel
j vertical

3 Adult to check

4 a regular b irregular c irregular d regular

5 semicircle, circle

6 Adult to check

7 Adult to check

8 a right angled b isosceles c equilateral d scalene

9 Equilateral triangle: All sides the same length, All angles the same size
Isosceles triangle: Two sides the same length, Two angles the same size
Scalene triangle: All sides different lengths, All angles different sizes
Right-angled triangle: One right angle

10 a parallelogram
b rhombus
c square
d kite
e trapezium
f rectangle

11 a parallelogram, rhombus, square, trapezium, rectangle
b square, trapezium, rectangle

12 a 8
b 9
c 7
d 12
e 10
f 4
g 5
h 6
i 7
j 3

13 a slide b turn c flip

14 a

b

c

15 a

b

c

16 a

b

c

3. 2D SHAPES CONTINUED

17 Adult to check

18 For example: square, rectangle, equilateral triangle, right-angled triangle, hexagon

19 For example: circle, oval, pentagon, octagon

20 Adult to check

21 half, reflection (or mirror image)

22 Adult to check

23 a 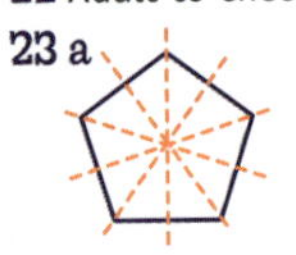 b 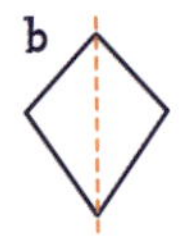 c 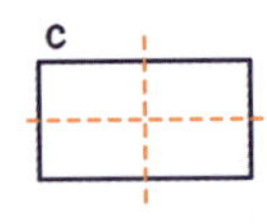 d 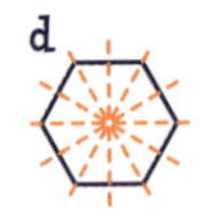 e 

24 a 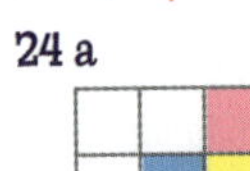 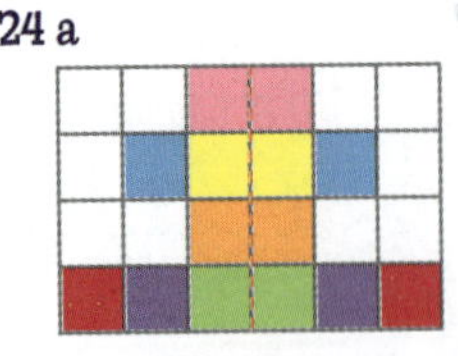 b 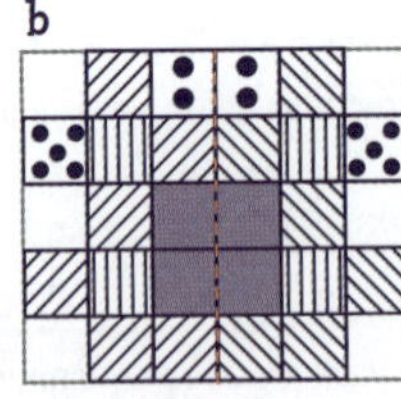 c 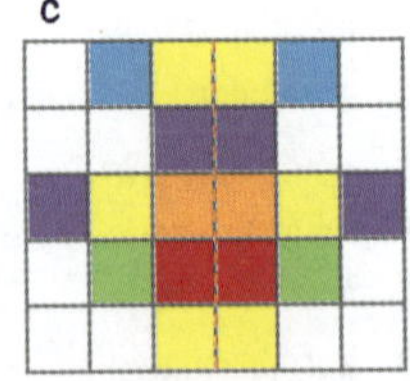

4. 3D OBJECTS

Prisms

Page 72 – Your Turn

Adult to check

Page 73 – Practice

1 a 8, 12, 6 b 10, 15, 7 c 12, 18, 8 d 16, 24, 10 e 5, 6, 5

2 a hexagonal prism b triangular prism c rectangular prism

Pyramids

Page 74 – Your Turn

2 a hexagonal pyramid b square pyramid c pentagonal pyramid

Page 75 – Practice

1 a 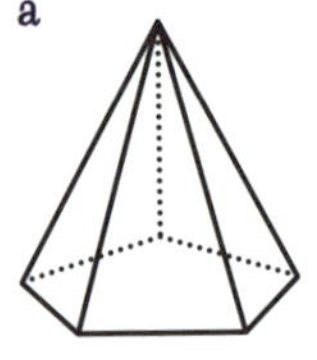 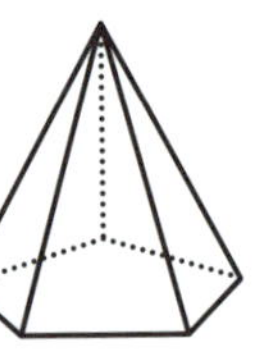 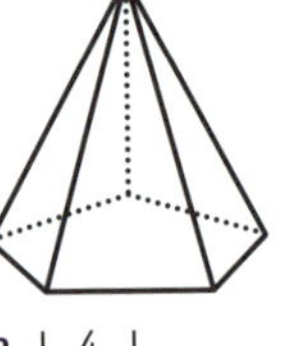  b 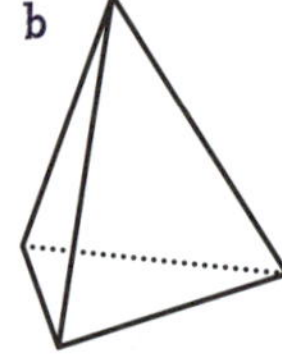 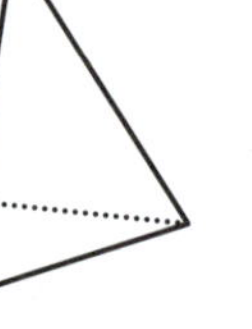 c 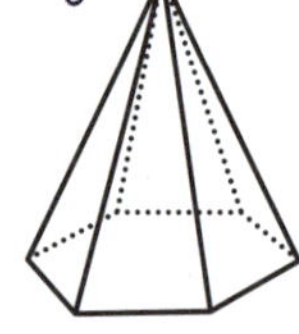

2 a 1, 4, 1 b 1, 3, 1 c 1, 5, 1 d 1, 6, 1 e 1, 8, 1

3 a triangular pyramid b hexagonal pyramid c pentagonal pyramid d octagonal pyramid

4. 3D OBJECTS CONTINUED

Curved Surfaces

Page 76 – Your Turn

1 a sphere b cylinder c cone d sphere e cylinder f cone g cone

Page 77 – Practice

1 a A D E J (balloons, Earth, basketball, marble)
b F G I (field hat, party hat, loud hailer)
c B C H (cans, log, watering can)

2

	Solid	Top View	Front View	Side View	Cross-section
a					
b					
c					

3 a cone
b cylinder

Cross-sections

Page 78 – Your Turn

a 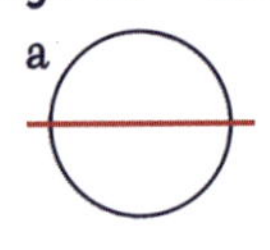 b 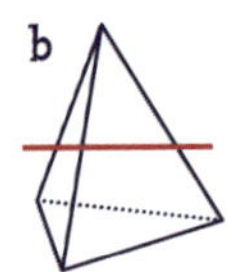

Page 79 – Practice

1 a 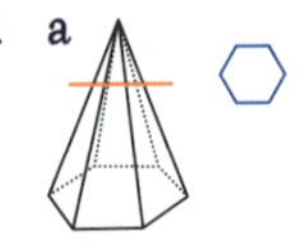

b  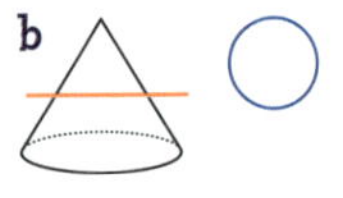 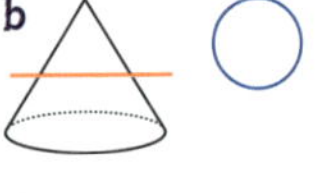

c 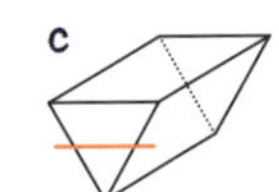 

d 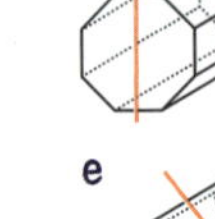 

e 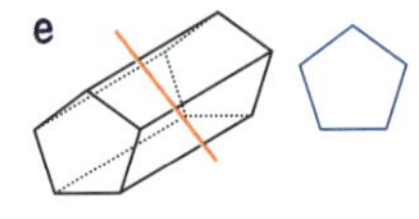

2

	Object	Cross-section: red line	Cross-section: blue line
a			
b			
c			
d			

  ISBN: 9781925726176

Nets of Prisms and Pyramids

Page 80 – Your Turn

a

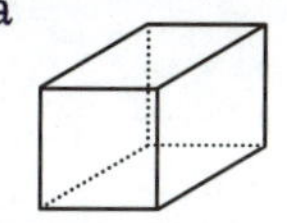

b

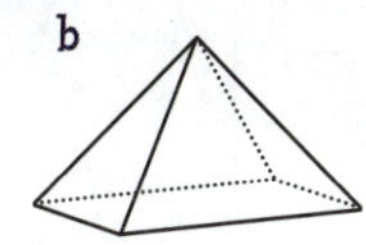

Page 81 – Practice

1 a pentagonal pyramid
b cube
c cone
d cylinder
e rectangular prism

2 a

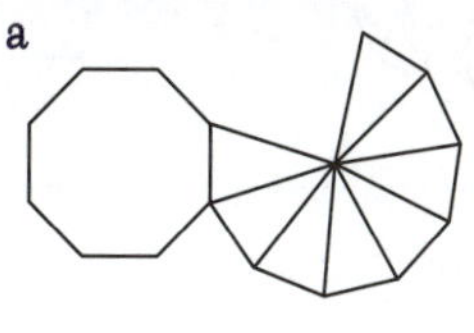

b

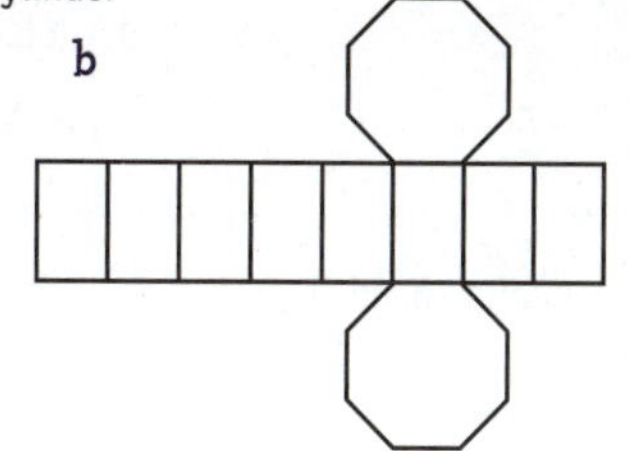

3 a pentagonal pyramid
b cube
c triangular prism
d octagonal pyramid

Views

Page 82 – Your Turn

a

b

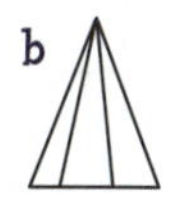

Page 83 – Practice

1 a b

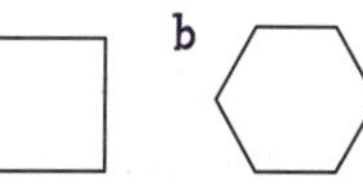

c

d

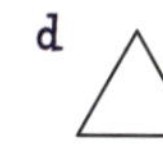

e

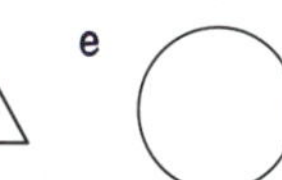

2

	Object	Top view	Front view	Side view
a				
b				
c				
d				
e				
f				
g				

3D Objects Review Page 84

1 a

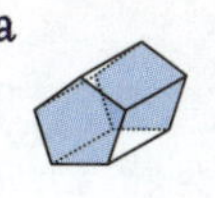

b

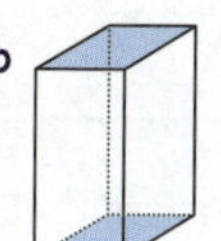

c

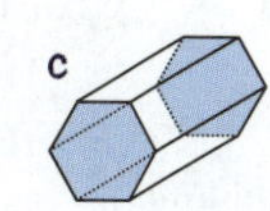

d

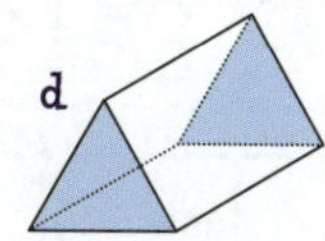

2 Adult to check

3 a

b

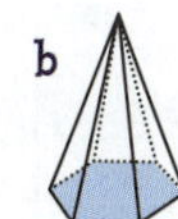

c

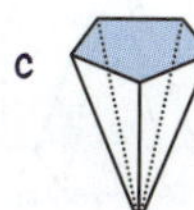

d

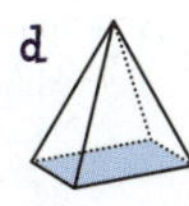

4 Adult to check

5

a cube	b triangular prism 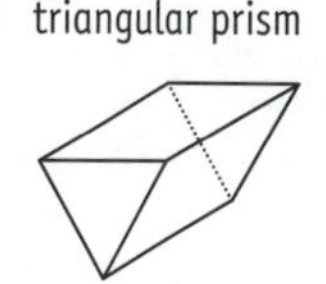	c rectangular prism 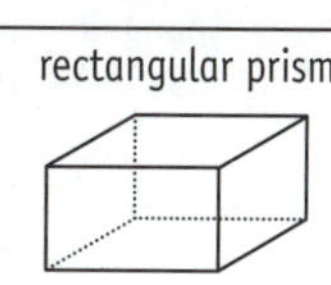
d rectangular pyramid 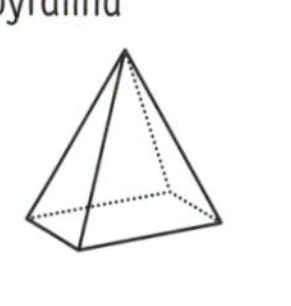	f pentagonal pyramid 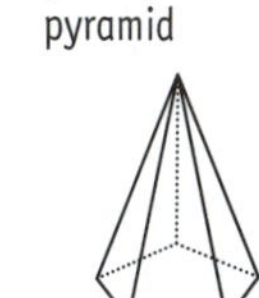	h triangular pyramid 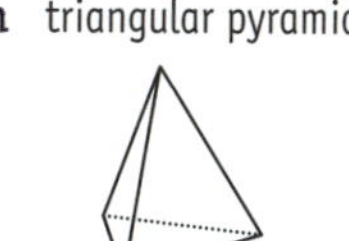
e octagonal prism 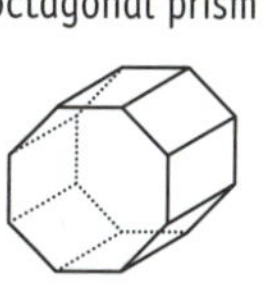	g hexagonal prism 	i octagonal pyramid 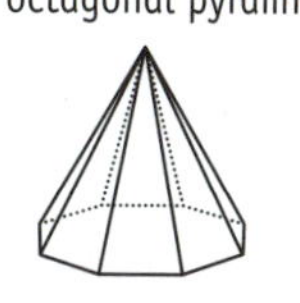

6

	Name	Diagram	No. vertices	No. edges	No. faces
a	triangular prism		6	9	5
b	rectangular prism		8	12	6
c	octagonal prism		26	24	10
d	pentagonal prism		10	15	7
e	hexagonal prism		12	18	8
f	cube		8	12	6

4. 3D OBJECTS CONTINUED

7

	Name	Diagram	No. vertices	No. edges	No. faces
a	pentagonal pyramid		6	10	6
b	triangular pyramid		4	6	4
c	square pyramid		5	8	5
d	octagonal pyramid		9	16	9

8 a sphere
b cone
c cylinder

9

	3D object	Top view	Front view	Side view
a				
b				
c				

10 Adult to check

11 a cone
b cylinder

12

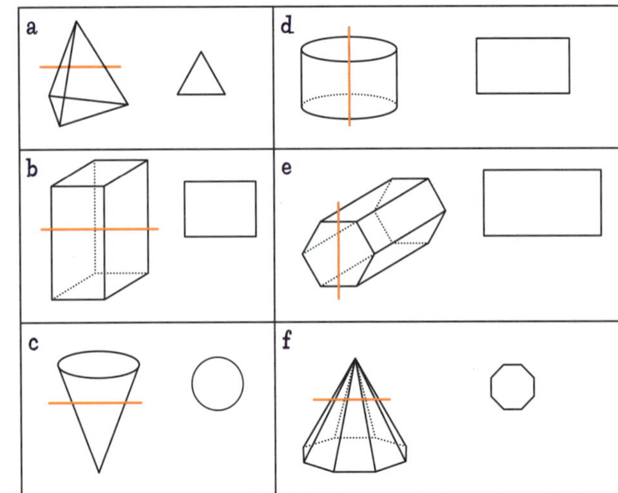

13

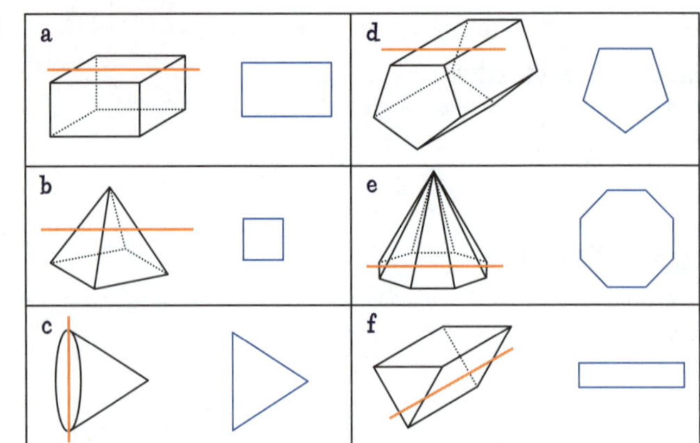

14 a hexagonal pyramid
b cube
c triangular prism
d pentagonal pyramid
e rectangular prism
f cylinder
g cone

15

	Object	Top view	Front view	Side view
a				
b				
c				
d				
e				
f				
g				
h				

16 a

b 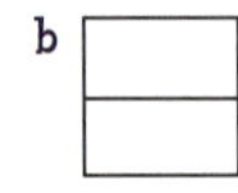

c 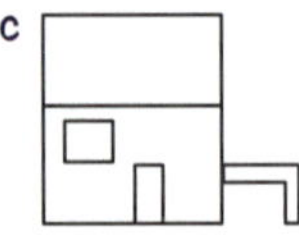

  ISBN: 9781925726176

5. AREA

The Square Centimetre

Page 90 – Your Turn

a 8 cm^2
b 18 cm^2

Page 91 – Practice

1 a 10 cm^2 c 12 cm^2 e 12 cm^2 g 11 cm^2
b $8\frac{1}{2}$ cm^2 d $9\frac{1}{2}$ cm^2 f 14 cm^2

2 a 7 c $3\frac{1}{2}$ cm^2 e 5
b 4 and 6 d 7 f $4\frac{1}{2}$

3 Adult to check

The Square Metre

Page 92 – Your Turn

a 12 m^2
b 4 m^2

Page 93 – Practice

1 Circle: door, screen, bed, fence
2 a 4 m^2 b 18 m^2 c 102 m^2 d 57 m^2 e 89 m^2
3 a 12 m^2 b 24 m^2 c 12 m^2 d 9 m^2 e 2 m^2
4 No
5 a cm^2 b m^2 c cm^2 d cm^2 e m^2

Square Kilometres

Page 94 – Your Turn

Adult to check

Page 95 – Practice

1 a 12 c 8 e 7 g 6
b $9\frac{1}{2}$ d 10 f 8

2 a ACT
b Western Australia
c Western Australia
d ACT, Tas., Vic., NSW, SA, NT, Qld, WA

Hectares

Page 96 – Your Turn

Adult to check

Page 97 – Practice

1 a 17 ha c 304 ha e 1 ha g 757 ha
b 28 ha d 100 ha f 349 ha

2 a 16 hectares c 41 hectares e 324 hectares g 890 hectares
b 22 hectares d 109 hectares f 603 hectares

3 a Great Barrier Reef Marine Park, Kakadu National Park, Uluru–Kata Tjuta National Park, Booderee National Park, Murramarang National Park
b 4112 ha
c 32.46 million ha

Area – Using Multiplication

Page 98 – Your Turn

a Area = length × width
= 2 m × 4 m
= 8 m^2

Page 99 – Practice

1 a Area = length × width
= 4 m × 5 m
= 20 m^2
b Area = length × width
= 1 m × 6 m
= 6 m^2
c Area = length × width
= 3 cm × 4 cm
= 12 cm^2

2 a Area = l × w
= 7 cm × 7 cm
= 49 cm^2
b Area = l × w
= 6 cm × 3 cm
= 18 cm^2
c Area = l × w
= 2 m × 5 m
= 10 m^2
d Area = l × w
= 6 m × 6 m
= 36 m^2
e Area = l × w
= 1 cm × 1 cm
= 1 cm^2

Perimeter

Page 100 – Your Turn

a 14 cm
b P = 4 m + 1 m + 1 m + 1 m + 1 m + 1 m + 1 m + 1 m + 1 m + 1 m + 4 m
= 16 m

Page 101 – Practice

1 a P = 2 cm + 4 cm + 2 cm + 4 cm
= 12 cm
b P = 1 cm + 2 cm + 1 cm + 4 cm
= 8 cm
c P = 2 cm + 2 cm + 2 cm + 2 cm + 2 cm
= 10 cm
d P = 1.5 cm + 3 cm + 1.5 cm + 3 cm
= 9 cm
e P = 1.5 m + 1.5 m + 1.5 m + 1.5 m
= 6 m
f P = 2.5 m + 2.5 m + 2.5 m + 2.5 m + 2.5 m + 2.5 m
= 15 m
g P = 5 cm + 2 cm + 1 cm + 1 cm + 1 cm + 1 cm + 3 cm + 2 cm + 5 cm + 6 cm
= 27 cm
h P = 4 m + 7 m + 4 m + 1 m + 3 m + 1 m + 3 m + 3 m + 3 m + 1 m + 3 m + 1 m
= 34 m

Area Review Page 102

1 a 8 b 7 c 8 d $7\frac{1}{2}$
2 Adult to check
3 Tick: a c d g h
4 a 20 c 12 e 20 g 11
b 10 d 11 f 18 h 11
5 a D, G, H c A, E e 10 m^2
b A, E, F d F
6 a 24 m^2 c 5 m^2 e 3 m^2 g 6 m^2
b 9 m^2 d 9 m^2 f 9 m^2 h 6 m^2
7 68 m^2
8 Adult to check
9 1, km, km
10 a 800 894 km^2 b 574 099 km^2 c 62 161 km^2 d 354 753 km^2
11 WA, Qld, NT, SA, NSW, Vic., Tas., ACT
12 100, 100, 10 000, ha
13 a 9 ha c 31 ha e 104 ha g 703 ha
b 12 ha d 97 ha f 212 ha h 390 ha

ANSWERS

5. AREA CONTINUED

14 a 7 hectares   c 28 hectares   e 630 hectares   g 100 hectares
b 136 hectares   d 211 hectares   f 909 hectares   h 687 hectares

15 a Riverside, Splashville, Rainbow Valley, Sunshine, Beachview
b Riverside
c 112 ha
d 293 ha
e 527 ha

16 a Area = length × width
= 1 m × 1 m
= 1 m^2

b Area = length × width
= 1 cm × 7 cm
= 7 cm^2

c Area = length × width
= 3 m × 3 m
= 9 m^2

d Area = length × width
= 3 cm × 5 cm
= 15 cm^2

e Area = length × width
= 8 m × 2 m
= 16 m^2

f Area = length × width
= 5 cm × 5 cm
= 25 cm^2

g Area = length × width
= 4 cm × 3 cm
= 12 cm^2

i Area = length × width
= 6 cm × 6 cm
= 36 cm^2

17 a P = 3 m + 5 m + 3 m + 5 m
= 16 m

b P = 4 cm + 4 cm + 4 cm + 4 cm
= 16 cm

c P = 4 m + 4 m + 1 m + 3 m + 1 m + 3 m + 1 m + 3 m + 1 m + 7 m
= 38 m

d P = 2 m + 2 m + 2 m + 2 m
= 8 m

e P = 6 cm + 3 cm + 6 cm + 3 cm
= 18 cm

f P = 8 m + 6 m + 3 m + 3 m + 5 m + 1 m + 5 m + 1 m + 5 m + 1 m
= 38 m

g P = 15 cm + 2 cm + 3 cm + 5 cm + 3 cm + 2 cm + 1 cm + 2 cm + 3 cm + 7 cm
= 43 cm

6. VOLUME & CAPACITY

Volume

Page 108 – Your Turn

a 5
b 11

Page 109 – Practice

1 a 7 cm^3 b 7 cm^3 c 32 cm^3 d 32 cm^3
2 a 2, 3   b 4, 5
3 a 0   b 22
4 10 cm^3 + 7 cm^3 + 7 cm^3 + 32 cm^3 + 32 cm^3 = 88 cm^3

Capacity

Page 110 – Your Turn

Adult to check

Page 111 – Practice

1 a 1, 2, 4, 3, 5   b 5, 3, 1, 2, 4   c 4, 5, 3, 2, 1
2 a 15 L   b 13 L   c 14 L   d 7 L   e 14 L

6. VOLUME & CAPACITY CONTINUED

Litres

Page 112 – Your Turn

a 7 L   c 22 L   e 105 L   g 630 L
b 9 L   d 85 L   f 1243 L

Page 113 – Practice

1 a 15 L   b 24 L   c 98 L   d 103 L   e 200 L
2 Adult to check
3 Adult to check
4 a 10
b 27
5 a

b

c

6 a 10 L   c 4 L   e 3 L   g 6 L
b 1 L   d 3 L   f 1 L

7 a

b

c

8 a 4   b 2   c 3   d 2
9 a 2 L   b 5 L   c 4 L   d 1 L
10 a 20 containers   c 48 oranges
b 30 buckets of water   d 10 L of juice

Millilitres

Page 116 – Your Turn

a c e b d

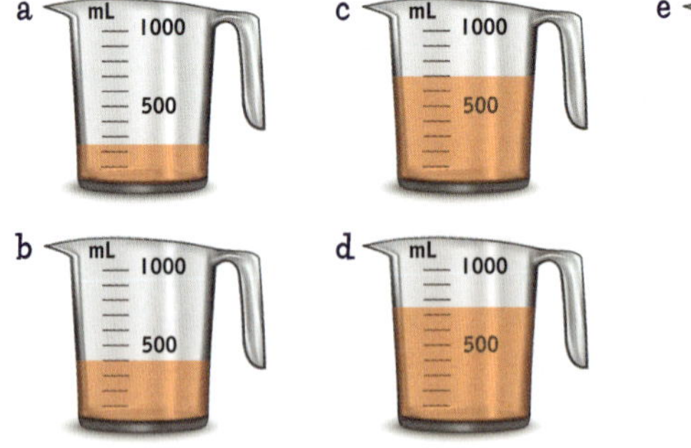

Page 117 – Practice

1 a L   c mL   e L   g mL
b mL   d L   f mL

2 a 1 L   c $9\frac{3}{4}$ L   e $8\frac{1}{2}$ L   g 10 L
b 2 L   d $1\frac{1}{2}$ L   f 6 L   h $1\frac{1}{4}$ L

3 a 4000 mL   c 3500 mL   e 1750 mL   g 6750 mL
b 3250 mL   d 8250 mL   f 2500 mL   h 8000 mL

4 a 3 L 800 mL   c 7 L 500 mL   e 3 L 800 mL   g 2 L 750 mL
b 1 L 425 mL   d 9 L 250 mL   f 1 L 100 mL   h 4 L 900 mL

5 a 9 L 365 mL   b 7 L 990 mL   c 15 L 100 mL
6 1, 6, 5, 3, 4, 2

Litres or Millilitres

Page 119 – Your Turn

1 a L b mL c mL d mL e L

Page 120 – Practice

1 a 575 mL b 450 mL c 1000 mL d 900 mL e 150 mL

2 a 8850 mL b 2025 mL c 1903 mL d 5600 mL e 7520 mL

3 a 7.265 L b 3.055 L c 9.0 L d 7.002 L e 10.420 L

4 a 6300 mL b 14 007 mL c 1250 mL d 4000 mL e 18 606 mL

5 a 5, 4, 2, 1, 3 b 5, 4, 3, 2, 1

Volume & Capacity Review Page 121

1 space

2 a 7 b 18 c 15 d 50

3 a a b d

4 a 3 cm^3 b 43 cm^3

5 liquid, hold, 1000

6 a left b left c right d right

7 a 1, 3, 5, 4, 2 b 2, 1, 3, 5, 4 c 5, 3, 2, 1, 4

8 a 8 b 6 c 21

9 a 8 L b 285 L c 20 L d 49 L e 1057 L f 10 L

10 Adult to check

11 Adult to check

12 a

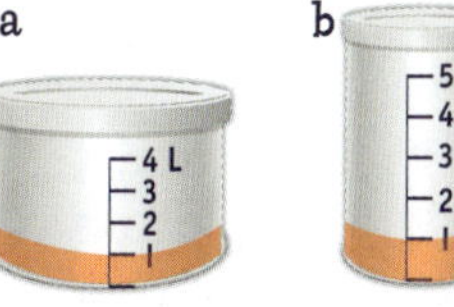

b

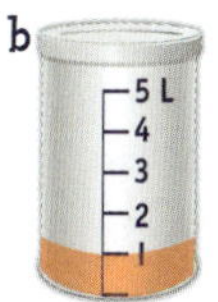

c

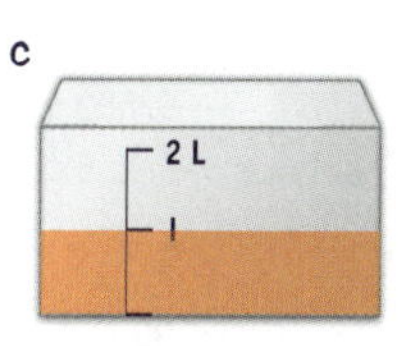

d

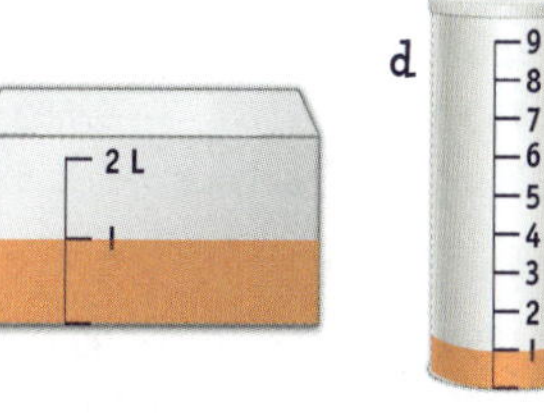

13 a 1 L b 2 L c 4 L d 5 L

14 a 13 L b 11 times c 11 buckets of water

15 a 1 b $\frac{3}{4}$ c 500 d 250

16 a b c d

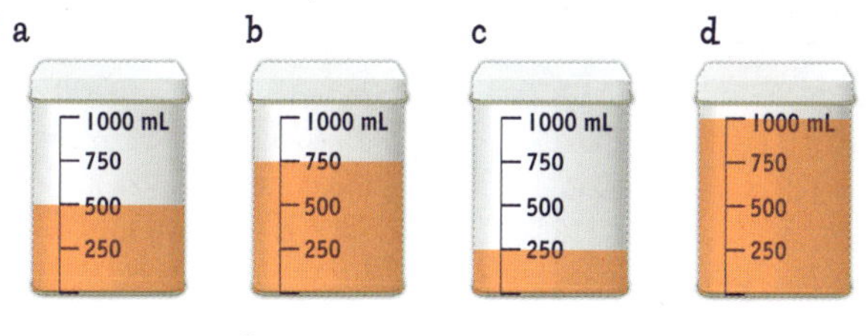

17 a b c d

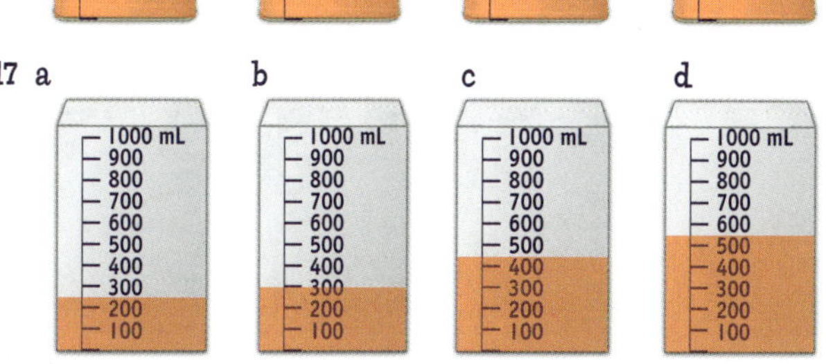

18 a mL b L c mL d mL e mL f L

19 a $3\frac{1}{2}$ L b 2 L c $1\frac{1}{2}$ L d $8\frac{3}{4}$ L e $9\frac{1}{4}$ L f $4\frac{3}{4}$ L

20 a 8000 mL b 3250 mL c 6500 mL d 2750 mL e 9000 mL f 10 250 mL

21 a 3 L 200 mL b 8 L 400 mL c 9 L 830 mL d 6 L 840 mL e 6 L 750 mL f 1 L 875 mL

22 a 6 L 780 mL b 14 L 730 mL c 14 L 700 mL d 10 L 70 mL e 13 L 158 mL f 9 L 609 mL

23 a 5, 4, 2, 3, 1 b 1, 4, 3, 5, 2 c 4, 2, 3, 5, 1

24 a 250 mL b 580 mL c 120 mL d 50 mL e 2000 mL f 98 mL

25 a 8535 mL b 6530 mL c 1680 mL d 5055 mL e 22 100 mL f 11 025 mL

26 a 7.680 L b 8.003 L c 13.303 L d 16.001 L e 3.0 L f 19.258 L

27 a 5204 mL b 9237 mL c 3750 mL d 12 250 mL e 6000 mL f 8520 mL

28 a 1, 2, 5, 4, 3 b 2, 1, 5, 3, 4 c 1, 2, 3, 5, 4 d 2, 4, 5, 3, 1

7. MASS

Kilograms

Page 128 – Your Turn

1 a a watermelon b a small child c 1-litre carton of milk d a full backpack e a net of oranges

Page 129 – Practice

1 a ✗ b ✗ c ✗ d ✓ e ✓ f ✗ g ✗

2 a Potatoes b Cherries c 21 kg

3 a 5 kg b 20 kg c 4 kg d 60 kg e 8 kg

4 a 3 b 10 c 2 d 5

5 Adult to check

Grams

Page 130 – Your Turn

Circle: b c d

Page 131 – Practice

1 a 4 g b 64 g c 105 g d 770 g

3 1 Chick peas 945 g
2 Soup 800 g
3 Carrots 345 g
4 Beans 275 g
5 Peas 250 g
6 Corn 125 g
7 Jam 120 g
8 Salmon 100g

4 a Carrots, Soup, Chickpeas b Salmon, Peas, Corn, Jam, Beans

5 a 750 g b 880 g c 655 g d 200 g e 875 g f 55 g g 725 g

6 a 4 b 8

Parts of a Kilogram

Page 132 – Your Turn

a 1500 g b 7000 g c 4250 g d 250 g e 5750 g f 2000 g g 3500 g h 9250 g

Page 133 – Practice

1 a 10 b 5 c 4 d 2 e 1

7. MASS CONTINUED

2 a $\frac{1}{4}$ kg = 250 g
$1\frac{1}{2}$ kg = 1500 g
3 kg = 3000 g
5 kg = 5000 g
$\frac{1}{2}$ kg = 500 g

d $2\frac{3}{4}$ kg = 2750 g
10 kg = 10 000 g
1 kg = 1000 g
$\frac{3}{4}$ kg = 750 g
$7\frac{1}{2}$ kg = 7500 g

b 0.25 kg = 250 g
2.75 kg = 2750 g
17.31 kg = 17 310 g
7.5 kg = 7500 g
0.5 kg = 500 g

e 0.75 kg = 750 g
8.5 kg = 8500 g
12.82 kg = 12 820 g
2.28 kg = 2280 g
4.71 kg = 4710 g

c 2 kg 300 g = 2300 g
4 kg 30 g = 4030 g
4 kg 300 g = 4300 g
2 kg 30 g = 2030 g
2 kg 3 g = 2003 g

f 8 kg 400 g = 8400 g
8 kg 40 g = 8040 g
8 kg 4 g = 8004 g
3 kg 800 g = 3800 g
3 kg 8 g = 3008 g

Comparing and Ordering Mass

Page 134 – Your Turn

a 3, 5, 2, 1, 4　b 3, 4, 1, 2, 5

Page 135 – Practice

1 a 4, 2, 1, 3, 5　b 2, 5, 3, 4, 1　c 1, 4, 5, 3, 2
2 a 9 kg　b 1 kg　c 3 kg
3 a 3　b 3
4 a 8 kg　b 10 kg　c 18 kg
5 2 paint cans

Reading Scales

Page 136 – Your Turn

a 3 kg　b 850 g　c 5 kg

Page 137 – Practice

1 a 3 kg　b 5 kg　c $4\frac{1}{2}$ kg　d 250 g　e 850 g　f 550 g　g 625 g
2 5, 1, 7, 8, 6, 2, 4, 3
3 a 2, 3, 4　b 1, 5, 6, 7, 8
4 a

b

c

d

The Tonne

Page 138 – Your Turn

a 250　b 750　c 3000

Page 139 – Practice

1 a 5　b 20　c 10　d 2
2 a 9730 kg
b 4.5 tonnes
c 4000 kg
d 8100 kg
e 3.12 tonnes
f 7 tonnes
g 6.125 tonnes
h 7845 kilograms
i 12 900 kilograms
j 2.4 tonnes
k 16.04 tonnes
l 16 250 kg
m 17 095 kg
n 21.09 tonnes
o 1003 kilograms
p 20 750 kg
q 14.001 tonnes
r 33 589 kg
s 17 010 kg

Gross Mass and Net Mass

Page 140 – Your Turn

a 515 g

Page 141 – Practice

1 a 880 g, 910 g, 880 g　b 560 g, 575 g, 560 g
2 a 625 g　b 605 g
3 a 425 g　b 405 g
4 35 g + 840 g = 875 g

Mass Review Page 142

1 1000, 1000
2 a 3 kg　b 18 kg　c 46 kg　d 89 kg　e 207 kg　f 16 kg　g 17 kg　h 109 kg　i 72 kg　j 80 kg
3 a bag of flour　b net of apples　c a baby　d a bicycle　e a woman　f a carton of juice
4 a ✗　b ✗　c ✔　d ✗　e ✗　f ✗　g ✗　h ✗　i ✔　j ✔　k ✗　l ✗
5 a 5, 2, 4, 3, 1　b 1, 4, 2, 5, 3　c 1, 2, 3, 5, 4
6 grams, 1000
7 a 43 g　b 2 g　c 903 g　d 15 g　e 71 g　f 59 g
8 a ✗　b ✔　c ✔　d ✗　e ✗　f ✔　g ✗　h ✗　i ✗　j ✔　k ✗　l ✔
9 A 2, B 3, C 6, D 1, E 4, F 7, G 8, H 5
10 a 750 g　b 700 g　c 350 g　d 850 g　e 600 g　f 255 g　g 50 g　h 500 g
11 a $3\frac{1}{2}$　b $1\frac{1}{4}$　c 4　d $7\frac{3}{4}$　e $5\frac{1}{2}$　f $6\frac{3}{4}$
12 a 1750 g　b 6000 g　c 2250 g　d 10 500 g　e 3750 g　f 7500 g　g 3500 g　h 8000 g　i 6750 g　j 1250 g
13 a 5　b 20　c 2　d 4　e 10
14 a 750 g
$\frac{1}{4}$ kg
1500 g
$2\frac{1}{2}$ kg
6000 g
$3\frac{3}{4}$ kg
10 kg

b 500 g
3.75 kg
15 420 g
0.2 kg
1750 g
400 g
14.125 kg

c 3200 g
1 kg 8 g
24 kg 530 g
5755 g
8 kg 40 g
7010 g
15 005 g

d 200 g
17.25 kg
6300 g
10 890 g
2.672 kg
2400 g
3070 g

15 a 3, 1, 5, 2, 4　b 1, 4, 3, 2, 5
16 a

b

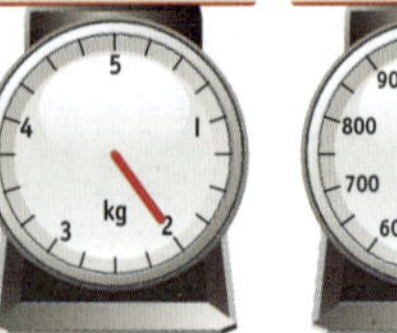

c

d

17 a $3\frac{1}{2}$ kg　b 650 g　c 150 g　d 500 g
18 1000
19 Adult to check
20 a 250　b 500　c 750　d 1000
21 a 5　b 2　c 10　d 4　e 20　f 100
22 a 2.4 t　b 8462 kg　c 10 500 kg　d 6.475 t　e 9603 kg　f 14.02 t　g 5004 kg　h 17 550 kg
23 625, 645, 625
24 655 g

CATCH UP MATHS YEAR 5 BOOK B © PASCAL PRESS ISBN: 9781925726176

8. TIME

O'clock and Half Past

Page 148 – Your Turn

a 

b 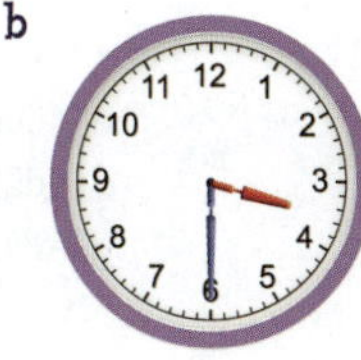

c 

Page 149 – Practice

a 

half past eight

8:30

eight thirty

d 

half past three

3:30

three thirty

g   

nine o'clock

9:00

nine o'clock

b  

eleven o'clock

eleven o'clock

e 

half past four

4:30

four thirty

c 

ten o'clock

10:00

ten o'clock

f    

half past twelve

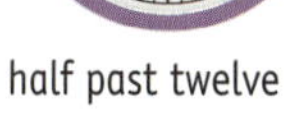

12:30

twelve thirty

2 a   

two o'clock

2:00

two o'clock

b   

half past eight

eight thirty

c   

three o'clock

three o'clock

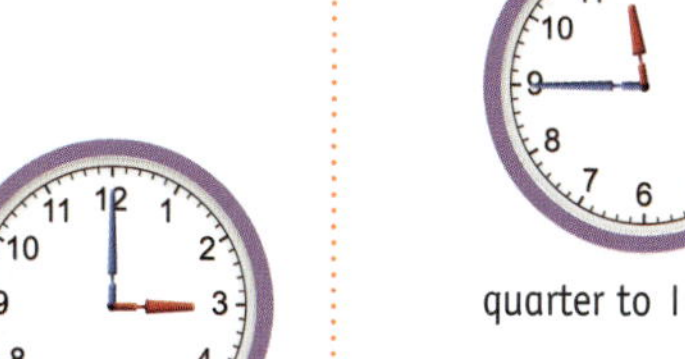

Quarter Past and Quarter To

Page 150 – Your Turn

a

b 11:15

c 

Page 151 – Practice

1 a 10:45

ten forty-five

quarter to 11

b 2:45

two forty-five

quarter to 3

c 

twelve fifteen

quarter past twelve

2 a 

quarter past seven

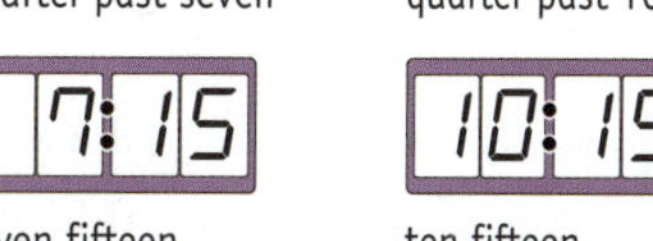

seven fifteen

d 

quarter past 10

ten fifteen

g 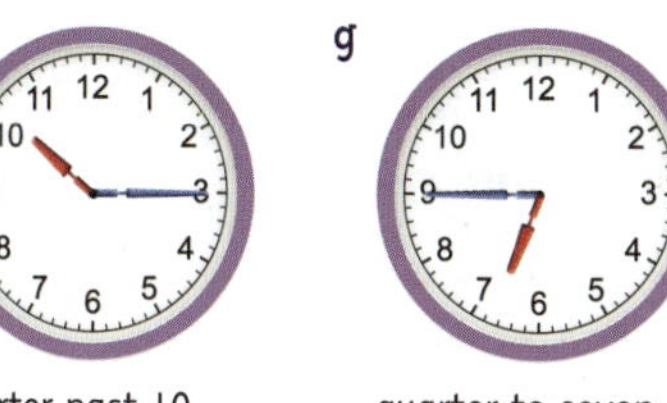

quarter to seven

six forty-five

b

quarter past three

3:15

three fifteen

e    

quarter past four

four fifteen

c

quarter to 12

11:45

eleven forty-five

f  

quarter to five

4:45

four forty-five

8. TIME CONTINUED

Intervals

Page 152 – Your Turn

B 14 to 11, ten forty six
C 11 to 8, seven forty-nine

Page 153 – Practice

1

	Digital time	How we read it	What it means
a	8:23	eight twenty-three	23 minutes past 8
b	11:44	eleven forty-four	16 minutes to 12
c	4:38	four thirty-eight	22 minutes to 5
d	2:18	two eighteen	18 minutes past 2
e	4:56	four fifty-six	4 minutes to 5

2 a

c

b

d

3 a 7:57 b 9:26 c 4:00 d 6:49
4 a 4:16 b 8:27 c 2:06 d 5:44
5 a 7:11 b 10:27 c 4:46 d 1:25

Converting Times

Page 154 – Your Turn

1 a 480 b 540 c 3

Page 155 – Practice

1 a 5 b 4 c 6 d 2 e 9 f 7 g 12 h 10 i 8
2 a 180 b 360 c 120 d 480 e 720 f 960 g 1080 h 840 i 600
3 a 2 b 5 c 9 d 11 e 2 f 70 g 20
4 a 300 b 420 c 240 d 480 e 660 f 1320 g 1800

24-hour Time

Page 156 – Your Turn

1 a before b before c after d after e before

Page 157 – Practice

1 a 07:30 b 08:00 c 06:14 d 23:27 e 01:31 f 02:22 g 00:53 h 19:43 i 07:12 j 16:55 k 17:46
2 a pm b am c pm d pm e am f am g am h pm i am j am k am
3 a 19:56 b 15:37 c 16:35 d 23:25 e 13:12 f 01:49 g 02:14 h 05:23 i 05:05 j 09:43 k 11:28
4 a False b False c False d True e True f False
5 Adult to check

Calendars

Page 158 – Your Turn

1 a Tuesday, 30th July
b Monday, 15th July
c 4th July and 18th July
d Friday, 26th July
e Adult to check
f Adult to check
g Adult to check

Page 160 – Practice

1 a 90 or 91 (leap year)
b 92
c 92
2 January, March, May, July, August, October, December
3 April, June, September, November
4 Adult to check
5 a Wednesday b Friday c Thursday d Wednesday e Saturday
6 a Friday b Saturday c Thursday d Sunday e Monday
7 a 5, 4 b 4, 5 c 5, 5 d 22, 23
8 Adult to check
9 Adult to check

Timetables

Page 162 – Your Turn

1 a 2 b 7:01 c 46 minutes

Page 163 – Practice

1 a Monday, 16th January
b 6
c 5:30 am
d 9
e 30 minutes
f 15, 5:00 am, 6:00 am
g 30 minutes
2 a 7:00 am b 5:30 am c 7:00 am
3 a 2 b 2 c 1 d 1
4 a No. This timetable is for weekends and public holidays.
b 9
c 40 minutes
d 12:36
e 8:30 am
f 6:30
g 7:05
5 a 7 minutes b 14 minutes c 25 minutes d 32 minutes e 25 minutes
6 a Monday, Wednesday, Friday
b Wednesday 2:15
c 20 minutes
d Thursday 1:15
7 a 11:15 b 11:20 c 10:15 d 10:45 e 11:15
8 a 6 b 6 c 4 d 5 e 6

Timelines

Page 167 – Your Turn

Adult to check

Page 168 – Practice

1 a *The Devil's Rain* b 1977 c Kelly Preston d 3 e 1991 f 2020

2 a 1978 b 2007 c 2007 d 1989

3 Adult to check

4 a 1900
b 1912
c *Wonder of the Seas*
d 1970s
e 1972
f Norwegian Capricorn Lines's
g TSS *Mardi Gras*

5 Adult to check

Time Review Page 170

1 a cross out 'small'
b cross out 'big'

2 a

two o'clock

two o'clock

b

half past four

four thirty

c
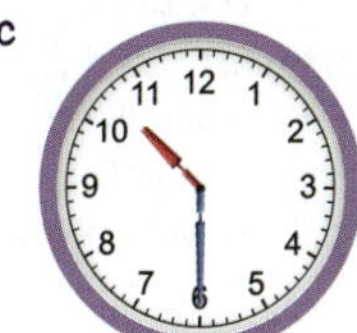
half past ten

ten thirty

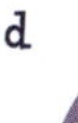
d

half past twelve

twelve thirty

3 a

half past eleven

eleven thirty

b

half past six

six thirty

b

five o'clock

five o'clock

a

seven o'clock

seven o'clock

4 a quarter to 6
b one fifteen
c quarter to 12
d three forty-five
e quarter past 3
f twelve fifteen

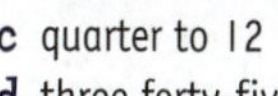

5 a

five fifteen

quarter past five

b
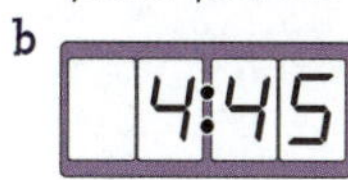
four forty-five

quarter to 5

c
2:15
two fifteen

quarter past two

d

eleven forty-five

quarter to 12

6 a
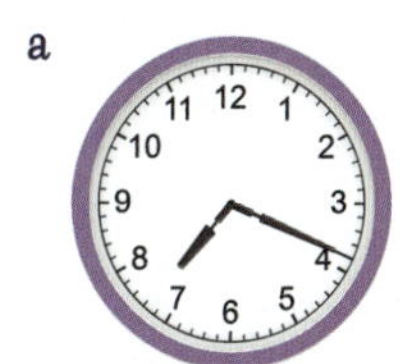
c

e

b

d

7

	Digital time	How we read it	What it means
a	11:29	eleven twenty-nine	29 minutes past 11
b	9:49	nine forty-nine	11 minutes to 10
c	11:34	eleven thirty-four	26 minutes to 12
d	1:17	one seventeen	17 minutes past 1
e	4:11	four eleven	11 minutes past 4

8 a 4 b 12 c 8 d 9 e 110 f 20

9 a 120 b 600 c 300 d 180 e 1800 f 2400

10 a 8 b 6 c 1 d 3 e 120 f 100

11 a 120 b 360 c 720 d 240 e 1920 f 900

12 a 05:09 b 23:53 c 10:44 d 12:18 e 16:37 f 08:35 g 01:52 h 18:23 i 18:46 j 15:46 k 02:19 l 07:58 m 08:05 n 21:21 o 00:17

13 Adult to check

14 Adult to check

8. TIME CONTINUED

15 a Tuesday b Thursday c Saturday d Sunday

16 a 4, 5 b 4, 4 c 5, 4 d 20, 23 e 10, 8

17 Adult to check

18 a 30 b 30 c 31 d 29 e 92 f 92 g 14 h 366 i 7

19 a 7 b 7 c 7 d 7 e 6 f 6

20 3

21 Tuesday 7 pm

22 a Bngid b Ari c Matt

23 a 8:15 am
b 60 minutes
c Circuit, Sunday, Ari

24 a 4 b 3 c 1 d 3 e 1 f 3

25 a 1863, London
b First international game held in Glasgow, Scotland
c 1904, France
d First successful international competition in Belgium at the Olympics
e 1930, Montevideo, Uruguay
f 1958
g 24
h 32
i 1992
j 208
k Adult to check

9. POSITION

Directions

Page 177 – Your Turn

1 a 4 left, 4 down, 2 left, 4 up, 4 left, 2 down
b 2 down, 1 left, 1 up, 2 left, 3 down, 1 left, 3 up, 2 left, 1 up, 1 left, 4 down, 2 left, 1 up, 1 left, 1 up

Page 178 – Practice

1 a Walk down the hallway past all the bedrooms and bathrooms. When you are past the kitchen on the right, the dining area is there on the right.
b As you leave the bathroom, turn right down the hallway and continue past all the bedrooms and the garage. When you get to the end, you have reacched the entry.

2 Sample answers:
a Between her parents' bedroom and bathroom 1
b Between Roisin's bedroom and the kitchen
c Opposite the family room and the kitchen
d Between bathroom 2 and the outside wall

3 a Exit onto Bell Street, turn right onto White Road, turn left onto John Street, then continue straight past Fort Street and Mary Street. Nina's house is on the left.
b Exit onto John Street and turn right. Take the next right onto White Road then left onto Bell Street. Gabby's house is on the right.
c Exit onto John Street and turn left. Continue straight past Fort Street and turn left into Mary Street, and Nina's house is on the left.

4 a Exit onto Bell Street, turn right onto White Road, turn left onto John Street, then continue straight past Fort Street and turn left onto Mary Street. Follow it around and the shops are on the left.
b Exit onto John Street, and turn right onto Mary Street. Follow it around and the shops are on the left.
c Exit onto John Street, and turn left. Continue straight past Fort Street and turn left onto Mary Street. Follow it around and the shops are on the left.

9. POSITION CONTINUED

5 a Hosain b Michael c Tiffany d Mrs White e Zac f Lachlan g Mia

6 a Third row from the top, on the left
b Second row from teh bottom, on the left
c Bottom row, second from the right
d Second row from the top, on the left
e Top row, second from the right

7 Adult to check

8 a Turn right onto Grey Road, continue past Green Road and Blue Street, and the park is on the left.
b Turn left onto Grey Road, turn right onto Blue Street and turn left onto Rainbow Road. Continue straight and you will arrive at the library.
c Turn right onto Green Road, then left onto Grey Road. The take away is on the right, on the corner with Navy Road.

The Compass

Page 182 – Your Turn

1 a b c

2

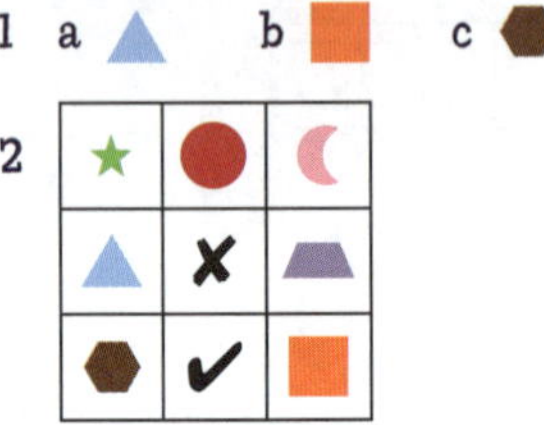

Page 183 – Practice

1

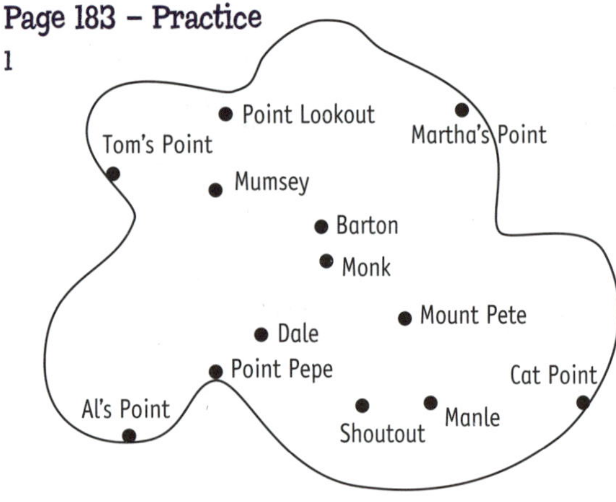

2 a giraffes b llamas c zebras

3 a south-west b north-east c east

4 a south-east b south c north-east d south-west e east f north g north-west

5 **Funaround Fun Park**

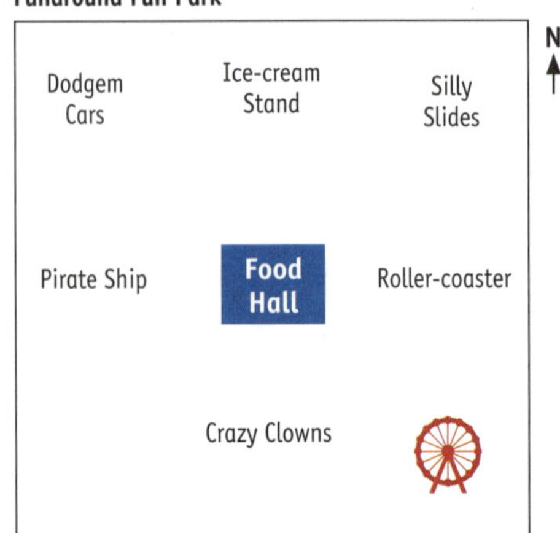

CATCH UP MATHS YEAR 5 BOOK B © PASCAL PRESS ISBN: 9781925726176

Grid References

Page 185 – Your Turn

1 a (D,3)
  b (A,3)

2

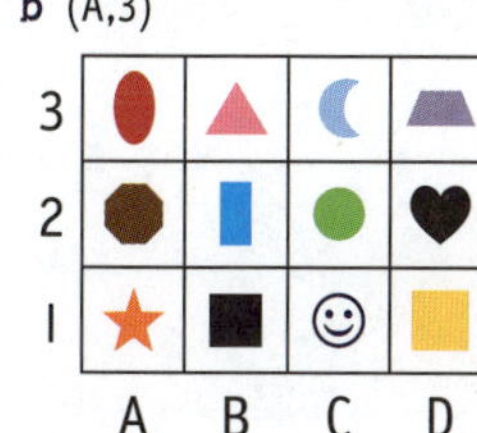

Page 186 – Practice

1 a (5,H) c (4,F) e (3,C) g (5,A)
  b (1,D) d (4,B) f (2,F)

2

H
G
F
E
D
C
B
A
1 2 3 4 5

3 a (2,8)
  b (5,8)
  c (1,7)
  d (3,5)
  e (1,4)
  f (3,4)
  g (5,3)
  h (1,1)

4 Adult to check

Coordinates

Page 187 – Your Turn

1 a (B,1) b (C,2) c (E,4)

2

5
4
3
2
1
0
A B C D E F

Page 188 – Practice

1 a (10,4) c (10,0) e (1,1) g (6,1)
  b (3,2) d (5,0) f (7,3)

2

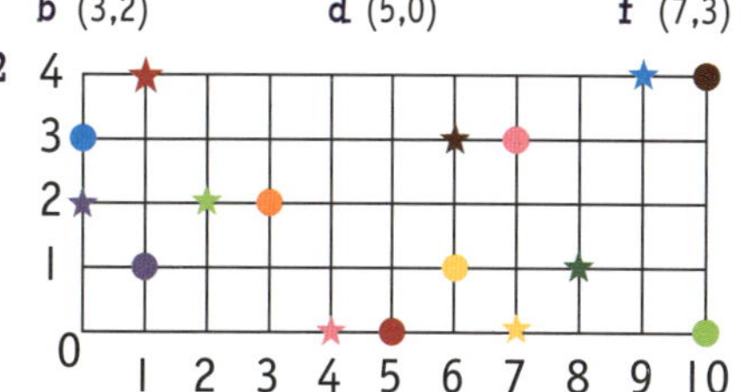

3 a (5,7), (8,7), (5,5), (8,5)
  b (0,2), (1,4), (2,4), (3,2)
  c (0,5), (3,5), (1, 7), (4,7)
  d (9,7), (10,7), (9,6), (10,6)
  e (4,3), (5,4), (6,4), (7,3), (6,2), (5,2)

4 a trapezium b rectangle c pentagon d parallelogram

Legends

Page 189 – Your Turn

1 a 1 b 1

2 14

Page 190 – Practice

1 a Picnic tables b Ticket booth c Donut Dive

2 a b c d A e

3 Adult to check

Position Review Page 191

1 a 3 left, 1 down, 2 left, 3 down, 2 right, 2 up, 2 right, 2 down
  b 6 right, 2 down, 1 right, 1 down, 2 left, 2 up, 2 left, 2 down, 4 left, 1 down
  c 4 down, 1 left
  d 3 left, 2 down, 7 right, 2 down, 5 left

2 a Walk past the office on the right and the garage on the left. Take the next left into the home theatre.
  b Walk past the office on the right and the garage and home theatre on the left. Continue past the bathroom and laundry on the right, and after that the area on the right is the family area.
  c Walk past the office on the right and the garage and home theatre on the left. Continue past the bathroom and the next door on the right is the laundry.

3 a Between the kitchen and the family area
  b Between the laundry and the office, opposite the home theatre
  c The first door on the left from the entry
  d Between the dining area and the home theatre

4 a Sam b Kira c Ravi d Joe e Lev f Zola

5 a Middle row, on the left
  b Back row, on the left
  c In the middle of the front row
  d Back row, second from the left
  e Middle row, second from the right
  f Back row, on the right

6 a North c East e South g West
  b North-east d South-east f South-west h North-west

7 a red b blue c purple d orange

8

9 a Flying Swings e Tickets/Entry
  b Magic Hill f Spinning Spider
  c Crazy Mirrors g Dodgem Cars
  d Slides h Roller-coaster

10 a East c North e South-east g North-east
  b South-west d North-west f South h West

11 a (B,4) (B,6) (C,7) (D,4) (D,6) c (E,4) (F,4) (G,3) (F,2) (E,2) (D,3)
  b (H,7) (I,6) (H,4) (G,6) d (M,7) (P,7) (Q,5) (M,5)

12 a parallelogram c square e rectangle
  b parallelogram d triangle

13 a (D,3) b (B,3)

14 a Brisbane c Perth e Melbourne g Adelaide
  b Hobart d Darwin f Sydney

15 a Library c Hospital e Bank g Cinema
  b School d Food f Shop h Park

16 a  b  c 

Catch-Up Maths Book 5B

ISBN: 9781925726176

Published by Pascal Press
PO Box 250
Glebe NSW 2037
www.pascalpress.com.au
contact@pascalpress.com.au

Design: Janice Bowles
Author: Deborah Frendo-Toman
Publisher: Lynn Dickinson
Typesetter: Ruth Schultz
Editor: Vaishali Batra

Printed in China by 1010 Printing International Ltd.